Daniel W. Cunningham

**Mathematical Logic**

## Also of Interest

*Advanced Mathematics*
*An Invitation in Preparation for Graduate School*
Patrick Guidotti, 2022
ISBN 978-3-11-078085-7, e-ISBN (PDF) 978-3-11-078092-5

*Abstract Algebra*
*An Introduction with Applications*
Derek J. S. Robinson, 2022
ISBN 978-3-11-068610-4, e-ISBN (PDF) 978-3-11-069116-0

*A Primer in Combinatorics*
Alexander Kheyfits, 2021
ISBN 978-3-11-075117-8, e-ISBN (PDF) 978-3-11-075118-5

*Brownian Motion*
*A Guide to Random Processes and Stochastic Calculus*
René L. Schilling, 2021
ISBN 978-3-11-074125-4, e-ISBN (PDF) 978-3-11-074127-8

*General Topology*
*An Introduction*
Tom Richmond, 2020
ISBN 978-3-11-068656-2, e-ISBN (PDF) 978-3-11-068657-9

*Partial Differential Equations*
*An Unhurried Introduction*
Vladimir A. Tolstykh, 2020
ISBN 978-3-11-067724-9, e-ISBN (PDF) 978-3-11-067725-6

Daniel W. Cunningham

# Mathematical Logic

An Introduction

DE GRUYTER

**Mathematics Subject Classification 2020**
03-01, 03B05, 03B10, 03B22, 03B25, 03D10, 03D20

**Author**
Dr. Daniel W. Cunningham
Department of Mathematics
Californial State University
5245 North Backer Avenue M/S PB108
Fresno CA 93740
USA
dwc17@csufresno.edu

ISBN 978-3-11-078201-1
e-ISBN (PDF) 978-3-11-078207-3
e-ISBN (EPUB) 978-3-11-078219-6

**Library of Congress Control Number: 2023931072**

**Bibliographic information published by the Deutsche Nationalbibliothek**
The Deutsche Nationalbibliothek lists this publication in the Deutsche Nationalbibliografie; detailed bibliographic data are available on the Internet at http://dnb.dnb.de.


Cover image: cicerocastro / iStock / Getty Images Plus
Typesetting: VTeX UAB, Lithuania
Printing and binding: CPI books GmbH, Leck

www.degruyter.com

# Preface

Mathematical logic is a particular branch of mathematics that applies mathematical tools to investigate the nature of mathematics. Consequently, in this book our attention will be focused on the language of mathematics. This will be done by discussing formal languages, first-order logic, model-theoretic semantics, deductive systems, and their mathematical properties and relations. Such a focus radiates a bright and direct light on mathematics itself.

Aristotle was the first formal logician who identified several key principles of correct reasoning, namely, the syllogisms. On the other hand, modern mathematical logic is based on the late nineteenth- and early twentieth-century innovative work of Boole, Frege, Peano, Russell, Whitehead, Hilbert, Skolem, Gödel, Tarski, Cantor, and their followers. The material presented in this text is the result of these pioneers in mathematical logic.

This textbook delivers an upper-division undergraduate course in mathematical logic. This subject typically attracts students with various backgrounds, some of whom may be quite familiar with logical notation and mathematical proof while others may not be as familiar. The book strives to address the needs of such a course. My primary goal was to write a book that would be accessible to all readers having a fundamental background in mathematics. Thus, I have made an effort to write clear and complete proofs throughout the text. In addition, these proofs favor detail over brevity. This approach should be to the benefit of all students, readers, and instructors.

## Topics covered

The book presents the fundamental topics in mathematical logic that will lead to the statements and coherent proofs of Gödel's completeness and incompleteness theorems. The basics of logic and elementary set theory are first discussed in Chapter 1. Since students, typically, are acquainted with these topics, one should not necessarily begin the book by starting with this chapter. However, Section 1.1.5 and Theorem 1.1.27 should definitely be discussed. In particular, Theorem 1.1.27 is a recursion theorem that justifies many of the key definitions and proofs that are presented in the text. Few books in mathematical logic explicitly state and prove this often applied result. Such books usually justify their definitions and proofs with the expression "by recursion;" however, we in such cases will cite and apply Theorem 1.1.27. Understanding the statement of this theorem is more important than reading and understanding its proof.

Chapter 2 introduces the syntax and semantics of propositional logic. The chapter also carefully discusses an induction principle that is illustrated by correctly proving fundamental results about the language of propositional logic. This is followed by establishing the completeness of the logical connectives, the compactness theorem, and the deduction theorem. We also state and prove the associated soundness and completeness

https://doi.org/10.1515/9783110782073-201

theorems. Most students who are familiar with propositional logic have not yet seen a mathematical development of this logic. Therefore, these topics offer an important prerequisite for the development of first-order logic.

Chapter 3 discusses the syntax and semantics of first-order languages. First-order logic is quite a bit more subtle than propositional logic, although they do share some common characteristics. The chapter also introduces structures, which can be viewed as vehicles for interpreting a given first-order language. Tarski's definition of satisfaction gives a precise meaning that yields a method for interpreting a first-order language in a given structure. This is then followed by the notion of a deduction (formal proof) in a first-order language.

The main goal of Chapter 4 is to present and prove the soundness and completeness theorems of first-order logic. For each of these proofs, I have respectively isolated the technical lemmas (Sections 4.1.1 and 4.2.1) that support the proofs of these important theorems. The compactness theorem is then presented with a proof. The chapter ends with several applications of the soundness, completeness, and compactness theorems.

In Gödel's proof of the incompleteness theorem, he encodes formulas into natural numbers using (primitive) recursive functions. In preparation for the proof of Gödel's theorem, Chapter 5 covers (primitive) recursive functions and relations. Since Gödel's encoding techniques created a link between logic and computing, we begin the chapter with an introduction to abstract computing machines and partial recursive functions. Today, computability theory in mathematical logic is closely related to the theory of computation in computer science.

In Chapter 6 the focus is on the language of elementary number theory $\mathcal{L}$ and the standard model $\mathcal{N}$ for number theory. The chapter begins with the question: Is there a decidable set of $\mathcal{L}$-sentences, that hold in $\mathcal{N}$, from which one can deduce all the sentences that are true in $\mathcal{N}$? This is followed by an introduction to the $\Omega$-axioms. These axioms allow one to deduce some of the basic statements that are true in the standard model. Then representable relations and functions are discussed. Eventually it is established that a function is representable if and only if the function is recursive. Then a technique is presented that allows one to perform a Gödel encoding of all of the formulas in the language $\mathcal{L}$. This is followed by a proof of the fixed-point lemma and two results of Gödel, namely, the first and second incompleteness theorems. These two theorems are among the most important results in mathematical logic.

## How to use the book

It is strongly recommended that the reader be familiar with the basics of sets, functions, relations, logic, and mathematical induction. These topics are typically introduced in a "techniques of proof" course (for example, see [1]). As the emphasis will be on theorems and their proofs, the reader should be comfortable reading and writing mathematical proofs.

If time is short or an instructor would like to end a one-semester course by covering Gödel's incompleteness theorems, certain topics can be bypassed. In particular, the following sections can be omitted without loss of continuity:
- 3.2.5 Classes of structures
- 3.2.6 Homomorphisms
- 4.3.1 Nonstandard models
- 4.3.4 Prenex normal form
- 5.1 The informal concept
- 5.2.1 Turing machines
- 5.2.2 Register machines

Furthermore, an instructor could focus on the statements of the technical lemmas in Sections 4.1.1 and 4.2.1 rather than on the proofs of these lemmas. These proofs could then be given as assigned reading. These technical lemmas support the respective proofs of the soundness and completeness theorems in Chapter 4. Similarly, in Sections 5.3 and 5.4 one could focus attention on the results rather than on the proofs. Of course, the material in Chapter 6 is more interesting than the proofs presented in these two sections. Perhaps, after seeing the theorems presented in Chapter 6, one would be more interested in the meticulous proofs presented in Sections 5.3 and 5.4.

Exercises are given at the end of each section in a chapter. An exercise marked with an asterisk $*$ is one that is cited, or referenced, elsewhere in the book. Suggestions are also provided for those exercises that a newcomer to upper-division mathematics may find more challenging.

# Acknowledgments

This book began as a set of lecture notes for an undergraduate course in mathematical logic that I taught at SUNY Buffalo State. The textbook used for this course was *A Mathematical Introduction to Logic* by Herbert B. Enderton [2]. As an undergraduate at UCLA, I also took courses in mathematical logic that used Enderton's book. Needless to say, my respect for Enderton's book has influenced the topics and presentation in my own book.

Sherman Page deserves to be recognized for his meticulous copy-editing and review of the manuscript. Sherman found many typos and errors that I had overlooked. I am extremely grateful for his salient comments and corrections. I am also grateful to Steven Elliot, editor at De Gruyter, for his guidance and enthusiasm. Thank you, Marianne Foley, for your support and tolerance.

https://doi.org/10.1515/9783110782073-202

# Contents

# 1 Basic set theory and basic logic

## 1.1 Basic set theory

### Fundamental definitions of set theory

A set is a collection of objects which are called its elements. As usual, we write "$t \in A$" to say that $t$ is a member of $A$, and we write "$t \notin A$"' to say that $t$ is not a member of $A$. We write $A = B$ to mean that the sets $A$ and $B$ are equal, that is, they have the same elements. For sets $A$ and $B$ we write $A \subseteq B$ to mean to mean that set $A$ is a subset of set $B$, that is, every element of $A$ is also an element of $B$.

**Remark.** Recall the following:
1. $A \subseteq B$ means that for all $x$, if $x \in A$, then $x \in B$.
2. $A = B$ means that for all $x$, $x \in A$ if and only if $x \in B$.
3. One special set is the empty set $\varnothing$, which has no members at all. Note that $\varnothing \subseteq A$ for any set $A$.
4. For any object $x$, the set $\{x\}$ is called a singleton because it has only one element, namely, $x$.

We now identify some important sets that often appear in mathematics:
1. $\mathbb{N} = \{x : x \text{ is a natural number}\} = \{0, 1, 2, 3, \ldots\}$,
2. $\mathbb{Z} = \{x : x \text{ is an integer}\} = \{\ldots, -3, -2, -1, 0, 1, 2, 3, \ldots\}$,
3. $\mathbb{Q} = \{x : x \text{ is a rational number}\}$; for example, $-2, \frac{2}{3} \in \mathbb{Q}$,
4. $\mathbb{R} = \{x : x \text{ is a real number}\}$; for example, $\frac{2}{3}, \pi \in \mathbb{R}$.

Given a property $P(x)$ we can form the set of just those elements in a set $A$ that make $P(x)$ true, that is, we can form the set $\{x \in A : P(x)\}$. For example, let $\mathbb{N}$ be the set of natural numbers. Suppose we want to collect just those elements in $\mathbb{N}$ that are odd. We can easily describe this set by $\{n \in \mathbb{N} : n \text{ is odd}\}$, that is, "the set of all $n \in \mathbb{N}$ such that $n$ is odd." Thus, $\{n \in \mathbb{N} : n \text{ is odd}\} = \{1, 3, 5, 7, \ldots\}$.

**Definition 1.1.1.** Given two sets $A$ and $B$ we define the following:
1. $A \cap B = \{x : x \in A \text{ and } x \in B\}$ is the *intersection* of $A$ and $B$,
2. $A \cup B = \{x : x \in A \text{ or } x \in B\}$ is the *union* of $A$ and $B$,
3. $A \setminus B = \{x : x \in A \text{ and } x \notin B\}$ is the *set difference* of $A$ and $B$ (also stated in English as $A$ "minus" $B$),
4. $A$ and $B$ are *disjoint* if they have no elements in common, that is, $A \cap B = \varnothing$,
5. to add one extra object $t$ to a set $A$, we will write $A; t$ to denote the set $A \cup \{t\}$.

Consider a set $A$ whose members are themselves sets. The union of $A$, denoted by $\bigcup A$, is the set of objects that belong to some member of $A$, that is,

$$\bigcup A = \{x : x \text{ belongs to some member of } A\}.$$

https://doi.org/10.1515/9783110782073-001

When $A$ is nonempty, the intersection of $A$, denoted by $\bigcap A$, is the set of objects that belong to every member of $A$, that is,

$$\bigcap A = \{x : x \text{ belongs to every member of } A\}.$$

For example, if $A = \{\{0,1,5\},\{1,6\},\{1,5\}\}$, then

$$\bigcup A = \{0,1,5,6\} \quad \text{and} \quad \bigcap A = \{1\}.$$

In cases where we have a set $A_i$ for each $i \in I$, the set $\{A_i : i \in I\}$ is called an indexed family of sets. The union $\bigcup\{A_i : i \in I\}$ is usually denoted by $\bigcup_{i\in I} A_i$ or $\bigcup_i A_i$. The intersection $\bigcap\{A_i : i \in I\}$ is usually denoted by $\bigcap_{i\in I} A_i$ or $\bigcap_i A_i$. We will often be dealing with unions of the form $\bigcup_{n\in\mathbb{N}} A_n$ and intersections of the form $\bigcap_{n\in\mathbb{N}} A_n$.

More generally, a set having the form $\{x_i : i \in I\}$ is called an indexed set and each $i \in I$ is called an index. Every element in $\{x_i : i \in I\}$ has the form $x_i$ for some $i \in I$. Such sets appear frequently in mathematics and will also appear in this text.

**Definition 1.1.2.** Let $A$ be a set. The *power set* of $A$, denoted by $\mathcal{P}(A)$, is the set whose elements are all of the subsets of $A$, that is, $\mathcal{P}(A) = \{X : X \subseteq A\}$.

Thus, $X \in \mathcal{P}(A)$ if and only if $X \subseteq A$. If $A$ is a finite set with $n$ elements, then one can show that the set $\mathcal{P}(A)$ has $2^n$ elements. The set $A = \{1,2,3\}$ has three elements, so $\mathcal{P}(A)$ has eight elements, namely,

$$\mathcal{P}(A) = \{\varnothing, \{1\}, \{2\}, \{3\}, \{1,2\}, \{1,3\}, \{2,3\}, \{1,2,3\}\}.$$

### 1.1.1 Relations

**Definition 1.1.3.** An *ordered pair* has the form $\langle a,b\rangle$, where $a$ is referred to as the first component and $b$ is identified as the second component.

**Example 1.1.4.** The pair $\langle 2,3\rangle$ is an ordered pair, and so is $\langle 3,2\rangle$. Note that these are different ordered pairs, that is, $\langle 2,3\rangle \neq \langle 3,2\rangle$.

**Definition 1.1.5.** Given sets $A$ and $B$, the *Cartesian product* $A \times B$ is the set

$$A \times B = \{\langle a,b\rangle : a \in A \text{ and } b \in B\}.$$

In other words, $A \times B$ is the set of *all* ordered pairs with first component in $A$ and second component in $B$.

**Definition 1.1.6.** An *n-sequence* is an ordered list of the form $\langle a_1, a_2, \ldots, a_n\rangle$, where $n \geq 1$ is a natural number. The term $a_1$ is called the first component, $a_2$ is the second component, …, and $a_n$ is the $n$-th component. We say that $\langle a_1, a_2, \ldots, a_k\rangle$ is a *proper initial segment* of $\langle a_1, a_2, \ldots, a_n\rangle$ when $1 \leq k < n$.

Given an $n$-sequence $s = \langle a_1, a_2, \dots, a_n\rangle$ and an $m$-sequence $t = \langle b_1, b_2, \dots, b_m\rangle$, the *concatenation* of $s$ with $t$ is the $(n+m)$-sequence given by

$$s^\frown t = \langle a_1, a_2, \dots, a_n, b_1, b_2, \dots, b_m\rangle.$$

**Example 1.1.7.** The sequence $\langle 3, -1, 2, 2\rangle$ is a 4-sequence, and so is $\langle 2, -1, 3, 2\rangle$. Note that these are different sequences, that is, $\langle 3, -1, 2, 2\rangle \neq \langle 2, -1, 3, 2\rangle$.

We will just say that $\langle a_1, a_2, \dots, a_n\rangle$ is a sequence, when $n$ is understood.

**Definition 1.1.8.** Let $n \geq 1$ and let $A$ be a set. Then the set $A^n$ is defined to be

$$A^n = \{\langle a_1, a_2, \dots, a_n\rangle : a_1 \in A, a_2 \in A, \dots, a_n \in A\}.$$

In other words, $A^n$ is the set of *all* $n$-sequences whose components are all in $A$.

For the record, an $n$-sequence $\langle a_1, a_2, \dots, a_n\rangle$ is rigorously defined to be a function

$$f\colon \{1, 2, \dots, n\} \to \{a_1, a_2, \dots, a_n\}$$

such that $f(1) = a_1, f(2) = a_2, \dots, f(n) = a_n$. Thus, $\langle a, b\rangle = \langle c, d\rangle$ if and only if $a = c$ and $b = d$. Moreover, $\langle a_1, a_2, \dots, a_n\rangle = \langle b_1, b_2, \dots, b_m\rangle$ if and only if $n = m$ and $a_1 = b_1, \dots, a_n = b_m$.

**Definition 1.1.9.** A subset $R$ of $A^n$ is said to be an *$n$-place relation* on $A$. We will write $R(a_1, a_2, \dots, a_n)$ to indicate that $\langle a_1, a_2, \dots, a_n\rangle \in R$.

A 1-place relation on a set $A$ is simply a subset of $A$.

**Definition 1.1.10.** A 2-place relation $R$ on a set $A$ is often just called a relation on $A$. The *domain* of $R$, $\mathrm{dom}(R)$, is the set $\{x \in A : \langle x, y\rangle \in R \text{ for some } y\}$. The *range* of $R$, $\mathrm{ran}(R)$, is the set $\{y \in A : \langle x, y\rangle \in R \text{ for some } x\}$. The union of $\mathrm{dom}(R)$ and $\mathrm{ran}(R)$ is called the *field* of $R$, $\mathrm{fld}(R)$.

A relation $R$ on $A$ is a subset of $A \times A$, that is, $R \subseteq A \times A$. We shall customarily write $xRy$ or $R(x, y)$ to denote $\langle x, y\rangle \in R$.

The equality relation is reflexive, symmetric, and transitive. Many relations also have some of these properties. Relations that have all of these properties often appear in mathematics.

**Definition 1.1.11.** For a relation $\sim$ on $A$, we define the following:

$\sim$ is *reflexive* if and only if $x \sim x$ for all $x \in A$,
$\sim$ is *symmetric* if and only if whenever $x \sim y$, then also $y \sim x$,
$\sim$ is *transitive* if and only if whenever both $x \sim y$ and $y \sim z$, then $x \sim z$.

Finally, we say that a relation $\sim$ on $A$ is an *equivalence relation* if and only if $\sim$ is reflexive, symmetric, and transitive.

### 1.1.2 Equivalence classes and partitions

A partition is a way of breaking up a set into nonempty disjoint parts such that each element of the set is in one of the parts.

**Definition 1.1.12.** Let $A$ be a set. Let $P$ be a collection of nonempty subsets of $A$. We say that $P$ is a *partition* of $A$ if the following hold:
1. For every element $a \in A$ there is a set $S \in P$ such that $a \in S$.
2. For all $S, T \in P$, if $S \neq T$, then $S \cap T = \varnothing$.

**Example 1.1.13.** The set $P = \{S_1, S_2, S_3\}$ forms a *partition* of the set $\mathbb{Z}$ where

$$\begin{aligned} S_1 &= \{\dots, -6, -3, 0, 3, 6, \dots\}, \\ S_2 &= \{\dots, -5, -2, 1, 4, 7, \dots\}, \\ S_3 &= \{\dots, -4, -1, 2, 5, 8, \dots\}. \end{aligned}$$

That is, $P = \{S_1, S_2, S_3\}$ breaks $\mathbb{Z}$ up into three disjoint nonempty sets and every integer is in one of these sets.

**Definition 1.1.14.** Let $\sim$ be an equivalence relation on a set $A$ and let $a \in A$ be an element of $A$. The *equivalence class* of $a$, denoted by $[a]_\sim$, is defined by

$$[a]_\sim = \{b \in A : b \sim a\}.$$

We write $[a] = [a]_\sim$ when the relation $\sim$ is clearly understood. So if $a \in A$, then $a \in [a]$ as $a \sim a$.

**Theorem 1.1.15.** *Let $\sim$ be an equivalence relation on $A$. Then for all $a, b \in A$,*

$$a \sim b \quad \textit{if and only if} \quad [a] = [b].$$

*Proof.* Let $\sim$ be an equivalence relation on $A$. Let $a, b \in A$. We shall prove that

$$a \sim b \quad \text{if and only if} \quad [a] = [b].$$

($\Rightarrow$). Assume that $a \sim b$. We prove that $[a] = [b]$. First we prove that $[a] \subseteq [b]$. Let $x \in [a]$. We show that $x \in [b]$. Since $x \in [a]$ and $[a] = \{y \in A : y \sim a\}$, it follows that $x \sim a$. By assumption, we also have $a \sim b$. Hence, $x \sim a$ and $a \sim b$. Since $\sim$ is transitive, we conclude that $x \sim b$. Now, because $[b] = \{y \in A : y \sim b\}$, it follows that $x \in [b]$. Therefore, $[a] \subseteq [b]$. The proof that $[b] \subseteq [a]$ is very similar. So $[a] = [b]$.

($\Leftarrow$). Assume that $[a] = [b]$. We prove that $a \sim b$. Since $a \in [a]$ and $[a] = [b]$, it follows that $a \in [b]$. But $[b] = \{y \in A : y \sim b\}$. Therefore, $a \sim b$. □

**Corollary 1.1.16.** *Let $\sim$ be an equivalence relation on a set $A$. Then for all $a \in A$ and $b \in A$, $a \in [b]$ if and only if $[a] = [b]$.*

**Theorem 1.1.17** (Fundamental theorem on equivalence relations). *Whenever $\sim$ is an equivalence relation on a set $A$, the collection $P = \{[a] : a \in A\}$ is a partition of $A$. We denote this partition by $P = A/{\sim}$.*

*Proof.* Let $\sim$ be an equivalence relation on $A$. We prove that the set $P = \{[a] : a \in A\}$ is a partition of $A$, that is, we prove that:
(i) for every element $x \in A$ we have $x \in [x]$,
(ii) for all $x, y \in A$, if $[x] \cap [y] \neq \varnothing$, then $[x] = [y]$.

*Proof of* (i). Let $x \in A$. Clearly, $[x] \in P$ and $x \in [x]$.

*Proof of* (ii). Let $x, y \in A$. Thus, $[x] \in P$ and $[y] \in P$. We must prove that if $[x] \neq [y]$, then $[x] \cap [y] = \varnothing$. So assume $[x] \neq [y]$. Assume, for a contradiction, that $[x] \cap [y] \neq \varnothing$. Since $[x] \cap [y] \neq \varnothing$, there exists a $z \in A$ such that $z \in [x]$ and $z \in [y]$. However, since $[x] = \{b \in A : b \sim x\}$ and $[y] = \{b \in A : b \sim y\}$, it thus follows that $z \sim x$ and $z \sim y$. Because $\sim$ is symmetric, we conclude that $x \sim z$ and $z \sim y$. As $\sim$ is transitive, we further conclude that $x \sim y$. But Theorem 1.1.15 now implies that $[x] = [y]$, which contradicts our assumption that $[x] \neq [y]$. Therefore, $[x] \cap [y] = \varnothing$. □

### 1.1.3 Functions

One of the most important concepts in mathematics is the *function* concept. A function is a way of associating each element of a set $A$ with exactly one element in another set $B$. We will give a precise set-theoretic definition of a function using relations.

**Definition 1.1.18.** A relation $R$ is *single-valued* if for each $x \in \operatorname{dom}(R)$ there is exactly one $y$ such that $\langle x, y\rangle \in R$.

Thus, if $R$ is a single-valued relation, then whenever $\langle x, y\rangle \in R$ and $\langle x, z\rangle \in R$, we can conclude that $y = z$.

**Definition 1.1.19.** A *function* $f$ is any single-valued relation; in other words, for each $x \in \operatorname{dom}(f)$ there is only one $y$ such that $\langle x, y\rangle \in f$.

Let $A$ and $B$ be sets. A *function $f$ from $A$ to $B$* is a subset of $A \times B$ such that for each $x \in A$ there is exactly one $y \in B$ so that $\langle x, y\rangle \in f$. For example, let $A = \{a, b, c, d, e\}$ and $B = \{5, 6, 7, 8, 9\}$. Then

$$f = \{\langle a, 8\rangle, \langle b, 7\rangle, \langle c, 9\rangle, \langle d, 6\rangle, \langle e, 5\rangle\}$$

is a function from $A$ to $B$ because for each $x \in A$, there is exactly one $y \in B$ such that $\langle x, y\rangle \in f$. We now express this notion in terms of a formal definition.

**Definition 1.1.20.** Let $A$ and $B$ be sets and let $f$ be a relation from $A$ to $B$. Then $f$ is said to be a *function from $A$ to $B$* if the following two conditions hold:
(1) $\mathrm{dom}(f) = A$, that is, for each $x \in A$, there is a $y \in B$ such that $\langle x,y\rangle \in f$,
(2) $f$ is single-valued, that is, if $\langle x,y\rangle \in f$ and $\langle x,z\rangle \in f$, then $y = z$.

The set $A$ is the *domain* of $f$ and the set $B$ is called the *codomain* of $f$.

We write $f: A \to B$ to indicate that $f$ is a function from the set $A$ to the set $B$. Thus, for each $x \in A$, there is exactly one $y \in B$ such that $\langle x,y\rangle \in f$. This unique $y$ is called "the value of $f$ at $x$" and is denoted by $f(x)$. Therefore, $\langle x,y\rangle \in f$ if and only if $f(x) = y$. We will say that $x \in A$ is an *input* to the function $f$ and $f(x)$ is the resulting *output*. One can also say that the function $f$ maps $x$ to $f(x)$, denoted by $x \mapsto f(x)$.

Given a function $f: A \to B$, the set $\{f(x) : x \in A\}$ is called the *range* of $f$ and is denoted by $\mathrm{ran}(f)$. Clearly, $\mathrm{ran}(f)$ is a subset of $B$. The following lemma offers a useful tool for showing that two functions are equal.

**Lemma 1.1.21.** *Let $f$ and $g$ be functions such that* $\mathrm{dom}(f) = \mathrm{dom}(g)$. *Then $f = g$ if and only if $f(x) = g(x)$ for all $x$ in their common domain.*

**Definition 1.1.22.** For a natural number $n \geq 2$ and a function $f: X^n \to Y$ from the set of sequences $X^n$ to the set $Y$, we shall write $f(\langle x_1, x_2, \ldots, x_n\rangle) = f(x_1, x_2, \ldots, x_n)$ for every element of $\langle x_1, x_2, \ldots, x_n\rangle \in X^n$. We will say that $f$ is an $n$-place function or that $f$ has arity $n$.

### 1.1.4 Induction and recursion

Let $\mathbb{N} = \{0, 1, 2, 3, \ldots\}$ be the set of natural numbers. Often one can use proof by *mathematical induction* to establish statements of the form "for all $n \in \mathbb{N}$, $P(n)$."

**Principle of mathematical induction**
Let $S \subseteq \mathbb{N}$. If
1. $0 \in S$ and
2. for all $n \in \mathbb{N}$, if $n \in S$, then $n + 1 \in S$,

then $S = \mathbb{N}$.

**Proof by mathematical induction**
Let $P(n)$ be a statement concerning a natural number variable $n$. If
1. $P(0)$ is true and
2. for all $n \in \mathbb{N}$, if $P(n)$, then $P(n + 1)$,

then $P(n)$ is true for *all* natural numbers $n$.

**Principle of strong induction**

Let $P(n)$ be a statement about a natural number variable $n$. If
1. $P(0)$ is true and
2. the statements $P(0), P(1), \ldots, P(n-1)$ imply $P(n)$, for all $n \in \mathbb{N}$,

then $P(n)$ is true for *all* natural numbers $n$.

Induction is often applied in mathematical proofs. One can also define a function by induction (recursion). A recursively defined function is one that is defined in terms of "previously evaluated values of the function." A function $h$ on $\mathbb{N}$ is defined recursively if its value at 0 is first specified and then all of the remaining values are defined by using a value that has previously been evaluated. A proof of the following theorem is given in [3, Theorem 4.2.1].

**Theorem 1.1.23** (Recursion on $\mathbb{N}$). *Let $a \in A$ and let $g: A \to A$ be a function, where $A$ is a set. Then there exists a unique function $h: \mathbb{N} \to A$ such that:*
1. $h(0) = a$,
2. $h(n+1) = g(h(n))$, *for all* $n \in \mathbb{N}$.

### 1.1.5 Defining sets by recursion

In this section, we shall develop and state a generalization of the above Theorem 1.1.23. First we must talk about defining sets by recursion. A recursive definition of a set $C$ has the following form:
(a) Basis: Specify the "initial" elements of $C$.
(b) Induction: Give one or more operations for constructing "new" elements of $C$ from "old" elements of $C$.
(c) Closure: The set $C$ is the smallest set that contains the initial elements and is also closed under the operations (see Theorem 1.1.25).

Suppose that $U$ is a set and we are given a function $f: U^n \to U$. We say that a set $S \subseteq U$ is *closed* under $f$ if and only if whenever $x_1, x_2, \ldots, x_n \in S$ we have $f(x_1, x_2, \ldots, x_n) \in S$. For any $A \subseteq U$, let us define

$$f[A] = \{f(x_1, x_2, \ldots, x_n) : x_1, x_2, \ldots, x_n \in A\}.$$

Now let $\mathcal{F}$ be a set of functions $\ell$ which have the form $\ell: U^n \to U$, where $n$ is the arity of $\ell$. The functions in $\mathcal{F}$ can have different arity. For any $A \subseteq U$ let us define $\mathcal{F}[A] = \bigcup\{\ell[A] : \ell \in \mathcal{F}\}$. Given $B \subseteq U$, in the next theorem, we show how to construct the smallest set $C$ such that $B \subseteq C \subseteq U$ and $C$ is closed under every function in $\mathcal{F}$.

**Theorem 1.1.24.** *Let $U$ be a set and let $\mathcal{F}$ be a set of functions $\ell$ of the form $\ell: U^n \to U$, where $n$ is the arity of $\ell$. Now let $B \subseteq U$. Define, by recursion on $\mathbb{N}$, the following sets:*

(1) $C_0 = B$,
(2) $C_{n+1} = C_n \cup \mathcal{F}[C_n]$ *for all* $n \in \mathbb{N}$.

*Let* $C = \bigcup_{n\in\mathbb{N}} C_n$. *Then* $B \subseteq C \subseteq U$ *and* $C$ *is closed under all the functions in* $\mathcal{F}$.

*Proof.* See Exercise 5. □

The set $C$, defined in Theorem 1.1.24, is referred to as the set *generated from* $B$ *by the functions in* $\mathcal{F}$. One feature of defining the set $C$ by the above recursion is that it yields an induction principle that will be frequently applied in the coming pages.

**Theorem 1.1.25** (Induction principle). *Let* $U$ *be a set and let* $\mathcal{F}$ *be a set of functions* $\ell$ *which have the form* $\ell\colon U^n \to U$, *where* $n$ *is the arity of* $\ell$. *Suppose that* $C$ *is the set generated from* $B$ *by the functions in* $\mathcal{F}$. *If* $I \subseteq C$ *satisfies*
(a) $B \subseteq I$ *and*
(b) $I$ *is closed under all of the functions in* $\mathcal{F}$,

*then* $I = C$.

*Proof.* One can prove by induction on $n$ that $C_n \subseteq I$ using (1) and (2) of Theorem 1.1.24. Thus, $C \subseteq I$, and since $I \subseteq C$, it follows that $I = C$. □

Suppose that $U$ is a set and that $f$ is a function of the form $f\colon U^n \to U$. Let $C \subseteq U$ and suppose that $C$ is closed under $f$. Then the function $f_C\colon C^n \to C$ is defined by $f_C(x_1, x_2, \dots, x_n) = f(x_1, x_2, \dots, x_n)$ for all $x_1, x_2, \dots, x_n \in C$.

**Definition 1.1.26.** Suppose that $U$ is a set and $\mathcal{F}$ is a set of functions $\ell$ of the form $\ell\colon U^n \to U$, where $n$ is the arity of $\ell$. We shall say that $C$ is *freely generated* from $B$ by the functions in $\mathcal{F}$ if the following hold:
1. the set $C$ is generated from $B$ by the functions in $\mathcal{F}$,
2. $f_C$ is one-to-one for every $f \in \mathcal{F}$,
3. the range of $f_C$ and $B$ are disjoint, for all $f \in \mathcal{F}$,
4. the range of $f_C$ and the range of $g_C$ are disjoint for all distinct $f, g \in \mathcal{F}$.

The following theorem shows that if a set $C$ is freely generated from $B$, then a function $h$ defined on the initial elements $B$ can be extended to a function $\overline{h}$ defined on all of the elements in $C$. This theorem will justify many results to be covered in the text. The proof requires some special set-theoretic tools and can be summarized as follows: The intersection of all the approximations to $\overline{h}$ is in fact $\overline{h}$.

**Theorem 1.1.27** (Recursion theorem). *Let* $U$ *be a set and let* $\mathcal{F}$ *be a set of functions* $\ell$ *of the form* $\ell\colon U^n \to U$, *where* $n$ *is the arity of* $\ell$. *Let* $B \subseteq U$ *and let* $C$ *be freely generated from* $B$ *by the functions in* $\mathcal{F}$. *Let* $V$ *be a set. Assume that:*
(a) *we are given a function* $h\colon B \to V$, *and*
(b) *for each* $\ell \in \mathcal{F}$, *there is an associated function* $F_\ell\colon V^n \to V$ *of the same arity as* $\ell$.

*Then there is a unique function $\overline{h}\colon C \to V$ such that:*
(i) $\overline{h}(x) = h(x)$ *for all* $x \in B$,
(ii) *for each* $\ell \in \mathcal{F}$, *we have*

$$\overline{h}(\ell(x_1, x_2, \ldots, x_n)) = F_\ell(\overline{h}(x_1), \overline{h}(x_2), \ldots, \overline{h}(x_n))$$

*for all* $x_1, x_2, \ldots, x_n \in C$, *where* $n$ *is the arity of* $\ell$.

*Proof.* Let us call a relation $R \subseteq C \times V$ *suitable* if the following two conditions hold:
(1) $\langle b, h(b)\rangle \in R$, for all $b \in B$,
(2) for all $\ell \in \mathcal{F}$ of arity $n$, if $x_1, x_2, \ldots, x_n \in \operatorname{dom}(R)$, then for each $1 \le i \le n$, we have $\langle x_i, y_i\rangle \in R$ for some $y_i \in V$ and

$$\langle \ell(x_1, x_2, \ldots, x_n), F_\ell(y_1, y_2, \ldots, y_n)\rangle \in R.$$

Our goal is to construct a suitable relation that is also a function.

Every suitable relation is a subset of $C \times V$. Let $\mathcal{S} = \{R : R \text{ is suitable}\}$. Because $C \times V$ is a suitable relation, $\mathcal{S}$ is nonempty. Let $\overline{h} = \bigcap \mathcal{S}$. Clearly, $\overline{h} \subseteq C \times V$, so $\overline{h}$ is a relation. We now prove that $\overline{h}$ is suitable and that it is a function.

**Claim 1.** *The relation* $\overline{h}$ *is suitable.*

*Proof.* We need to show that $\overline{h}$ satisfies items (1) and (2). Let $b \in B$. Every relation in $\mathcal{S}$ is suitable. So $\langle b, h(b)\rangle$ is an element in every relation in $\mathcal{S}$. Thus, $\langle b, h(b)\rangle \in \bigcap \mathcal{S}$, that is, $\langle b, h(b)\rangle \in \overline{h}$ for every $b \in B$. To prove (2), let $\ell \in \mathcal{F}$ be of arity $n$ and let $x_1, x_2, \ldots, x_n \in \operatorname{dom}(\overline{h})$. Therefore, for each $1 \le i \le n$, $\langle x_i, y_i\rangle \in \overline{h}$ for some $y_i \in V$. As $\overline{h} = \bigcap \mathcal{S}$, it follows that each $\langle x_i, y_i\rangle$ belongs to every relation in $\mathcal{S}$. Let $R \in \mathcal{S}$. Since $\langle x_i, y_i\rangle \in R$ for all $1 \le i \le n$ and $R$ is suitable, item (2) implies that $\langle \ell(x_1, x_2, \ldots, x_n), F_\ell(y_1, y_2, \ldots, y_n)\rangle \in R$. Thus, $\langle \ell(x_1, x_2, \ldots, x_n), F_\ell(y_1, y_2, \ldots, y_n)\rangle$ belongs to every relation in $\mathcal{S}$ and therefore, $\langle \ell(x_1, x_2, \ldots, x_n), F_\ell(y_1, y_2, \ldots, y_n)\rangle \in \overline{h}$. Hence, $\overline{h}$ is suitable. (Claim 1) □

We now must prove that $\overline{h}$ is a function with domain $C$.

**Claim 2.** *The relation* $\overline{h}$ *is a function from* $C$ *to* $V$.

*Proof.* To prove that $\overline{h}$ is a function from $C$ to $V$, we must show that for each $x \in C$, there is exactly one $y \in V$ such that $\langle x, y\rangle \in \overline{h}$. Let $I \subseteq C$ be defined by

$$I = \{x \in C : \text{there is exactly one } y \in V \text{ such that } \langle x, y\rangle \in \overline{h}\}.$$

We shall prove that $I = C$ by applying Theorem 1.1.25. To do this, we must first prove that $B \subseteq I$. Let $b \in B$. We know that $\langle b, h(b)\rangle \in \overline{h}$, because $\overline{h}$ is suitable. To prove that $b \in I$, we need to show there is no $\langle b, y\rangle \in \overline{h}$ where $y \ne h(b)$. Suppose, for a contradiction, that $\langle b, y\rangle \in \overline{h}$, where $y \ne h(b)$. Consider the relation $R = \overline{h} \setminus \{\langle b, y\rangle\}$. So (▲) $\langle b, y\rangle \notin R$. Since $C$ is freely generated from $B$ by the functions in $\mathcal{F}$, it follows that whenever $\ell \in \mathcal{F}$, we have

$\ell(x_1, x_2, \ldots, x_n) \notin B$ for all $x_1, x_2, \ldots, x_n \in C$. As $\overline{h}$ is suitable, it thus follows that $R$ is also suitable. Hence, $R \in \mathcal{S}$. Since $\overline{h} = \bigcap \mathcal{S}$, we conclude that $\overline{h} \subseteq R$, and thus, $\langle b, y\rangle \in R$, which contradicts (▲). Thus, $h(b)$ is the only element in $V$ such that $\langle b, h(b)\rangle \in \overline{h}$. Therefore, $B \subseteq I$.

Now we show that $I$ is closed under all of the functions in $\mathcal{F}$. Let $g\colon U^k \to U$ be a function in $\mathcal{F}$ and let $z_1, z_2, \ldots, z_k \in I$. We need to show that $g(z_1, z_2, \ldots, z_k) \in I$. For each $1 \le i \le k$, since $z_i \in I$, there is a unique $w_i$ such that $\langle z_i, w_i\rangle \in \overline{h}$. Because $\overline{h}$ is suitable, we conclude that

$$\langle g(z_1, z_2, \ldots, z_k), F_g(w_1, w_2, \ldots, w_k)\rangle \in \overline{h}.$$

To show that $g(z_1, z_2, \ldots, z_k) \in I$, we must show there is no $\langle g(z_1, z_2, \ldots, z_k), y\rangle \in \overline{h}$ where $F_g(w_1, w_2, \ldots, w_k) \neq y$. Suppose, for a contradiction, that there exists a $y \in V$ such that (◀) $\langle g(z_1, z_2, \ldots, z_k), y\rangle \in \overline{h}$ where (▼) $F_g(w_1, w_2, \ldots, w_k) \neq y$. Consider the relation

$$R = \overline{h} \setminus \{\langle g(z_1, z_2, \ldots, z_k), y\rangle\}.$$

Clearly, (♦) $\langle g(z_1, z_2, \ldots, z_k), y\rangle \notin R$. We now show that $R$ is suitable. Let $b \in B$. Since $C$ is freely generated from $B$ by the functions in $\mathcal{F}$, it follows that $g(z_1, z_2, \ldots, z_k) \neq b$. Since $\overline{h}$ is suitable, it follows that $\langle b, h(b)\rangle \in R$. Thus, $R$ satisfies item (1) of the above definition of suitability. To verify that $R$ satisfies item (2), let $\ell \in \mathcal{F}$ be of arity $n$ and let $x_1, x_2, \ldots, x_n \in \operatorname{dom}(R)$. Since $R \subseteq \overline{h}$, we have $x_1, x_2, \ldots, x_n \in \operatorname{dom}(\overline{h})$. Since $\overline{h}$ is suitable, it follows that

$$\langle \ell(x_1, x_2, \ldots, x_n), F_\ell(y_1, y_2, \ldots, y_n)\rangle \in \overline{h}.$$

So if

$$\langle \ell(x_1, x_2, \ldots, x_n), F_\ell(y_1, y_2, \ldots, y_n)\rangle \neq \langle g(z_1, z_2, \ldots, z_k), y\rangle,$$

then $\langle \ell(x_1, x_2, \ldots, x_n), F_\ell(y_1, y_2, \ldots, y_n)\rangle \in R$. On the other hand, if

$$\langle \ell(x_1, x_2, \ldots, x_n), F_\ell(y_1, y_2, \ldots, y_n)\rangle = \langle g(z_1, z_2, \ldots, z_k), y\rangle, \quad \text{then}$$
$$(\clubsuit)\ \ell(x_1, x_2, \ldots, x_n) = g(z_1, z_2, \ldots, z_k) \quad \text{and} \quad (\blacktriangleright)\ F_\ell(y_1, y_2, \ldots, y_n) = y.$$

Because $C$ is freely generated from $B$ by the functions in $\mathcal{F}$, (♣) implies that $\ell = g$, $n = k$, and $x_1 = z_1, x_2 = z_2, \ldots, x_n = z_k$. Therefore, $x_1, x_2, \ldots, x_n \in I$ and thus, $y_1 = w_1, y_2 = w_2, \ldots, y_n = w_k$. Hence,

$$F_\ell(y_1, y_2, \ldots, y_n) = F_g(w_1, w_2, \ldots, w_k),$$

which contradicts (▼) and (▶). Thus, $R$ is suitable. So $R \in \mathcal{S}$. Since $\overline{h} = \bigcap \mathcal{S}$, we conclude that $\overline{h} \subseteq R$ so $\langle g(z_1, z_2, \ldots, z_k), y\rangle \in R$ (see (◀)), which contradicts (♦). Therefore,

$g(z_1, z_2, \ldots, z_k) \in I$. Theorem 1.1.25 now implies that $I = C$ and, as a result, $\overline{h}$ is a function from $C$ to $V$. (Claim 2) □

Since the function $\overline{h}$ is suitable, it satisfies conditions (i) and (ii) given in the statement of the theorem. To prove that $\overline{h}$ is unique, let $g: C \to V$ also satisfy properties (i) and (ii). Thus, $g$ is a suitable relation, so $g \in \mathcal{S}$. Since $\overline{h} = \bigcap \mathcal{S}$, we have $\overline{h} \subseteq g$. As $\overline{h}$ and $g$ are functions with domain $C$, Lemma 1.1.21 implies that $\overline{h} = g$. Hence, $\overline{h}$ is unique. (Theorem) □

The function $\overline{h}$ in Theorem 1.1.27, satisfying (1) and (2), is often said to be *defined by recursion*. If a function is defined by recursion, then proofs of statements about this function typically use "proof by induction."

The proof of Theorem 1.1.27 requires that $C$ be freely generated. To illustrate this, let $f: \mathbb{N} \to \mathbb{N}$ and $g: \mathbb{N} \to \mathbb{N}$ be defined by $f(n) = n^2$ and $g(n) = n^3$. Let $B = \{0, 1, 2\}$. Let $C$ be the set generated from $B$ by the functions in $\mathcal{F} = \{f, g\}$. The functions in $\mathcal{F}$ are one-to-one. However, their ranges are not disjoint and their ranges are not disjoint with $B$. So $C$ is not freely generated. Now let $F_f: \mathbb{R} \to \mathbb{R}$ be the identity function and let $F_g: \mathbb{R} \to \mathbb{R}$ be the cube root function. Let $h: \{0, 1, 2\} \to \mathbb{R}$ be such that $h(1) = 3$. If $h$ could be extended to a function $\overline{h}: C \to \mathbb{R}$ as in Theorem 1.1.27, then we would have

$$\overline{h}(f(1)) = F_f(\overline{h}(1)) = F_f(3) = 3,$$
$$\overline{h}(g(1)) = F_g(\overline{h}(1)) = F_g(3) = \sqrt[3]{3}.$$

Since $f(1) = g(1) = 1$, we conclude that $\overline{h}(1) = 3$ and $\overline{h}(1) = \sqrt[3]{3}$. Thus, no such function $\overline{h}$ exists.

### 1.1.6 Countable sets

Let $\mathbb{N} = \{0, 1, 2, 3, \ldots\}$ be the set of natural numbers. A set is countable if it has the same size as some subset of $\mathbb{N}$. In other words, a set is *countable* if there is a one-to-one correspondence between the set and a subset of $\mathbb{N}$. Our next definition expresses this concept in mathematical terms.

**Definition 1.1.28.** A set $A$ is *countable* if and only if there exists a one-to-one function $f: A \to \mathbb{N}$.

The following theorem will be used to prove that the set of all finite sequences of a countable set is also countable (see Theorem 1.1.30). A proof of this theorem is given in [1, Theorem 4.4.7].

**Theorem 1.1.29** (Fundamental theorem of arithmetic). *For every natural number $n > 1$, there exist distinct primes $p_1, p_2, \ldots, p_k$ together with natural numbers $a_1 \geq 1, a_2 \geq 1, \ldots, a_k \geq 1$ such that*

$$n = p_1^{a_1} p_2^{a_2} \cdots p_k^{a_k}.$$

*Furthermore, given any prime factorization into distinct primes,*

$$n = q_1^{b_1} q_2^{b_2} \cdots q_\ell^{b_\ell}$$

*we have $\ell = k$, the primes $q_i$ are the same as the primes $p_j$ (except for order), and the corresponding exponents are the same.*

**Theorem 1.1.30.** *Let $A$ be a set and let $S$ be the set of all finite sequences of elements of $A$. If $A$ is countable, then $S$ is countable.*

*Proof.* Let $A$ be a countable set and let $f: A \to \mathbb{N}$ be one-to-one. Let $S$ be the set of all finite sequences of elements of $A$. Thus,

$$S = \bigcup_{n \in \mathbb{N}} A^{n+1}.$$

Let $h: S \to \mathbb{N}$ be defined as follows: Let $\langle a_1, a_2, \ldots, a_m \rangle \in S$. Define

$$h(\langle a_1, a_2, \ldots, a_m \rangle) = 2^{f(a_1)+1} \cdot 3^{f(a_2)+1} \cdot 5^{f(a_3)+1} \cdots p_m^{f(a_m)+1},$$

where $p_m$ is the $m$-th prime. To prove that $h: S \to \mathbb{N}$ is one-to-one, let $\langle a_1, a_2, \ldots, a_m \rangle$ and $\langle b_1, b_2, \ldots, b_n \rangle$ be arbitrary elements of $S$ and assume that

$$h(\langle a_1, a_2, \ldots, a_m \rangle) = h(\langle b_1, b_2, \ldots, b_n \rangle).$$

Thus

$$2^{f(a_1)+1} \cdot 3^{f(a_2)+1} \cdot 5^{f(a_3)+1} \cdots p_m^{f(a_m)+1} = 2^{f(b_1)+1} \cdot 3^{f(b_2)+1} \cdot 5^{f(b_3)+1} \cdots p_n^{f(b_n)+1}.$$

By Theorem 1.1.29, we conclude that $m = n$ and

$$f(a_1) + 1 = f(b_1) + 1, f(a_2) + 1 = f(b_2) + 1, \ldots, f(a_m) + 1 = f(b_n) + 1.$$

Because $f$ is one-to-one, we see that

$$a_1 = b_1, a_2 = b_2, \ldots, a_m = b_n.$$

Hence,

$$\langle a_1, a_2, \ldots, a_m \rangle = \langle b_1, b_2, \ldots, b_n \rangle,$$

and thus $h: S \to \mathbb{N}$ is one-to-one. Therefore, $S$ is countable. □

**Definition 1.1.31.** Let $A$ be a nonempty set. We shall say that $A$ is *denumerable* if and only if there is an enumeration

$$a_1, a_2, a_3, \ldots, a_n, \ldots \tag{1.1}$$

of all of the elements in $A$, that is, every element in $A$ appears in the above list indexed by the positive natural numbers.

So a set is *denumerable* if we can list the elements of the set in the same way that we list the set of nonzero natural numbers, namely, $1, 2, 3, 4, 5, \ldots$. However, our definition of a denumerable set allows elements in the list (1.1) to be repeated, or not to be repeated.

We will show below that a countable set is denumerable. Using an enumeration of a set will allow us to construct new sets that will be useful in Chapters 2 and 4 (for example, see the proof of Theorem 2.4.2 on page 44).

**Theorem 1.1.32.** *Let $A$ be a nonempty countable set. Then there is a function $g: \mathbb{N} \to A$ that is onto $A$.*

*Proof.* Assume that $A$ is a nonempty countable set. Thus, there is a function $f: A \to \mathbb{N}$ that is one-to-one. We shall use the function $f$ to define our desired function $g: \mathbb{N} \to A$. Let $c \in A$ be some fixed element. Define $g$ as follows: For each $n \in \mathbb{N}$,

$$g(n) = \begin{cases} a, & \text{if } n \text{ is in the range of } f \text{ and } f(a) = n, \\ c, & \text{otherwise.} \end{cases}$$

Because $f: A \to \mathbb{N}$ is one-to-one, one can show that $g$ is a function and it is onto. □

**Corollary 1.1.33.** *Let $A$ be a nonempty countable set. Then $A$ is denumerable.*

*Proof.* Assume that $A$ is a nonempty countable set. Theorem 1.1.32 implies that there is a function $g: \mathbb{N} \to A$ that is onto. For each $n \geq 1$ let $a_n = g(n-1)$. Since $g$ is onto, it follows that the enumeration $a_1, a_2, a_3, \ldots, a_n, \ldots$ lists every element in $A$. □

### 1.1.7 Cardinality

The cardinality of a set is a measure of how many elements are in the set. For example, the set $A = \{1, 2, 3, \ldots, 10\}$ has 10 elements, so the cardinality of $A$ is 10, denoted by $|A| = 10$. The cardinality of an infinite set $X$ will also be denoted by $|X|$. In this section, we briefly discuss Georg Cantor's method for measuring the size of an infinite set without the use of numbers. There are two infinite sets where one of these sets has cardinality much larger than the other infinite set. Therefore, it is possible for one infinite set to have "many more" elements than another infinite set.

What does it mean to say that two sets have the same cardinality, that is, the same size? Cantor discovered a simple answer to this question.

**Definition 1.1.34.** For sets $A$ and $B$, we say that $A$ has the *same cardinality* as $B$, denoted by $|A| =_c |B|$, if there is a bijection $f: A \to B$.

The expression $|A| =_c |B|$ looks like an equation; however, the assertion $|A| =_c |B|$ should be viewed only as an abbreviation for the statement "$A$ has the same cardinality as $B$." In other words, $|A| =_c |B|$ means that "there is a function $f: A \to B$ that is one-to-one and onto $B$." The relationship $=_c$, given in Definition 1.1.34, is reflexive, symmetric, and transitive.

The following theorem is very useful for proving many results about cardinality. The theorem states that if there are functions $f: A \to B$ and $g: B \to A$ that are both one-to-one, then there exists a function $h: A \to B$ that is one-to-one *and* onto $B$.

**Theorem 1.1.35** (Schröder–Bernstein). *Let $A$ and $B$ be any two sets. Suppose that $|A| \leq_c |B|$ and $|B| \leq_c |A|$. Then $|A| =_c |B|$.*

### 1.1.8 The axiom of choice

Suppose that a set $S$ contains only nonempty sets. Is it possible to uniformly select exactly one element from each set in $S$? In other words, is there a function $F$ so that for each $A \in S$, we have $F(A) \in A$? The following set-theoretic principle will allow us to positively answer this question.

**Axiom of Choice.** Let $\mathcal{C}$ be a set of nonempty sets. Then there is a function $H: \mathcal{C} \to \bigcup \mathcal{C}$ such that $H(A) \in A$ for all $A \in \mathcal{C}$.

Cantor developed a theory of infinite cardinal numbers which, assuming the axiom of choice, allows one to measure the size of any infinite set. Thus, if $A$ is an infinite set, there is a cardinal number $\kappa$ such that $\kappa = |A|$. Moreover, there is an ordering on these cardinal numbers such that for any cardinal numbers $\lambda$ and $\kappa$, we have $\lambda \leq \kappa$ or $\kappa \leq \lambda$. Moreover, if $\lambda \leq \kappa$ and $\kappa \leq \lambda$, then $\lambda = \kappa$. The cardinality of any countable infinite set is denoted by the cardinal $\aleph_0$.

Zorn's lemma is an important theorem about sets that is normally used to prove the existence of a mathematical object when it cannot be explicitly identified. The lemma involves the concept of a chain. A *chain* is a collection of sets $C$ such that for all sets $x$ and $y$ in $C$, we have either $x \subseteq y$ or $y \subseteq x$.

**Zorn's Lemma 1.1.36.** *Let $\mathcal{F}$ be a set of sets. Suppose that for every chain $C \subseteq \mathcal{F}$, $\bigcup C$ is in $\mathcal{F}$. Then there exists an $M \in \mathcal{F}$ that is maximal, that is, $M$ is not the proper subset of any $A \in \mathcal{F}$.*

Surprisingly, Zorn's lemma is equivalent to the axiom of choice. There will be times when we may apply the axiom of choice or Zorn's lemma. However, when working with countable sets, Corollary 1.1.33 will allow us to avoid such applications.

**Exercises 1.1.**

1. Let $A$ be a set and let $\mathcal{F}$ be a nonempty set of sets. Prove the following:
   (1) $A \setminus \bigcup \mathcal{F} = \bigcap\{A \setminus C : C \in \mathcal{F}\}$,
   (2) $A \setminus \bigcap \mathcal{F} = \bigcup\{A \setminus C : C \in \mathcal{F}\}$.
2. Consider the function $h: \mathbb{N} \to \mathbb{N}$ defined by the recursion
   (a) $h(0) = 0$,
   (b) $h(n+1) = 5h(n) + 1$, for all $n \in \mathbb{N}$.
   Prove by induction that $h(n) = \frac{5^n - 1}{4}$ for all $n \in \mathbb{N}$.
3. Let $A$ and $B$ be countable sets. Since $A$ is countable, there is a one-to-one function $f: A \to \mathbb{N}$. Also, as $B$ is countable, there is a one-to-one function $g: B \to \mathbb{N}$.
   (a) Prove that the function $h: A \to \mathbb{N}$ defined by $h(a) = 2^{f(a)+1}$ is one-to-one.
   (b) Prove that the function $k: B \to \mathbb{N}$ defined by $k(b) = 3^{g(b)+1}$ is one-to-one.
   (c) Define the function $\ell: A \cup B \to \mathbb{N}$ by
   $$\ell(x) = \begin{cases} h(x), & \text{if } x \in A, \\ k(x), & \text{if } x \in B \setminus A, \end{cases}$$
   for each $x \in A \cup B$. Prove that $\ell$ is one-to-one. Thus, $A \cup B$ is countable.
4. Let $\ell: \mathbb{N} \times \mathbb{N} \to \mathbb{N}$ be defined by $\ell(x, y) = 2^{x+1} \cdot 3^{y+1}$ and let $g: \mathbb{N} \to \mathbb{N}$ be defined by $g(x) = 2^{x+1}$. Let $C$ be the set generated from $B = \{1, 2\}$ by the set $\mathcal{F} = \{\ell, g\}$.
   (a) Find $C_1$ and then find an element in $C_2$.
   (b) Show that $\ell_C$ and $g_C$ are one-to-one (use Theorem 1.1.29).
   (c) Show that the set $C$ is freely generated by $B$.
   (d) Let $h: B \to \mathbb{R}$ be defined by $h(1) = \pi$ and $h(2) = e$. Let $F_\ell: \mathbb{R} \times \mathbb{R} \to \mathbb{R}$ and $F_g: \mathbb{R} \to \mathbb{R}$ be defined by $F_\ell(x, y) = x + y$ and $F_g(x) = 5x$. Let $\overline{h}: C \to \mathbb{R}$ be the function given by Theorem 1.1.27. Find $\overline{h}(36)$, $\overline{h}(4)$, and $\overline{h}(72)$.

*5. Let $U$ be a set and let $\mathcal{F}$ be a set of functions $\ell$ of the form $\ell: U^n \to U$, where $n$ is the arity of $\ell$. Let $B \subseteq U$ and let $C$ be generated from $B$ by the functions in $\mathcal{F}$.
   (a) Prove that $C$ is closed under the functions in $\mathcal{F}$.
   (b) Let $B \subseteq D$, where $D$ is closed under the functions in $\mathcal{F}$. Prove that $C \subseteq D$.

6. Let $U$ be a set and let $\mathcal{F}$ be a set of functions $\ell$ of the form $\ell: U^n \to U$, where $n$ is the arity of $\ell$. Let $B \subseteq U$ and let $C$ be freely generated from $B$ by the functions in $\mathcal{F}$. Let $f, g \in \mathcal{F}$, $x_1, x_2, \ldots, x_n \in C$, and $z_1, z_2, \ldots, z_k \in C$. Suppose that $f(x_1, x_2, \ldots, x_n) = g(z_1, z_2, \ldots, z_k)$. Explain why one can conclude that $f = g$, $n = k$, and $x_1 = z_1, x_2 = z_2, \ldots, x_n = z_k$.
7. In the proof of Theorem 1.1.25, show that $C_n \subseteq I$ for all $n \in \mathbb{N}$.

*8. Let $U$ be a set and let $\mathcal{F}$ be a set of functions $\ell$ of the form $\ell: U^n \to U$, where $n$ is the arity of $\ell$. Let $B \subseteq B' \subseteq U$. Let $C$ be generated from $B$ by the functions in $\mathcal{F}$ and let $C'$ be generated from $B'$ by the functions in $\mathcal{F}$. Prove that $C \subseteq C'$.

*9. Let $U$ be a set and let $\mathcal{F}$ be a set of functions $\ell$ of the form $\ell: U^n \to U$, where $n$ is the arity of $\ell$. For each $\ell \in \mathcal{F}$, let $F_\ell: V^n \to V$ be of the same arity as $\ell$. Let $B \subseteq B' \subseteq U$. Let $C$ be freely generated from $B$ by the functions in $\mathcal{F}$ and let $C'$ be freely generated from $B'$ by the functions in $\mathcal{F}$. Let $h: B \to V$ and $g: B' \to V$ be such that $h(b) = g(b)$ for all $b \in B$. Show that $\overline{h}(x) = \overline{g}(x)$ for all $x \in C$.

*10. Let $U$ be a set and let $\mathcal{F}$ be a set of functions $\ell$ of the form $\ell: U^n \to U$, where $n$ is the arity of $\ell$. For each $\ell \in \mathcal{F}$, let $F_\ell: V^n \to V$ be of the same arity as $\ell$. Let $B \subseteq U$, let $h: B \to V$, and let $C$ be freely generated from $B$ by the functions in $\mathcal{F}$ as in Theorem 1.1.24, that is, $C = \bigcup_{n\in\mathbb{N}} C_n$, where:
(1) $C_0 = B$,
(2) $C_{n+1} = C_n \cup \mathcal{F}[C_n]$ for all $n \in \mathbb{N}$.
Suppose that:
- the function $h$ is one-to-one;
- for each $\ell \in \mathcal{F}$, $F_\ell$ is one-to-one and the functions $h, F_\ell$ have disjoint ranges;
- $F_\ell$ and $F_{\ell'}$ have disjoint ranges whenever $\ell, \ell' \in \mathcal{F}$ are distinct.

Let $\overline{h}: C \to V$ be as in Theorem 1.1.27. Prove by induction on $n$ that $\overline{h}$ is one-to-one on $C_n$. Then conclude that $\overline{h}$ is one-to-one.

**Exercise Notes:** For Exercise 8, let $I = C \cap C'$. Using Theorem 1.1.25, prove that $I = C$. For Exercise 9, $C \subseteq C'$ by Exercise 8. Let $I = \{x \in C : \overline{h}(x) = \overline{g}(x)\}$. Using Theorem 1.1.25, prove that $I = C$. For Exercise 10, if $c \in C_{n+1}$, let $k$ be the least such that $c \in C_k$. If $k > 0$, then $c$ is in the range of a function in $\mathcal{F}$ restricted to $C_{k-1}$.

## 1.2 Basic logic

### 1.2.1 Propositions and logical connectives

A *proposition* is a declarative sentence that is either true or false, but not both. When discussing the logic of propositional statements in this section, we shall use symbols to represent these statements. Capital letters, for instance, $P$, $Q$, $R$, are used to symbolize propositional statements which may be called *propositional components*. Using the five logical connectives $\wedge, \vee, \neg, \to, \leftrightarrow$ together with the components, we can form new logical sentences called *compound sentences*. For example:

1. $P \wedge Q$ (means "$P$ and $Q$" and is called a *conjunction*),
2. $P \vee Q$ (means "$P$ or $Q$" and is called a *disjunction*),
3. $\neg P$ (means "not $P$" and is called a *negation*),
4. $P \to Q$ (means "if $P$, then $Q$" and is called a *conditional*),
5. $P \leftrightarrow Q$ (means "$P$ if and only if $Q$" and is called a *biconditional*).

Using the propositional components and the logical connectives, one can construct more complicated sentences, for example,

$$((P \wedge (\neg Q)) \vee (S \to (\neg R))).$$

We will more formally investigate propositional logic in Chapter 2.

### 1.2.2 Truth tables and truth functions

Given a collection of propositional components, say $P$, $Q$, and $R$, we can assign truth values to these components. For example, we can assign the truth values $T, F, T$ to $P, Q, R$, respectively, where $T$ means "true" while $F$ means "false." The truth value of a sentence of propositional logic can be evaluated from the truth values assigned to its components. We shall explain what this "means" by using truth tables.

The logical connectives $\wedge, \vee, \neg$ yield the natural truth values given by Table 1.1.

**Table 1.1:** Truth tables for conjunction, disjunction, and negation.

| (1) Conjunction | | | (2) Disjunction | | | (3) Negation | |
|---|---|---|---|---|---|---|---|
| $P$ | $Q$ | $P \wedge Q$ | $P$ | $Q$ | $P \vee Q$ | $P$ | $\neg P$ |
| $T$ | $T$ | $T$ | $T$ | $T$ | $T$ | $T$ | $F$ |
| $T$ | $F$ | $F$ | $T$ | $F$ | $T$ | $F$ | $T$ |
| $F$ | $T$ | $F$ | $F$ | $T$ | $T$ | | |
| $F$ | $F$ | $F$ | $F$ | $F$ | $F$ | | |

Table 1.1(1) has four rows (not including the header). The columns beneath $P$ and $Q$ list all the possible pairs of truth values that can be assigned to the components $P$ and $Q$. For each such pair, the corresponding truth value for $P \wedge Q$ appears to the right. For example, consider the third pair of truth values in this table, $FT$. Thus, if the propositional components $P$ and $Q$ are assigned the respective truth values $F$ and $T$, we see that the truth value of $P \wedge Q$ is $F$.

Table 1.1(2) shows that if $P$ and $Q$ are assigned the respective truth values $T$ and $F$, then the truth value of $P \vee Q$ is $T$. Moreover, when $P$ and $Q$ are assigned the truth values $T$ and $T$, the truth value of $P \vee Q$ is also $T$. In mathematics, the logical connective "or" has the same meaning as "and/or," that is, $P \vee Q$ is true if and only if only $P$ is true, only $Q$ is true, or both $P$ and $Q$ are true. Table 1.1(3) shows that the negation of a statement reverses the truth value of the statement.

A *truth function* accepts truth values as input and yields a unique truth value as output. Let $V = \{T, F\}$. The above three truth tables yield the corresponding truth functions $F_\wedge\colon V^2 \to V$, $F_\vee\colon V^2 \to V$, and $F_\neg\colon V \to V$, defined by

$$F_\wedge(x,y) = \begin{cases} T, & \text{if } x = T \text{ and } y = T, \\ F, & \text{otherwise,} \end{cases} \tag{1.2}$$

$$F_\vee(x,y) = \begin{cases} T, & \text{if } x = T \text{ or } y = T, \\ F, & \text{otherwise,} \end{cases} \tag{1.3}$$

$$F_\neg(x) = \begin{cases} T, & \text{if } x = F, \\ F, & \text{otherwise.} \end{cases} \tag{1.4}$$

The standard truth tables for the conditional and biconditional connectives are given in Table 1.2, which states that when $P$ and $Q$ are assigned the respective truth values $T$ and $F$, then the truth value of $P \to Q$ is $F$; otherwise, it is $T$. In particular, when $P$ is false, we shall say that $P \to Q$ is *vacuously* true. Table 1.2(5) shows that $P \leftrightarrow Q$ is true when $P$ and $Q$ are assigned the same truth value; when $P$ and $Q$ have different truth values, then the biconditional is false.

**Table 1.2:** Truth tables for the conditional and biconditional.

| **(4) Conditional** | | | **(5) Biconditional** | | |
|---|---|---|---|---|---|
| $P$ | $Q$ | $P \to Q$ | $P$ | $Q$ | $P \leftrightarrow Q$ |
| $T$ | $T$ | $T$ | $T$ | $T$ | $T$ |
| $T$ | $F$ | $F$ | $T$ | $F$ | $F$ |
| $F$ | $T$ | $T$ | $F$ | $T$ | $F$ |
| $F$ | $F$ | $T$ | $F$ | $F$ | $T$ |

Let $V = \{T, F\}$. The conditional and biconditional truth tables yield the two truth functions $F_{\to}\colon V^2 \to V$ and $F_{\leftrightarrow}\colon V^2 \to V$, defined by

$$F_{\to}(x, y) = \begin{cases} F, & \text{if } x = T \text{ and } y = F, \\ T, & \text{otherwise,} \end{cases} \tag{1.5}$$

$$F_{\leftrightarrow}(x, y) = \begin{cases} T, & \text{if } x = y, \\ F, & \text{otherwise.} \end{cases} \tag{1.6}$$

Using the truth tables for the sentences $P \wedge Q$, $P \vee Q$, $\neg P$, $P \to Q$, and $P \leftrightarrow Q$, one can build truth tables for more complicated compound sentences. Given a compound sentence, the "outside" connective is the "last connective that one needs to evaluate." After the outside connective has been determined, one can break up the sentence into its "parts." For example, in the compound sentence $\neg P \wedge (Q \vee P)$ we see that $\wedge$ is the outside connective with two parts $\neg P$ and $Q \vee P$.

**Problem 1.2.1.** Construct a truth table for the sentence $\neg P \to (Q \wedge P)$.

**Solution.** The two components $P$ and $Q$ will each need a column in our truth table. Since there are two components, there are four truth assignments for $P$ and $Q$. We will enter these combinations in the two leftmost columns in the same order as in Table 1.1(1). The outside connective of the propositional sentence $\neg P \to (Q \wedge P)$ is $\to$. We can break this sentence into the two parts $\neg P$ and $Q \wedge P$. So these parts will also need a column in our truth table. As we can break the sentences $\neg P$ and $Q \wedge P$ only into components (namely, $P$ and $Q$), we obtain the following truth table:

| | $P$ | $Q$ | $\neg P$ | $Q \wedge P$ | $\neg P \to (Q \wedge P)$ |
|---|---|---|---|---|---|
| | $T$ | $T$ | $F$ | $T$ | $T$ |
| | $T$ | $F$ | $F$ | $F$ | $T$ |
| | $F$ | $T$ | $T$ | $F$ | $F$ |
| | $F$ | $F$ | $T$ | $F$ | $F$ |
| STEP # | 1 | 1 | 2 | 3 | 4 |

We will now describe in steps how to obtain the truth values in the above table.

STEP 1: Specify all of the truth values that can be assigned to the components.

STEP 2: In each row, use the truth value assigned to the component $P$ to obtain the corresponding truth value for $\neg P$, using Table 1.1(3).

STEP 3: In each row, use the truth values assigned to $Q$ and $P$ to determine the corresponding truth value in the column under $Q \wedge P$ via Table 1.1(1).

STEP 4: In each row, use the truth values assigned to $\neg P$ and $Q \wedge P$ to evaluate the matching truth value for the final column under the sentence $\neg P \to (Q \wedge P)$, employing Table 1.2(4).

### 1.2.3 Predicates and quantifiers

Variables are used throughout mathematics and logic to represent unspecified values. They are used when one is interested in "properties" that may be true or false depending on the values represented by the variables. A *predicate* is just a statement that asserts that certain variables satisfy a property. For example, "$x$ is an irrational number" is a predicate. We can symbolize this predicate as $Ix$, where $I$ is called a *predicate symbol*. Of course, the truth or falsity of the expression $Ix$ can be evaluated only when a value for $x$ is given. For example, if $x$ is given the value $\sqrt{2}$, then $Ix$ would be true, whereas if $x$ is given the value 2, then $Ix$ would be false.

When our attention is focused on just the elements in a particular set, we will refer to that set as our *universe of discourse*. For example, if we were just talking about real numbers, then our universe of discourse would be the set of real numbers $\mathbb{R}$. Every statement made in a specific universe of discourse applies only to the elements in that universe.

Given a predicate $Px$ and variable $x$, we may want to assert that *every* element $x$ in the universe of discourse satisfies $Px$. We may also want to express the fact that *at least one* element $x$ in the universe makes $Px$ true. We can then form logical sentences using the quantifiers $\forall$ and $\exists$. The quantifier $\forall$ means "for all" and is called the *universal quantifier*. The quantifier $\exists$ means "there exists" and it is referred to as the *existential quantifier*. For example, we can form the sentences:

1. $\forall xPx$ (which means "for all $x$, $Px$"),
2. $\exists xPx$ (which means "there exists an $x$ such that $Px$").

A statement of the form $\forall x Px$ is called a *universal statement*. Any statement having the form $\exists x Px$ is called an *existential statement*. Quantifiers offer a valuable tool for clear thinking in mathematics, where many ideas begin with the expression "for every" or "there exists." Of course, the truth or falsity of a quantified statement depends on the particular universe of discourse.

Let $x$ be a variable that appears in a predicate $Px$. In the statements $\forall x Px$ and $\exists x Px$, we say that $x$ is a *bound variable* because $x$ is *bound* by a quantifier. In other words, when every occurrence of a variable in a statement is attached to a quantifier, then that variable is called a bound variable. If a variable appears in a statement and it is not bound by a quantifier, then the variable is said to be a *free variable*. Whenever a variable is free, substitution may take place, that is, one can replace a free variable with any particular value from the universe of discourse—perhaps 1 or 2. For example, the assertion $\forall x(Px \to x = y)$ has the one free variable $y$. Therefore, we can perform a substitution to obtain $\forall x(Px \to x = 1)$. In a given context, if all of the free variables in a statement are replaced with values from the universe of discourse, then one can determine the truth or falsity of the resulting statement.

**Problem 1.2.2.** Consider the predicates (properties) $Px$, $Ox$, and $Dxy$, where the variables $x$ and $y$ are intended to represent natural numbers:
1. $Px$ represents the statement "$x$ is a prime number,"
2. $Ox$ represents the statement "$x$ is an odd natural number,"
3. $Dxy$ represents the statement "$x$ evenly divides $y$."

Using the above predicates, determine whether or not the following logical formulas are true or false, where the universe of discourse is the set of natural numbers:
1. $\forall x(Px \to Ox)$,
2. $\forall y(y \geq 2 \to \exists x(Px \land Dxy))$.

**Solution.** The expression $\forall x(Px \to Ox)$ states that "all prime numbers are odd," which is clearly false. On the other hand, item 2 states that "every natural number greater than or equal to 2 is divisible by a prime number" and this is true.

We will formally investigate predicates and quantifiers in Chapter 3. The understanding of the logic of quantifiers is (one can argue) a requisite tool for the study of mathematics and logic. To help students learn the language of quantifier logic, Jon Barwise and John Etchemendy created an innovative software program called Tarski's World. This program presents a visual display of geometric shapes sitting on a grid, referred to as a world (or universe). The shapes have a variety of colors and positions on the grid. The user can create logic formulas and then determine whether or not these formulas are true or false in the world. Tarski's World is named after Alfred Tarski, an early pioneer in mathematical logic. We end this section with our own version of Tarski's World. In the following problem, we are given a Tarskian World and some

English statements. We will be given some predicates and be asked to translate these statements into logical form.

**Problem 1.2.3.** Consider the following Tarskian World, where some of the individuals are labeled with a name. The universe consists of all the objects in this Tarskian world.

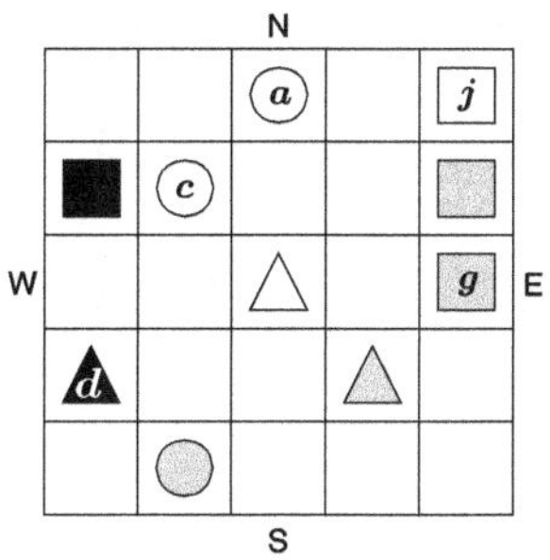


Define the following predicates (properties):
- $Tx$ means "$x$ is a triangle." $Cx$ means "$x$ is a circle." $Sx$ means "$x$ is a square."
- $Ix$ means "$x$ is white." $Gx$ means "$x$ is gray." $Bx$ means "$x$ is black."
- $Nxy$ means "$x$ is on the northern side of $y$."
- $Wxy$ means "$x$ is on the western side of $y$."
- $Kxy$ means "$x$ has the same color as $y$."

The constants $a, c, d, g, j$ are the names of five individuals in the above world. Using the given predicates, write each of the following statements in logical form, looking for possible hidden quantifiers and logical connectives.
1. There is a black square.
2. Every circle is white.
3. There are no black circles.
4. $a$ is north of $c$.
5. $a$ is not north of $j$.
6. Every circle is north of $d$.
7. Some circle has the same color as $d$.
8. $d$ is west of every circle.

**Solution.** Statements 2, 6, and 7 are the only ones that are false in the given Tarskian world. We express the above sentences in the following logical forms:
1. The sentence means that "for some $x$, $x$ is black and $x$ is a square." In logical form, we have $\exists x(Bx \wedge Sx)$.
2. The sentence means that "for all $x$, if $x$ is a circle, then $x$ is white." In logical form, $\forall x(Cx \to Ix)$.

3. The sentence can be stated in two equivalent ways. First, the sentence means that "it is not the case that some circle is black," that is, ¬(some circle is black). In logical form, we obtain $\neg\exists x(Cx \wedge Bx)$. Second, the sentence also means that "every circle is not black" and we get $\forall x(Cx \rightarrow \neg Bx)$.
4. In logical form, the sentence becomes $Nac$.
5. The logical form of this sentence is $\neg Naj$.
6. Rephrasing, we obtain "for all $x$, if $x$ is a circle, then $x$ is north of $d$." In logical form, we have $\forall x(Cx \rightarrow Nxd)$.
7. The sentence asserts that "for some $x$, $x$ is a circle and $x$ has the same color as $d$." In logical form, we have $\exists x(Cx \wedge Kxd)$.
8. Stated more clearly, we obtain "for all $x$, if $x$ is a circle, then $d$ is west of $x$." In logical form, we have $\forall x(Cx \rightarrow Wdx)$.

There will be cases when we will have to prove that there is exactly one value that satisfies a property. There is another quantifier that is sometimes used. It is called the *uniqueness quantifier*. This quantifier is written as $\exists!xPx$, which means that "there exists a unique $x$ satisfying $Px$," whereas $\exists xPx$ simply asserts that "at least one $x$ satisfies $Px$."

The quantifier $\exists!$ can be expressed in terms of the other quantifiers $\exists$ and $\forall$. In particular, the statement $\exists!xPx$ is equivalent to

$$\exists xPx \wedge \forall x\forall y((Px \wedge Py) \rightarrow x = y),$$

because the above statement means that "there is an $x$ such that $Px$ holds, and any individuals $x$ and $y$ that satisfy $Px$ and $Py$ must be the same individual."

#### Quantifier negation laws

We now introduce logic laws that involve the negation of quantifiers. Let $Px$ be any predicate. The statement $\forall xPx$ means that "for every $x$, $Px$ is true." Thus, the assertion $\neg\forall xPx$ means that "it is not the case that every $x$ makes $Px$ true." Therefore, $\neg\forall xPx$ means there is an $x$ that does not make $Px$ true, which can be expressed as $\exists x\neg Px$. This reasoning is reversible, as we will now show. The assertion $\exists x\neg Px$ means that "there is an $x$ that makes $Px$ false." Hence, $Px$ is not true for every $x$, that is, $\neg\forall xPx$. Therefore, $\neg\forall xPx$ and $\exists x\neg Px$ are logically equivalent. Similar reasoning shows that $\neg\exists xPx$ and $\forall x\neg Px$ are also equivalent. We now formally present these important logic laws that connect quantifiers with negation.

**Quantifier Negation Laws 1.2.4.** The following logical equivalences hold for any predicate $Px$:
1. $\neg\forall xPx \Leftrightarrow \exists x\neg Px$,
2. $\neg\exists xPx \Leftrightarrow \forall x\neg Px$,
3. $\neg\forall x\neg Px \Leftrightarrow \exists xPx$,
4. $\neg\exists x\neg Px \Leftrightarrow \forall xPx$.

We will be using the symbols $\Leftrightarrow$ and $\Rightarrow$ to abbreviate two English expressions. The symbol $\Leftrightarrow$ denotes the phrase "is equivalent to" and $\Rightarrow$ denotes the word "implies."

### Quantifier interchange laws

*Adjacent quantifiers* have four forms: $\exists x \exists y$, $\forall x \forall y$, $\forall x \exists y$, and $\exists x \forall y$. How should one interpret statements that contain adjacent quantifiers? If a statement contains adjacent quantifiers, one should address the quantifiers, one at a time, in the order in which they are presented. This is illustrated in our solutions of the following three problems.

**Problem 1.2.5.** Let the universe of discourse be a group of people and let $Lxy$ mean "$x$ likes $y$." What do the following formulas mean in English?
1. $\exists x \exists y Lxy$,
2. $\exists y \exists x Lxy$.

**Solution.** Note that "$x$ likes $y$" also means that "$y$ is liked by $x$." We will now translate each of these formulas from "left to right" as follows:
1. $\exists x \exists y Lxy$ means "there is a person $x$ such that $\exists y Lxy$," that is, "there is a person $x$ who likes some person $y$." Therefore, $\exists x \exists y Lxy$ means that "someone likes someone."
2. $\exists y \exists x Lxy$ states that "there is a person $y$ such that $\exists x Lxy$," that is, "there is a person $y$ who is liked by some person $x$." Thus, $\exists y \exists x Lxy$ means that "someone is liked by someone."

Hence, the statements $\exists x \exists y Lxy$ and $\exists y \exists x Lxy$ mean the same thing.

**Problem 1.2.6.** Let the universe of discourse be a group of people and let $Lxy$ mean "$x$ likes $y$." What do the following formulas mean in English?
1. $\forall x \forall y Lxy$,
2. $\forall y \forall x Lxy$.

**Solution.** We will work again from "left to right" as follows:
1. $\forall x \forall y Lxy$ means "for every person $x$, we have $\forall y Lxy$," that is, "for every person $x$, $x$ likes every person $y$"; hence, $\forall x \forall y Lxy$ means that "everyone likes everyone";
2. $\forall y \forall x Lxy$ means that "for each person $y$, we have $\forall x Lxy$," that is, "for each person $y$, $y$ is liked by every person $x$"; thus, $\forall y \forall x Lxy$ means "everyone is liked by everyone."

So the statements $\forall x \forall y Lxy$ and $\forall y \forall x Lxy$ mean the same thing.

Adjacent quantifiers of a different type are referred to as *mixed quantifiers*.

**Problem 1.2.7.** Let the universe of discourse be a group of people and let $Lxy$ mean "$x$ likes $y$." What do the following mixed quantifier formulas mean in English?
1. $\forall x \exists y Lxy$,
2. $\exists y \forall x Lxy$.

**Solution.** We will translate the formulas as follows:
1. $\forall x \exists y Lxy$ asserts that "for every person $x$, we have $\exists y Lxy$," that is, "for every person $x$, there is a person $y$ such that $x$ likes $y$." Thus, $\forall x \exists y Lxy$ means that "everyone likes someone."
2. $\exists y \forall x Lxy$ states that "there is a person $y$ such that $\forall x Lxy$," that is, "there is a person $y$ who is liked by every person $x$." In other words, $\exists y \forall x Lxy$ means "someone is liked by everyone."

We conclude that the mixed quantifier statements $\forall x \exists y Lxy$ and $\exists y \forall x Lxy$ are not logically equivalent, that is, they do *not* mean the same thing.

To clarify the conclusion that we obtained in our solution of Problem 1.2.7, consider the universe of discourse $U = \{a, b, c, d\}$, which consists of just four individuals with names as given. Figure 1.1 presents a world in which $\forall x \exists y Lxy$ is true, where we portray the property $Lxy$ using the "arrow notation" $x \xrightarrow{\text{likes}} y$. In this world, "everyone likes someone."

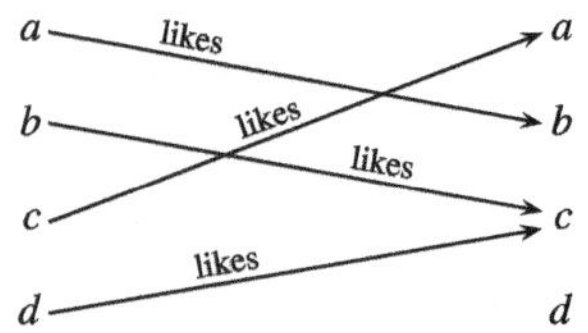


**Figure 1.1:** A world where $\forall x \exists y Lxy$ is true, because everyone likes someone.

Let us focus our attention on Figure 1.1. Clearly, the statement $\forall x \exists y Lxy$ is true in the world depicted in this figure. Moreover, note that $\exists y \forall x Lxy$ is actually false in this world. This is the case because there is no individual whom everyone likes. Thus, $\forall x \exists y Lxy$ is true and $\exists y \forall x Lxy$ is false in the world of Figure 1.1. We can now conclude that $\forall x \exists y Lxy$ and $\exists y \forall x Lxy$ do not mean the same thing.

Our solution to Problem 1.2.5 shows that assertions $\exists x \exists y Lxy$ and $\exists y \exists x Lxy$ both mean "someone likes someone." This supports the equivalence:

$$\exists x \exists y Lxy \Leftrightarrow \exists y \exists x Lxy.$$

Similarly, Problem 1.2.6 confirms the equivalence:

$$\forall x \forall y Lxy \Leftrightarrow \forall y \forall x Lxy.$$

Therefore, interchanging adjacent quantifiers of the same kind does not change the meaning. Problem 1.2.7, however, demonstrates that the two statements $\forall y \exists x Lxy$ and $\exists x \forall y Lxy$ are not equivalent. We conclude this discussion with a summary of the above observations:

- *Adjacent quantifiers of the same type are interchangeable.*
- *Adjacent quantifiers of a different type may not be interchangeable.*

**Quantifier Interchange Laws 1.2.8.** For every predicate $Pxy$, the following three statements are valid:

1. $\exists x \exists y Pxy \Leftrightarrow \exists y \exists x Pxy$,
2. $\forall x \forall y Pxy \Leftrightarrow \forall y \forall x Pxy$,
3. $\exists x \forall y Pxy \Rightarrow \forall y \exists x Pxy$.

It should be noted that the implication in item 3 cannot, in general, be reversed.

**Quantifier distribution laws**

A quantifier can sometimes "distribute" over a conjunction or a disjunction. The quantifier distribution laws, given below, express relationships that hold between a quantifier and the two logical connectives $\vee$ and $\wedge$. Namely, the existential quantifier distributes over disjunction (see 1.2.9(1)), and the universal quantifier distributes over conjunction (see 1.2.9(2)). The following four quantifier distribution laws can be useful when proving certain set identities.

**Quantifier Distribution Laws 1.2.9.** For any predicates $Px$ and $Qx$, we have the following distribution laws:

1. $\exists x Px \vee \exists x Qx \Leftrightarrow \exists x(Px \vee Qx)$,
2. $\forall x Px \wedge \forall x Qx \Leftrightarrow \forall x(Px \wedge Qx)$.

If $R$ is a predicate that does not involve the variable $x$, then we have:

3. $R \wedge \exists x Qx \Leftrightarrow \exists x(R \wedge Qx)$,
4. $R \vee \forall x Qx \Leftrightarrow \forall x(R \vee Qx)$.

**Exercises 1.2.**

1. Let $C(x)$ represent the predicate "$x$ is in the class" and let $Mx$ represent "$x$ is a mathematics major." Let the universe of discourse be the set of all students. Analyze the logical form of the following sentences.
   (a) Everyone in the class is a mathematics major.
   (b) Someone in the class is a mathematics major.
   (c) No one in the class is a mathematics major.
2. Determine whether the following statements are true or false in the universe $\mathbb{R}$:
   (a) $\forall x(x^2 + 1 > 0)$,
   (b) $\forall x(x^2 + x \geq 0)$,
   (c) $\forall x(x > \frac{1}{2} \rightarrow \frac{1}{x} < 3)$,
   (d) $\exists x(\frac{1}{x-1} = 3)$,
   (e) $\exists x(\frac{1}{x-1} = 0)$.
3. Given the properties and the Tarskian world in Problem 1.2.3 on page 21, determine the truth or falsity of the following statements:

(a) $\forall x(Ix \rightarrow (Tx \vee Sx))$,
(b) $\forall x(Bx \rightarrow (Tx \vee Sx))$,
(c) $\exists y(Cy \wedge Nyd)$,
(d) $\exists y(Cy \wedge Ndy)$,
(e) $\exists y(Cy \wedge (Nyd \wedge Ndy))$.

4. Using the Tarskian predicates given in Problem 1.2.3, translate the following English sentences into logical sentences where $b$ and $d$ are names of individuals in some Tarskian world:
   (a) Something is white.
   (b) Some circle is white.
   (c) All squares are black.
   (d) No squares are black.
   (e) All triangles are west of $d$.
   (f) A triangle is west of $d$.
   (g) Some triangle is north of $d$.
   (h) Some triangle is not gray.
   (i) Every triangle is north of $b$.
   (j) No square has the same color as $b$.

# 2 Propositional logic

In this chapter we will utilize mathematical tools, namely induction and recursion, to formally investigate an important form of logic called propositional logic. In particular, we will establish a number of significant theorems concerning the syntax and semantics of propositional logic. An informal introduction to this topic is given in Section 1.2.1.

## 2.1 The language

The syntax of the language of propositional logic is specified by first identifying the symbols (the alphabet) of the language and then defining expressions to be any finite string of these symbols. Some expressions are meaningful while others are not. To single out the meaningful expressions requires a recursive definition (see Section 1.1.5). The meaningful expressions in propositional logic will be called *well-formed formulas* (wffs).

**Definition 2.1.1.** Our official language for propositional logic, denoted by $\mathcal{L}$, consists of the distinct symbols in the set

$$\mathcal{L} = \{ (, ), \neg, \wedge, \vee, \rightarrow, \leftrightarrow, \mathbf{A}_1, \mathbf{A}_2, \ldots, \mathbf{A}_n, \ldots \},$$

where the symbols in the language $\mathcal{L}$ are described in the following table:

| Symbol | Name | Remark |
|---|---|---|
| ( | left parenthesis | punctuation |
| ) | right parenthesis | punctuation |
| $\neg$ | negation | English: not |
| $\wedge$ | conjunction | English: and |
| $\vee$ | disjunction | English: or |
| $\rightarrow$ | conditional | English: if _, then _ |
| $\leftrightarrow$ | biconditional | English: if and only if |
| $\mathbf{A}_1$ | first sentence symbol | |
| $\mathbf{A}_2$ | second sentence symbol | |
| $\vdots$ | $\vdots$ | |
| $\mathbf{A}_n$ | $n$-th sentence symbol | |
| $\vdots$ | $\vdots$ | |

**Remark.** Several remarks are in order:

1. The distinct symbols $\neg, \wedge, \vee, \rightarrow, \leftrightarrow$ are called *logical connectives.*
2. The infinite number of distinct symbols $\mathbf{A}_1, \mathbf{A}_2, \ldots, \mathbf{A}_n, \ldots$ are called *sentence symbols*. Each $\mathbf{A}_i$ stands for a particular sentence.

   Recall that $\langle a_1, a_2, \ldots, a_n \rangle$ denotes a sequence (see Definition 1.1.6).

https://doi.org/10.1515/9783110782073-002

**Definition 2.1.2.** An $\mathcal{L}$-*expression* is a finite string of symbols from $\mathcal{L}$ obtained by omitting the sequence brackets $\langle\,\rangle$ and the commas from a finite sequence of symbols from $\mathcal{L}$. If $\alpha$ and $\beta$ are $\mathcal{L}$-expressions, then $\alpha\beta$ is the $\mathcal{L}$-expression consisting of the symbols in the expression $\alpha$ followed by the symbols in the expression $\beta$.

**Example 2.1.3.** Clearly, $\langle (, \neg, \mathbf{A}_1, ) \rangle$ is a finite sequence of symbols from $\mathcal{L}$, and after removing the sequence brackets and commas we obtain the $\mathcal{L}$-expression $(\neg\mathbf{A}_1)$. Let $\alpha = (\neg\mathbf{A}_1)$ and $\beta = \mathbf{A}_1$. Then $\alpha$ and $\beta$ are $\mathcal{L}$-expressions. In addition, we can write $(\alpha \to \beta)$ to denote the $\mathcal{L}$-expression $((\neg\mathbf{A}_1) \to \mathbf{A}_1)$. Also, note that $\neg \to)(\mathbf{A}_8$ is an $\mathcal{L}$-expression. Moreover, $\alpha\beta$ is the expression $(\neg\mathbf{A}_1)\mathbf{A}_1$.

There is a one-to-one correspondence between each $\mathcal{L}$-expression $\alpha$ and a finite sequence of symbols in $\mathcal{L}$ denoted by $\langle\alpha^*\rangle$, where $\alpha^*$ denotes the result of putting commas between all of the symbols in $\alpha$. So an $\mathcal{L}$-expression $\alpha$ is a *proper initial segment* of the $\mathcal{L}$-expression $\beta$ when $\langle\alpha^*\rangle$ is a proper initial segment of $\langle\beta^*\rangle$ (see Definition 1.1.6).

### Grammatically correct expressions

Is the English expression "then work not. is, Sally work is Bill if at at" a grammatically correct sentence? No! English expressions that are not grammatically correct make no sense. This observation motivates the following question: What does it mean for an $\mathcal{L}$-expression to be grammatically correct? Surely, the $\mathcal{L}$-expression $(\mathbf{A}_2 \to \mathbf{A}_5)$ is a meaningful expression. However, the expression $\to \mathbf{A}_2)(\mathbf{A}_5$ appears to be nonsensical. We know that English has "correct grammar." Can we give the language $\mathcal{L}$ correct grammar? The answer is yes.

We want to define the notion of a *well-formed formula* (*wff*) in the language $\mathcal{L}$. The wffs will be the grammatically correct $\mathcal{L}$-expressions. Our definition will have the following consequences:

(a) Every sentence symbol is a wff.
(b) If $\alpha$ and $\beta$ are wffs, then so are $(\neg\alpha)$, $(\alpha \wedge \beta)$, $(\alpha \vee \beta)$, $(\alpha \to \beta)$, and $(\alpha \leftrightarrow \beta)$.
(c) Every wff is obtained by applying (a) or (b).

We will often use lower-case Greek characters to represent wffs and upper-case Greek characters to represent sets of wffs.

Items (a)–(c) above declare that every sentence symbol is a wff and every other wff is built from other wffs by using the logical connectives and parentheses in particular ways. For example, $\mathbf{A}_{1123}$, $(\mathbf{A}_2 \to (\neg\mathbf{A}_1))$, and $(((\neg\mathbf{A}_1) \wedge (\mathbf{A}_1 \to \mathbf{A}_7)) \to \mathbf{A}_7)$ are all wffs, but $X_3$, $(\mathbf{A}_5)$, $()\neg\mathbf{A}_{41}$, $\mathbf{A}_5 \to \mathbf{A}_7)$, and $(\mathbf{A}_2 \vee (\neg\mathbf{A}_1)$ are not wffs.

In order to define wffs we will need the following five *formula building functions*: Let $\alpha$ and $\beta$ be $\mathcal{L}$-expressions. Then

$$\begin{aligned}
\mathcal{E}_\neg(\alpha) &= (\neg\alpha),\\
\mathcal{E}_\wedge(\alpha,\beta) &= (\alpha \wedge \beta),\\
\mathcal{E}_\vee(\alpha,\beta) &= (\alpha \vee \beta),
\end{aligned} \tag{2.1}$$

$$\mathcal{E}_{\rightarrow}(\alpha,\beta) = (\alpha \rightarrow \beta),$$
$$\mathcal{E}_{\leftrightarrow}(\alpha,\beta) = (\alpha \leftrightarrow \beta).$$

Using the above operations, our next definition is an application of Theorem 1.1.24.

**Definition 2.1.4.** Let $\mathcal{F} = \{\mathcal{E}_{\neg}, \mathcal{E}_{\wedge}, \mathcal{E}_{\vee}, \mathcal{E}_{\rightarrow}, \mathcal{E}_{\leftrightarrow}\}$ and let $\mathcal{S} \subseteq \{\mathbf{A}_1, \mathbf{A}_2, \mathbf{A}_3, \dots\}$ be a set of sentence symbols in the language $\mathcal{L}$. Let $\overline{\mathcal{S}}$ be the set generated from $\mathcal{S}$ by the functions in $\mathcal{F}$. An $\mathcal{L}$-expression $\alpha$ is a *well-formed formula (wff or formula) with sentence symbols in* $\mathcal{S}$ if and only if $\alpha \in \overline{\mathcal{S}}$. The set $\overline{\mathcal{S}}$ is the set of wffs that can be built from the symbols in S using the five formula building functions in (2.1). When $\mathcal{S}$ is the set of all sentence symbols, we shall just say that $\alpha$ is a *well-formed formula (wff or formula).*

For example, if $S = \{\mathbf{A}_3, \mathbf{A}_7\}$, then $\overline{\mathcal{S}}$ consists of the wffs whose only sentence symbols are $\mathbf{A}_3$ and/or $\mathbf{A}_7$.

**Remark.** Let $\mathcal{S}$ and $\mathcal{T}$ be sets of sentence symbols. Exercise 8 on page 37 implies that if $\mathcal{S} \subseteq \mathcal{T}$, then $\overline{\mathcal{S}} \subseteq \overline{\mathcal{T}}$.

Definition 2.1.4 asserts that a wff is an expression that can be built from sentence symbols by applying the *formula building functions* a finite number of times (see Section 1.1.5). Thus, a wff is constructed by using the sentence symbols as building blocks and the formula building functions as mortar. This leads to the following definition, which gives a road map for the construction of a particular wff.

**Definition 2.1.5.** A *construction sequence* is a finite sequence $\langle \varepsilon_1, \varepsilon_2, \dots, \varepsilon_n \rangle$ of $\mathcal{L}$-expressions such that for each $i \leq n$ one of the following holds:

1. $\varepsilon_i$ is a sentence symbol,
2. $\varepsilon_i = \mathcal{E}_{\neg}(\varepsilon_j)$ for some $j < i$,
3. $\varepsilon_i = \mathcal{E}_{\boxdot}(\varepsilon_j, \varepsilon_k)$ for some $j < i$ and $k < i$, where $\boxdot$ represents any of the connectives $\wedge, \vee, \rightarrow, \leftrightarrow$.

An $\mathcal{L}$-expression $\alpha$ is a wff whenever there is a construction sequence $\langle \varepsilon_1, \varepsilon_2, \dots, \varepsilon_n \rangle$ such that $\alpha = \varepsilon_n$, that is, the sequence ends in $\alpha$.

**Problem 2.1.6.** Let $\alpha$ denote the wff $((\mathbf{A}_1 \wedge \mathbf{A}_{10}) \rightarrow ((\neg \mathbf{A}_3) \vee \mathbf{A}_8))$. Identify a construction sequence that ends with $\alpha$.

**Solution.** The following "tree" gives a picture of how the above wff $\alpha$ is constructed from its sentence symbols using the formula building functions:

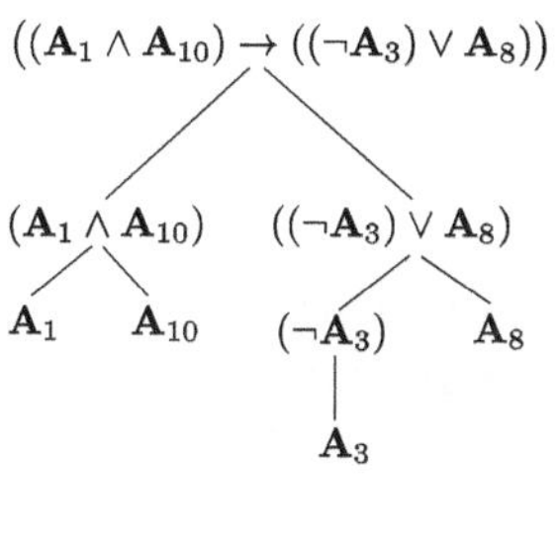

Thus we have the following construction sequence that ends in $\alpha$:

$$\langle \mathbf{A}_1, \mathbf{A}_{10}, (\mathbf{A}_1 \wedge \mathbf{A}_{10}), \mathbf{A}_3, (\neg\mathbf{A}_3), \mathbf{A}_8, ((\neg\mathbf{A}_3) \vee \mathbf{A}_8), \alpha\rangle.$$

### Induction on wffs principle

To construct a wff, one starts with the sentence symbols and then applies the formula building functions. Consequently, Theorem 1.1.25 yields the following *induction on wffs principle* and associated proof strategy.

**Wff Induction Principle.** Let $\mathbb{S}(\alpha)$ be a statement about a wff $\alpha$. If

1. $\mathbb{S}(\mathbf{A}_i)$ is true for every sentence symbol $\mathbf{A}_i$ and
2. for all wffs $\alpha$ and $\beta$, if $\mathbb{S}(\alpha)$ and $\mathbb{S}(\beta)$, then $\mathbb{S}((\neg\alpha))$ and $\mathbb{S}((\alpha \boxdot \beta))$ (here $\boxdot$ represents each of the connectives $\wedge, \vee, \rightarrow, \leftrightarrow$),

then $\mathbb{S}(\alpha)$ is true for *all* wffs $\alpha$.

**Proof Strategy.** In order to prove a statement "for all wffs $\alpha$, $\mathbb{S}(\alpha)$" by induction, use the following diagram:

*Base step:* Prove $\mathbb{S}(\mathbf{A}_i)$ for all sentence symbols $\mathbf{A}_i$.
*Inductive step:* Let $\alpha$ and $\beta$ be arbitrary wffs.
Assume $\mathbb{S}(\alpha)$ and $\mathbb{S}(\beta)$.
Prove $\mathbb{S}((\neg\alpha))$ and $\mathbb{S}((\alpha \boxdot \beta))$.

### Applications of the induction principle

**Theorem 2.1.7.** *Let $\alpha$ be any wff. The number of left parentheses equals the number of right parentheses in $\alpha$.*

*Proof.* We shall use proof by *induction on wffs*.

*Base step:* Let $\alpha = \mathbf{A}_i$ be a sentence symbol. Then $\mathbf{A}_i$ has zero left parentheses and the same number of right parentheses.

*Inductive step:* Let $\alpha$ and $\beta$ be arbitrary wffs. Assume the induction hypothesis

the number of left parentheses in $\alpha$ equals the number of right parentheses,
the number of left parentheses in $\beta$ equals the number of right parentheses. (IH)

We must prove that the same holds for each of the following:

$$(\neg\alpha), (\alpha \wedge \beta), (\alpha \vee \beta), (\alpha \rightarrow \beta), (\alpha \leftrightarrow \beta).$$

From the induction hypothesis (IH), it immediately follows that the number of left parentheses in $(\neg\alpha)$ equals the number of right parentheses. A similar argument applies for the rest of the binary connectives. □

**Theorem 2.1.8.** *A proper initial segment of a wff has more left parentheses than right parentheses. Thus, no proper initial segment of a wff is itself a wff.*

*Proof.* We shall use proof by *induction on wffs*.

*Base step:* Let $\alpha = \mathbf{A}_i$ be a sentence symbol. Since $\mathbf{A}_i$ has no proper initial segments, the result follows vacuously.

*Inductive step:* Let $\alpha$ and $\beta$ be arbitrary wffs. Assume the induction hypothesis

$$\begin{array}{l}\text{a proper initial segment of } \alpha \text{ has more left parentheses than right parentheses,}\\ \text{a proper initial segment of } \beta \text{ has more left parentheses than right parentheses.}\end{array} \tag{IH}$$

We must prove that any proper initial segment of each of the following has more left parentheses than right parentheses:

$$(\neg\alpha), (\alpha \wedge \beta), (\alpha \vee \beta), (\alpha \to \beta), (\alpha \leftrightarrow \beta).$$

Let us first consider the wff $(\alpha \vee \beta)$. Any proper initial segment of $(\alpha \vee \beta)$ has one of the following forms:

1. (,
2. $(\alpha_0$, where $\alpha_0$ is a proper initial segment of $\alpha$,
3. $(\alpha$,
4. $(\alpha\vee$,
5. $(\alpha \vee \beta_0$, where $\beta_0$ is a proper initial segment of $\beta$,
6. $(\alpha \vee \beta$.

Case 1 is clear. In cases 3, 4, and 6, Theorem 2.1.7 implies that these proper initial segments have more left parentheses than right ones. For cases 2 and 5, the induction hypothesis implies that these two proper initial segments have more left parentheses than right parentheses. A similar argument applies for the other logical connectives. Theorem 2.1.7 now implies that a proper initial segment of a wff is not a wff. □

**Exercises 2.1.**

1. Let **B** and **S** be sentence symbols that represent the sentences "Bill is at work" and "Sally is at work," respectively. Translate the following English sentences into wffs:
   (a) Bill is at work and Sally is not at work.
   (b) If Bill is at work, then Sally is not at work.
   (c) Bill is not at work if and only if Sally is at work.
2. Let $\alpha$ be the wff $((\neg\mathbf{A}_1) \to ((\mathbf{A}_4 \vee (\mathbf{A}_3 \leftrightarrow \mathbf{A}_2)) \wedge \mathbf{A}_3))$. Identify a construction sequence that ends with $\alpha$.
3. Given any wff $\alpha$, let

$$s(\alpha) = \text{the number of occurrences of sentence symbols in } \alpha,$$

$$c(\alpha) = \text{the number of occurrences of binary connectives in } \alpha.$$

Prove by induction on $\alpha$ the following statement: For all wffs $\alpha$, $s(\alpha) = c(\alpha) + 1$.
4. In the inductive step in the proof of Theorem 2.1.8, using the induction hypothesis, show that any proper initial segment of $(\neg\alpha)$ is not a wff.
5. Prove by induction on $\alpha$ the following statement: For every wff $\alpha$ there is a construction sequence that ends with $\alpha$ (see Definition 1.1.6).
6. Let $\alpha_1, \alpha_2, \ldots, \alpha_n$ be wffs, where $n > 1$. Show that $\alpha_1\alpha_2\cdots\alpha_n$ is not a wff.

## 2.2 Truth assignments

The set of truth values $\{F, T\}$ consists of two distinct items:

$F$, called *falsity*,
$T$, called *truth*.

Clearly, there are an infinite number of sentence symbols. Suppose that we have truth values assigned to each of these sentence symbols. Can we then identify a function that will evaluate the truth value of *all* the wffs? If so, is there only one such function? In order to address these questions, we first need to establish a unique readability theorem, that is, we need to show that one can read a wff without ambiguity.

**Theorem 2.2.1** (Unique readability). *Let* $\mathcal{F} = \{\mathcal{E}_\neg, \mathcal{E}_\wedge, \mathcal{E}_\vee, \mathcal{E}_\rightarrow, \mathcal{E}_\leftrightarrow\}$ *and let* $B = \{\mathbf{A}_1, \mathbf{A}_2, \mathbf{A}_3, \ldots\}$. *When restricted to the set of wffs, we have the following:*
(a) *The range of each operation in* $\mathcal{F}$ *is disjoint from* $B$.
(b) *Any two distinct operations in* $\mathcal{F}$ *have disjoint ranges.*
(c) *Every operation in* $\mathcal{F}$ *is one-to-one.*

*In other words, the set of all wffs is freely generated from the set of sentence symbols by the five operations* $\mathcal{E}_\wedge, \mathcal{E}_\vee, \mathcal{E}_\neg, \mathcal{E}_\rightarrow, \mathcal{E}_\leftrightarrow$.

*Proof.* Let $\mathcal{F} = \{\mathcal{E}_\wedge, \mathcal{E}_\vee, \mathcal{E}_\neg, \mathcal{E}_\rightarrow, \mathcal{E}_\leftrightarrow\}$ and let $B = \{\mathbf{A}_1, \mathbf{A}_2, \mathbf{A}_3, \ldots\}$.
(a) We show that the range of $\mathcal{E}_\wedge$ is disjoint from $B$. Let $\alpha$ and $\beta$ be wffs. Clearly, $(\alpha \wedge \beta) \neq \mathbf{A}_i$, because $\mathbf{A}_i$ has no parentheses. Hence, the range of $\mathcal{E}_\wedge$ is disjoint from $B$. A similar argument applies for the other operations in $\mathcal{F}$.
(b) We need to show that any two distinct operations in $\mathcal{F}$ have disjoint ranges. Let $\alpha$, $\beta$, $\gamma$, and $\sigma$ be wffs. Suppose that $(\alpha \wedge \beta) = (\gamma \vee \sigma)$. Then, by dropping the first parenthesis in each expression, the resulting expressions are equal, that is,

$$\alpha \wedge \beta) = \gamma \vee \sigma). \tag{2.2}$$

Since $\alpha$ and $\gamma$ are wffs, Theorem 2.1.8 implies that $\alpha$ and $\gamma$ cannot be proper initial segments of the other. Thus, (2.2) implies that $\wedge\beta) = \vee\sigma)$. So $\wedge = \vee$, which is a contradiction. So $\mathcal{E}_\wedge$ and $\mathcal{E}_\vee$ have disjoint ranges. Similar reasoning applies in the other cases.

(c) Finally, we must show that each operation in $\mathcal{F}$ is one-to-one. So let $\alpha$, $\beta$, $\gamma$, and $\sigma$ be wffs. Suppose that $(\alpha \wedge \beta) = (\gamma \wedge \sigma)$. Reasoning as in part (b), we see that $\alpha = \gamma$ and thus, $\beta = \sigma$. Thus, $\mathcal{E}_\wedge$ is one-to-one on the wffs. Similar reasoning also shows that the other operations are one-to-one. □

The unique readability theorem will allow us to positively address the questions posed at the beginning of this section (see Theorem 2.2.4 below).

**Definition 2.2.2.** Let $\mathcal{S}$ be a set of sentence symbols. A function $v\colon \mathcal{S} \to \{F, T\}$ is called a *truth assignment* for $\mathcal{S}$.

**Example 2.2.3.** Consider the set $\mathcal{S} = \{\mathbf{A}_3, \mathbf{A}_6, \mathbf{A}_8\}$ of sentence symbols. Then the function $v\colon \mathcal{S} \to \{F, T\}$ defined by

$$v(\mathbf{A}_3) = T, \quad v(\mathbf{A}_6) = F, \quad v(\mathbf{A}_8) = F$$

is a truth assignment for $\mathcal{S}$.

Let $\mathcal{S}$ be a set of sentence symbols and let $v\colon \mathcal{S} \to \{F, T\}$ be a truth assignment for $\mathcal{S}$. Let $\mathcal{F} = \{\mathcal{E}_\wedge, \mathcal{E}_\vee, \mathcal{E}_\neg, \mathcal{E}_\to, \mathcal{E}_\leftrightarrow\}$. Theorem 2.2.1 implies that the set $\overline{\mathcal{S}}$ of all wffs generated by $\mathcal{S}$ from the functions in $\mathcal{F}$ is freely generated. For each function in $\mathcal{F}$ we also have the corresponding truth functions $F_\wedge$, $F_\vee$, $F_\neg$, $F_\to$, $F_\leftrightarrow$ introduced in Section 1.2.2 (see (1.2)–(1.6)). Theorem 1.1.27 now implies the following result.

**Theorem 2.2.4.** *Let $\mathcal{S}$ be a set of sentence symbols and let $v\colon \mathcal{S} \to \{F, T\}$ be a truth assignment for $\mathcal{S}$. Then there is a unique function $\overline{v}\colon \overline{\mathcal{S}} \to \{F, T\}$ satisfying the following:*

(i) *For every $\mathbf{A} \in \mathcal{S}$, $\overline{v}(\mathbf{A}) = v(\mathbf{A})$.*

(ii) *For every $\alpha, \beta \in \overline{\mathcal{S}}$, we have:*

(1) $\overline{v}((\neg\alpha)) = \begin{cases} T, & \text{if } \overline{v}(\alpha) = F, \\ F, & \text{if } \overline{v}(\alpha) = T, \end{cases}$

(2) $\overline{v}((\alpha \wedge \beta)) = \begin{cases} T, & \text{if } \overline{v}(\alpha) = T \text{ and } \overline{v}(\beta) = T, \\ F, & \text{otherwise}, \end{cases}$

(3) $\overline{v}((\alpha \vee \beta)) = \begin{cases} T, & \text{if } \overline{v}(\alpha) = T \text{ or } \overline{v}(\beta) = T \text{ (or both)}, \\ F, & \text{otherwise}, \end{cases}$

(4) $\overline{v}((\alpha \to \beta)) = \begin{cases} F, & \text{if } \overline{v}(\alpha) = T \text{ and } \overline{v}(\beta) = F, \\ T, & \text{otherwise}, \end{cases}$

(5) $\overline{v}((\alpha \leftrightarrow \beta)) = \begin{cases} T, & \text{if } \overline{v}(\alpha) = \overline{v}(\beta), \\ F, & \text{otherwise}. \end{cases}$

Conditions (1)–(5) of the above theorem are given in tabular form below:

| $\alpha$ | $\beta$ | $(\neg\alpha)$ | $(\alpha\wedge\beta)$ | $(\alpha\vee\beta)$ | $(\alpha\rightarrow\beta)$ | $(\alpha\leftrightarrow\beta)$ |
|---|---|---|---|---|---|---|
| $T$ | $T$ | $F$ | $T$ | $T$ | $T$ | $T$ |
| $T$ | $F$ | $F$ | $F$ | $T$ | $F$ | $F$ |
| $F$ | $T$ | $T$ | $F$ | $T$ | $T$ | $F$ |
| $F$ | $F$ | $T$ | $F$ | $F$ | $T$ | $T$ |

**Remark 2.2.5** (On Theorem 2.2.4). Let $\mathcal{S}$ be a set of sentence symbols and $v: \mathcal{S} \to \{F, T\}$ be a truth assignment for $\mathcal{S}$. Let $\overline{v}: \overline{\mathcal{S}} \to \{F, T\}$ be as in Theorem 2.2.4. For all $\alpha, \beta \in \overline{\mathcal{S}}$ (iff means "if and only if"),

(1) $\overline{v}((\neg\alpha)) = T$ iff $\overline{v}(\alpha) = F$,
(2) $\overline{v}((\alpha\wedge\beta)) = T$ iff $\overline{v}(\alpha) = T$ and $\overline{v}(\beta) = T$,
(3) $\overline{v}((\alpha\vee\beta)) = T$ iff $\overline{v}(\alpha) = T$ or $\overline{v}(\beta) = T$,
(4) $\overline{v}((\alpha\rightarrow\beta)) = T$ iff if $\overline{v}(\alpha) = T$, then $\overline{v}(\beta) = T$,
(5) $\overline{v}((\alpha\leftrightarrow\beta)) = T$ iff $\overline{v}(\alpha) = \overline{v}(\beta)$.

**Example 2.2.6.** Consider the set $\mathcal{S} = \{\mathbf{A}_3, \mathbf{A}_6, \mathbf{A}_8\}$ of three sentence symbols and let $v: \mathcal{S} \to \{F, T\}$ be defined by

$$v(\mathbf{A}_3) = T, \quad v(\mathbf{A}_6) = F, \quad v(\mathbf{A}_8) = F.$$

Let $\alpha$, $\beta$, and $\gamma$ be defined by

$$\alpha = (\mathbf{A}_3 \to (\neg\mathbf{A}_6)),$$
$$\beta = (\mathbf{A}_3 \to (\mathbf{A}_6 \vee \mathbf{A}_8)),$$
$$\gamma = ((\mathbf{A}_3 \to (\neg\mathbf{A}_6)) \leftrightarrow (\mathbf{A}_3 \to (\mathbf{A}_6 \vee \mathbf{A}_8))).$$

So $\alpha, \beta, \gamma \in \overline{\mathcal{S}}$. What is $\overline{v}(\alpha), \overline{v}(\beta), \overline{v}(\gamma)$?

**Definition 2.2.7.** Let $\varphi$ be a wff. Let $v$ be a truth assignment defined on the sentence symbols appearing in $\varphi$. Then $v$ *satisfies* $\varphi$ if and only if $\overline{v}(\varphi) = T$.

**Example 2.2.8.** Let $\mathcal{S} = \{\mathbf{A}_3, \mathbf{A}_6, \mathbf{A}_8\}$, let $v: \mathcal{S} \to \{F, T\}$, and let $\alpha, \beta, \gamma$ all be as in the above Example 2.2.6. Does $v$ satisfy $\alpha$? Answer the same question for $\beta$ and $\gamma$.

**Definition 2.2.9.** Let $\Sigma$ be a set of wffs and let $v$ be a truth assignment defined on the sentence symbols appearing in any wff from $\Sigma$. Then $v$ *satisfies* $\Sigma$ if and only if $\overline{v}(\varphi) = T$ for every $\varphi \in \Sigma$. We shall say that $\Sigma$ is *satisfiable* if there is a truth assignment that satisfies $\Sigma$.

Given that $\Sigma$ is a set of wffs and $\tau$ is a wff, our next definition will address the following question: When can we view $\tau$ as a conclusion that follows from $\Sigma$, where $\Sigma$ is regarded as a set of hypotheses?

**Definition 2.2.10.** Let $\Sigma$ be a set of wffs and let $\tau$ be a wff. Then $\Sigma$ *tautologically implies* $\tau$ (written $\Sigma \models \tau$) if and only if for every truth assignment $v$ defined on the sentence symbols that appear in wffs in $\Sigma$ and in $\tau$, if $v$ satisfies $\Sigma$, then $v$ satisfies $\tau$.

Exercise 9 on page 15 implies the following theorem.

**Theorem 2.2.11.** *Let $v_1 : \mathcal{S} \to \{F, T\}$ and $v_2 : \mathcal{T} \to \{F, T\}$ be two truth assignments, where $\mathcal{S}$ and $\mathcal{T}$ are sets of sentence symbols such that $\mathcal{S} \subseteq \mathcal{T}$. If $v_1(\mathbf{A}_i) = v_2(\mathbf{A}_i)$ for all $\mathbf{A}_i \in \mathcal{S}$, then $\overline{v}_1(\alpha) = \overline{v}_2(\alpha)$ for all $\alpha \in \overline{\mathcal{S}}$.*

**Corollary 2.2.12.** *Let $\Sigma$ be a set of wffs and let $\tau$ be a wff. Let*

$$\mathcal{S} = \{\mathbf{A}_i : \mathbf{A}_i \textit{ appears in a wff in } \Sigma \textit{ or in } \tau\}$$

*and let $\mathcal{T}$ be a set of sentence symbols such that $\mathcal{S} \subseteq \mathcal{T}$. Then $\Sigma \models \tau$ if and only if every truth assignment for $\mathcal{T}$ that satisfies every member in $\Sigma$ also satisfies $\tau$.*

Definition 2.2.10 can be used to identify valid forms of reasoning, or inference rules. For example, let $\Sigma = \{(\alpha \to \beta), \alpha\}$ and let $v$ be a truth assignment that satisfies $\Sigma$. Thus, $\overline{v}((\alpha \to \beta)) = T$ and $\overline{v}(\alpha) = T$. So, by Remark 2.2.5(4), we conclude that $\overline{v}(\beta) = T$. Therefore, $\Sigma \models \beta$. This particular tautological implication justifies the formal inference rule called *modus ponens*, represented by the diagram

$$\begin{array}{l} \alpha \to \beta \\ \underline{\alpha \qquad\quad} \\ \therefore \beta \end{array},$$

which asserts that if $(\alpha \to \beta)$ and $\alpha$ are true, we therefore can conclude that $\beta$ is true. In other words, modus ponens offers an argument that is truth preserving.

**Remark 2.2.13.** A few special cases concerning Definitions 2.2.7–2.2.10 deserve some comments.

(a) If $\Sigma$ is the empty set $\varnothing$, then every truth assignment satisfies $\Sigma$.
(b) It follows from (a) that $\varnothing \models \tau$ if and only if every truth assignment satisfies $\tau$. In this case, we say that $\tau$ is a *tautology* (written $\models \tau$). In other words, $\tau$ is a *tautology* if and only if for every truth assignment $v$ containing the sentence symbols in $\tau$, we have $\overline{v}(\tau) = T$.
(c) If there is no truth assignment that satisfies $\Sigma$, then for any $\tau$ it is vacuously true that $\Sigma \models \tau$. This can occur if $\Sigma$ contains a contradiction; for example, if $(\mathbf{A}_i \wedge (\neg \mathbf{A}_i)) \in \Sigma$.
(d) If $\Sigma$ is a singleton $\{\sigma\}$, then we write $\sigma \models \tau$ in place of $\{\sigma\} \models \tau$.

**Problem 2.2.14.** Let $\alpha$ and $\beta$ be wffs. Show that $(\alpha \to (\beta \to \alpha))$ is a tautology.

**Solution.** We apply Remark 2.2.13(b). Let $v$ be a truth assignment that contains the sentence symbols occurring in $\alpha$ and $\beta$. We must show that $\overline{v}((\alpha \to (\beta \to \alpha))) = T$. Suppose, for a contradiction, that $\overline{v}((\alpha \to (\beta \to \alpha))) = F$. By Theorem 2.2.4(4), (▲) $\overline{v}(\alpha) = T$ and $\overline{v}((\beta \to \alpha)) = F$. Since $\overline{v}((\beta \to \alpha)) = F$, we conclude that $\overline{v}(\alpha) = F$, which contradicts (▲). So $\overline{v}((\alpha \to (\beta \to \alpha))) = T$. Thus, $(\alpha \to (\beta \to \alpha))$ is a tautology.

**Definition 2.2.15.** Let $\sigma$ and $\tau$ be wffs. Then $\sigma$ and $\tau$ are said to be *tautologically equivalent* (written $\sigma \vDash\dashv \tau$) if $\sigma \vDash \tau$ and $\tau \vDash \sigma$.

**Remark 2.2.16** (On Definition 2.2.15). Let $\sigma$ and $\tau$ be wffs. Then one can show that $\sigma \vDash\dashv \tau$ if and only if for every truth assignment $v$ containing the sentence symbols in $\sigma$ and $\tau$, we have $\overline{v}(\sigma) = \overline{v}(\tau)$.

### 2.2.1 Some tautologies

For the remainder of this chapter, we will let **A**, **B**, **C**, ..., **Z** denote arbitrary sentence symbols. We note the following.

1. *Two simple tautologies:*
   $(\mathbf{A} \to \mathbf{A})$
   $(\mathbf{A} \vee (\neg \mathbf{A}))$
2. *De Morgan's laws:*
   $((\neg(\mathbf{A} \vee \mathbf{B})) \leftrightarrow ((\neg \mathbf{A}) \wedge (\neg \mathbf{B})))$
   $((\neg(\mathbf{A} \wedge \mathbf{B})) \leftrightarrow ((\neg \mathbf{A}) \vee (\neg \mathbf{B})))$
3. *Commutative laws:*
   $((\mathbf{A} \wedge \mathbf{B}) \leftrightarrow (\mathbf{B} \wedge \mathbf{A}))$
   $((\mathbf{A} \vee \mathbf{B}) \leftrightarrow (\mathbf{B} \vee \mathbf{A}))$
4. *Associative laws:*
   $((\mathbf{A} \vee (\mathbf{B} \vee \mathbf{C})) \leftrightarrow ((\mathbf{A} \vee \mathbf{B}) \vee \mathbf{C}))$
   $((\mathbf{A} \wedge (\mathbf{B} \wedge \mathbf{C})) \leftrightarrow ((\mathbf{A} \wedge \mathbf{B}) \wedge \mathbf{C}))$
5. *Distribution laws:*
   $((\mathbf{A} \wedge (\mathbf{B} \vee \mathbf{C})) \leftrightarrow ((\mathbf{A} \wedge \mathbf{B}) \vee (\mathbf{A} \wedge \mathbf{C})))$
   $((\mathbf{A} \vee (\mathbf{B} \wedge \mathbf{C})) \leftrightarrow ((\mathbf{A} \vee \mathbf{B}) \wedge (\mathbf{A} \vee \mathbf{C})))$
6. *Conditional laws:*
   $((\mathbf{A} \to \mathbf{B}) \leftrightarrow ((\neg \mathbf{A}) \vee \mathbf{B}))$
   $((\mathbf{A} \to \mathbf{B}) \leftrightarrow (\neg(\mathbf{A} \wedge (\neg \mathbf{B}))))$
   $(((\neg \mathbf{A}) \to \mathbf{B}) \leftrightarrow (\mathbf{A} \vee \mathbf{B}))$
   $((\neg(\mathbf{A} \to (\neg \mathbf{B}))) \leftrightarrow (\mathbf{A} \wedge \mathbf{B}))$
7. *Double negation law:* $((\neg(\neg \mathbf{A})) \leftrightarrow \mathbf{A})$
8. *Contrapositive law:* $((\mathbf{A} \to \mathbf{B}) \leftrightarrow ((\neg \mathbf{B}) \to (\neg \mathbf{A})))$
9. *Biconditional law:* $((\mathbf{A} \leftrightarrow \mathbf{B}) \leftrightarrow ((\mathbf{A} \to \mathbf{B}) \wedge (\mathbf{B} \to \mathbf{A})))$
10. *Exportation law:* $(((\mathbf{A} \wedge \mathbf{B}) \to \mathbf{C}) \leftrightarrow (\mathbf{A} \to (\mathbf{B} \to \mathbf{C})))$

### 2.2.2 Omitting parentheses

We now describe some conventions that will allow us to reduce (without any ambiguity) the number of parentheses used in our wffs. These conventions will give us a "shortcut" in how we can correctly express a propositional statement.

1. The outermost parentheses need not be explicitly mentioned. For example, we can write $\mathbf{A} \wedge \mathbf{B}$ to denote $(\mathbf{A} \wedge \mathbf{B})$.
2. The negation symbol shall apply to as little as possible. For example, we can write $\neg\mathbf{A} \wedge \mathbf{B}$ to denote $(\neg\mathbf{A}) \wedge \mathbf{B}$, that is, $((\neg\mathbf{A}) \wedge \mathbf{B})$.
3. The conjunction and disjunction symbols shall apply to as little as possible. For example, we will write
$$\mathbf{A} \wedge \mathbf{B} \to \neg\mathbf{C} \vee \mathbf{D} \text{ to denote } ((\mathbf{A} \wedge \mathbf{B}) \to ((\neg\mathbf{C}) \vee \mathbf{D})),$$
using conventions 1 and 2.
4. When one logical connective is used repeatedly, the parentheses are assumed to be grouped to the right. For example, we shall write
$$\alpha \wedge \beta \wedge \gamma \text{ to denote } \alpha \wedge (\beta \wedge \gamma) \text{ and}$$
$$\alpha \to \beta \to \gamma \text{ to denote } \alpha \to (\beta \to \gamma).$$
This convention allows us to observe that for $n \geq 2$,
$$\alpha_1 \wedge \alpha_2 \wedge \cdots \wedge \alpha_n \wedge \alpha_{n+1} \text{ denotes } \alpha_1 \wedge \alpha_2 \wedge \cdots \wedge (\alpha_n \wedge \alpha_{n+1}), \tag{2.3}$$
$$\alpha_1 \to \alpha_2 \to \cdots \to \alpha_n \to \alpha_{n+1} \text{ denotes } \alpha_1 \to \alpha_2 \to \cdots \to (\alpha_n \to \alpha_{n+1}). \tag{2.4}$$

In some of the following exercises, we will omit parentheses following these conventions.

**Exercises 2.2.**

1. Using Theorem 2.2.11, prove Corollary 2.2.12.

*2. Let $\Sigma$ be a set of wffs and let $\tau$ be a wff. Let $\tau$ be either in $\Sigma$ or a tautology. Show that $\Sigma \models \tau$.

3. Determine whether or not $((\mathbf{A} \to \mathbf{B}) \vee (\mathbf{B} \to \mathbf{A}))$ is a tautology.

*4. Let $\alpha$ and $\beta$ be wffs. Show that $\alpha \to \beta \to (\alpha \wedge \beta)$ is a tautology.

*5. Let $\alpha$ and $\beta$ be wffs. Show that $(\neg\alpha \to \beta) \to (\neg\alpha \to \neg\beta) \to \alpha$ is a tautology.

*6. Let $\Sigma$ be a set of wffs and let $\theta, \psi$ be wffs. Suppose that $\Sigma \models \theta$ and $\Sigma \models (\theta \to \psi)$. Show that $\Sigma \models \psi$.

7. Let $\Sigma_1$ and $\Sigma_2$ be sets of wffs and let $\alpha$ and $\beta$ be wffs. Suppose that $\Sigma_1 \models \alpha$ and $\Sigma_2 \models \beta$. Show that $\Sigma_1 \cup \Sigma_2 \models (\alpha \wedge \beta)$.

8. Let $\Sigma = \{\alpha_1, \alpha_2, \ldots, \alpha_n\}$ be a finite set of wffs and let $\tau$ be a wff. Prove that if $\Sigma \models \tau$, then $(\alpha_1 \wedge \alpha_2 \wedge \cdots \wedge \alpha_n) \to \tau$ is a tautology.

*9. Let $\Sigma$ be any set of wffs and let $\alpha$ and $\beta$ be wffs. Show that the following hold:
   (a) $\Sigma \cup \{\alpha\} \vDash \beta$ iff $\Sigma \vDash (\alpha \to \beta)$,
   (b) $\alpha \vDash\Dashv \beta$ iff $\vDash (\alpha \leftrightarrow \beta)$.

*10. Suppose that $(\alpha \leftrightarrow \beta)$ is a tautology. Show that $(\alpha \to \beta)$ is a tautology.

11. Suppose that either $\Sigma \vDash \alpha$ or $\Sigma \vDash \beta$. Prove that $\Sigma \vDash (\alpha \vee \beta)$.

12. Find $\Sigma, \alpha, \beta$ such that $\Sigma \vDash (\alpha \vee \beta)$, and yet $\Sigma \nvDash \alpha$ and $\Sigma \nvDash \beta$.

13. Let $\alpha, \beta$ be wffs. Suppose that $\alpha$ and $(\alpha \leftrightarrow \beta)$ are both tautologies. Show that $\beta$ is also a tautology.

*14. Let $\alpha, \alpha', \beta$, and $\beta'$ be wffs. Suppose that $\alpha \vDash\Dashv \alpha'$ and $\beta \vDash\Dashv \beta'$. Show that
   (a) $(\neg\alpha) \vDash\Dashv (\neg\alpha')$,
   (b) $(\alpha \wedge \beta) \vDash\Dashv (\alpha' \wedge \beta')$,
   (c) $(\alpha \vee \beta) \vDash\Dashv \neg((\neg\alpha') \wedge (\neg\beta'))$,
   (d) $(\alpha \wedge \beta) \vDash\Dashv \neg(\alpha' \to (\neg\beta'))$.

15. The sets of wffs below are either satisfiable or not. Determine which are satisfiable by finding a truth assignment that satisfies the given set $\Sigma$. If $\Sigma$ is not satisfiable, explain (that is, give an argument) why this is the case.
   (a) $\Sigma = \{\mathbf{A}, (\mathbf{A} \to \mathbf{B}), \neg\mathbf{B}\}$,
   (b) $\Sigma = \{(\mathbf{A} \to \mathbf{B}), \neg\mathbf{A}, \neg\mathbf{B}\}$,
   (c) $\Sigma = \{(\mathbf{A} \to \mathbf{B}), \neg\mathbf{A}, \mathbf{B}\}$,
   (d) $\Sigma = \{(\mathbf{A} \wedge \mathbf{B}), \neg\mathbf{A}, \neg\mathbf{B}\}$,
   (e) $\Sigma = \{(\mathbf{A} \wedge \neg\mathbf{A}), \neg\mathbf{C}, \mathbf{B}\}$,
   (f) $\Sigma = \{\neg(\mathbf{A} \leftrightarrow \mathbf{B}), \mathbf{A}, \neg\mathbf{B}\}$,
   (g) $\Sigma = \{(\mathbf{A} \leftrightarrow \mathbf{B}), \neg\mathbf{A}, \neg\mathbf{B}\}$,
   (h) $\Sigma = \{\mathbf{A} \wedge \mathbf{B}, \neg(\mathbf{C} \wedge \mathbf{D}), \mathbf{D} \to \neg\mathbf{A}, \mathbf{C} \to \neg\mathbf{B}\}$,
   (i) $\Sigma = \{\mathbf{A} \wedge \mathbf{B}, \neg\mathbf{C} \wedge \mathbf{D}, \mathbf{D} \to \neg\mathbf{A}, \mathbf{C} \to \neg\mathbf{B}\}$.

*16. Let $\Sigma$ be a set of wffs and let $\tau$ be a wff. Prove that

$$\Sigma \nvDash \tau \quad \text{if and only if} \quad \Sigma \cup \{\neg\tau\} \text{ is satisfiable.}$$

17. Let $\Sigma$ be a set of wffs and let $\varphi$, $\psi$ be wffs. Let $v$ be a truth assignment that satisfies $\Sigma$. Prove the following statements:
   (a) If $(\varphi \wedge \psi) \in \Sigma$, then $\overline{v}(\varphi) = T$.
   (b) If $(\varphi \to \psi) \in \Sigma$ and $\neg\psi \in \Sigma$, then $\overline{v}(\neg\varphi) = T$.
   (c) If $(\psi \to \varphi) \in \Sigma$ and $\neg\varphi \in \Sigma$, then $\overline{v}(\psi) = F$.
   (d) If $\neg(\psi \to \varphi) \in \Sigma$, then $\overline{v}(\psi) = T$.

18. Determine whether or not the following statements are true. Explain (that is, give an argument) why each statement is true or explain why it is false.
   (a) Let $\Sigma = \{\mathbf{A}, (\mathbf{A} \to \mathbf{B})\}$. Then $\Sigma \vDash \mathbf{B}$.
   (b) Let $\Sigma = \{(\mathbf{A} \vee \mathbf{B}), \neg\mathbf{B}\}$. Then $\Sigma \vDash \mathbf{A}$.
   (c) Let $\Sigma = \{\mathbf{C} \to (\mathbf{A} \to \mathbf{B}), \mathbf{A}, \mathbf{C}\}$. Then $\Sigma \vDash (\mathbf{C} \to \mathbf{B})$.
   (d) Let $\Sigma = \{\mathbf{C} \to (\neg\mathbf{A} \to \mathbf{B}), \mathbf{C}\}$. Then $\Sigma \vDash (\neg\mathbf{B} \to \mathbf{A})$.
   (e) Let $\Sigma = \{(\mathbf{A} \leftrightarrow \mathbf{B}), \neg\mathbf{C}\}$. Then $\Sigma \vDash (\mathbf{A} \to \mathbf{B})$.

(f) Let $\Sigma = \{(\mathbf{A} \leftrightarrow \mathbf{B}), \neg\mathbf{C}\}$. Then $\Sigma \models \mathbf{A}$.
(g) Let $\Sigma = \{(\mathbf{A} \leftrightarrow \mathbf{B}), \neg\mathbf{C}, \neg\mathbf{B}\}$. Then $\Sigma \models \neg\mathbf{A}$.
(h) Let $\Sigma = \{(\mathbf{A} \vee \mathbf{B}), \neg\mathbf{C}\}$. Then $\Sigma \models \mathbf{A}$.
(i) Let $\Sigma = \{(\mathbf{A} \wedge \neg\mathbf{A}), \neg\mathbf{C}, \mathbf{B}\}$. Then $\Sigma \models \mathbf{D}$.

19. Let $\mathcal{S}$ be the set of all sentence symbols. For a set $\Gamma$ of wffs, define the truth assignment $v_\Gamma\colon \mathcal{S} \to \{F, T\}$ by

$$v_\Gamma(\mathbf{A}_i) = \begin{cases} T, & \text{if } \mathbf{A}_i \in \Gamma, \\ F, & \text{if } \mathbf{A}_i \notin \Gamma. \end{cases}$$

Justify your answers to the following questions:
(a) Let $\Gamma = \{(\mathbf{A} \wedge \mathbf{B}), \mathbf{A}, \neg\mathbf{B}\}$. Does $v_\Gamma$ satisfy $\Gamma$?
(b) Let $\Gamma = \{(\mathbf{A} \wedge \neg\mathbf{A}), \mathbf{C}, \mathbf{B}\}$. Does $v_\Gamma$ satisfy $\Gamma$?
(c) Let $\Gamma = \{\neg(\mathbf{A} \leftrightarrow \mathbf{B}), \mathbf{A}, \neg\mathbf{B}\}$. Does $v_\Gamma$ satisfy $\Gamma$?
(d) Let $\Gamma = \{(\mathbf{A} \leftrightarrow \mathbf{B}), \neg\mathbf{A}, \neg\mathbf{B}\}$. Does $v_\Gamma$ satisfy $\Gamma$?
(e) Let $\Gamma = \{\mathbf{A} \wedge \mathbf{B}, \neg(\mathbf{C} \wedge \mathbf{D}), \mathbf{D} \to \neg\mathbf{A}, \mathbf{C} \to \neg\mathbf{B}, \mathbf{A}, \mathbf{B}\}$. Does $v_\Gamma$ satisfy $\Gamma$?
(f) Let $\Gamma = \{\mathbf{D}, \neg\mathbf{C} \wedge \mathbf{D}, \mathbf{D} \to \neg\mathbf{A}, \mathbf{C} \to \neg\mathbf{B}\}$. Does $v_\Gamma$ satisfy $\Gamma$?

20. (Substitution) Let $\alpha_1, \alpha_2, \ldots, \alpha_n, \ldots$ be an infinite sequence of wffs. For each wff $\varphi$, let $\varphi^*$ be the result of replacing each occurrence of $\mathbf{A}_n$ in $\varphi$ with $\alpha_n$.
(a) Let $v$ be a truth assignment for the set $\mathcal{S}$ of all the sentence symbols. Define $u\colon \mathcal{S} \to \{F, T\}$ by $u(\mathbf{A}_n) = \overline{v}(\alpha_n)$. Prove that $\overline{u}(\varphi) = \overline{v}(\varphi^*)$ for all wffs $\varphi$.
(b) Prove that if $\varphi$ is a tautology, then so is $\varphi^*$.
(c) Prove that if $\psi \vDash\dashv \varphi$, then $\psi^* \vDash\dashv \varphi^*$.

**Exercise Notes:** For Exercise 14, apply Remark 2.2.16. For Exercise 19, note that if $\Gamma = \{(\mathbf{A} \wedge \mathbf{B}), \mathbf{A}, \neg\mathbf{B}\}$, then $\mathbf{A} \in \Gamma$ and $\mathbf{B} \notin \Gamma$. For Exercise 20(a), let $P(\varphi)$ be the statement $\overline{u}(\varphi) = \overline{v}(\varphi^*)$. Now prove "for all wffs $\varphi$, $P(\varphi)$" by induction on $\varphi$. (Question: Is $(\varphi \wedge \psi)^* = (\varphi^* \wedge \psi^*)$?)

## 2.3 Completeness of the logical connectives

In logic, a *truth function* is one that accepts truth values as input and produces a unique truth value as output. Let $V = \{T, F\}$. Thus, an $n$-place truth function has the form $H\colon V^n \to V$ for some $n \geq 1$. Truth functions are also called *Boolean functions*. Such functions are applied in a variety of different fields, namely, electrical engineering, computer science, game theory, and combinatorics.

Let $\alpha$ be a wff whose sentence symbols are among $\mathbf{A}_1, \mathbf{A}_2, \ldots, \mathbf{A}_n$. We can now define an $n$-place truth function $H_\alpha\colon V^n \to V$ by

$$H_\alpha(x_1, x_2, \ldots, x_n) = \overline{v}(\alpha), \text{ where } v(\mathbf{A}_1) = x_1,\ v(\mathbf{A}_2) = x_2,\ \ldots,\ v(\mathbf{A}_n) = x_n. \tag{2.5}$$

So, given any wff, we can use it to define a truth function. For example, let $\alpha$ be the wff $(\mathbf{A}_1 \vee \neg\mathbf{A}_2)$. Then we see that $H_\alpha \colon V^2 \to V$ satisfies

$$\begin{aligned} H_\alpha(T,T) &= T, \\ H_\alpha(T,F) &= T, \\ H_\alpha(F,T) &= F, \\ H_\alpha(F,F) &= T, \end{aligned}$$

which is essentially the truth table for $\alpha$. This invites an interesting question: Given any truth function $G$, is there a wff $\psi$ such that $G = H_\psi$? Before addressing this question, let us look at some examples.

**Example 2.3.1.** Suppose that $G\colon V^3 \to V$ always has output $F$. Let $\alpha$ be the wff $(\mathbf{A}_1 \wedge \neg\mathbf{A}_1)$. Since $\alpha$ is always false and its sentence symbols are among $\mathbf{A}_1, \mathbf{A}_2, \mathbf{A}_3$, by applying (2.5) we see that $H_\alpha(x_1, x_2, x_3) = \overline{v}(\alpha) = F$ for all $x_1, x_2, x_3 \in V$ (see Theorem 2.2.11). So $G = H_\alpha$.

**Example 2.3.2.** Let $G\colon V^3 \to V$ be such that

$$G(x_1, x_2, x_3) = \begin{cases} T, & \text{if } x_1 = T, x_2 = F, x_3 = T, \\ F, & \text{otherwise.} \end{cases}$$

Thus, $G(T,F,T) = T$ and this is the only input that produces the output value $T$. Let $\gamma$ be the wff $(\mathbf{A}_1 \wedge \neg\mathbf{A}_2 \wedge \mathbf{A}_3)$ which corresponds to $T,F,T$. Note that $\overline{v}(\gamma) = T$ if and only if $v(\mathbf{A}_1) = T$, $v(\mathbf{A}_2) = F$, $v(\mathbf{A}_3) = T$. Thus, by applying (2.5), we see that $G(x_1, x_2, x_3) = H_\gamma(x_1, x_2, x_3) = \overline{v}(\gamma)$ for all $x_1, x_2, x_3 \in V$. So $G = H_\gamma$.

In the above two examples, we were given relatively simple truth functions $G$ and we were able to find a wff $\psi$ such that $G = H_\psi$. Our next example gives a slightly more complicated truth function. These examples will provide a guide that will allow us to prove that for any truth function $G$ there is always a wff $\psi$ such that $G = H_\psi$.

**Example 2.3.3.** Suppose that truth function $G\colon V^3 \to V$ is such that

$$\begin{aligned} G(T,T,T) &= T, \\ G(T,T,F) &= F, \\ G(T,F,T) &= F, \\ G(T,F,F) &= T, \\ G(F,T,T) &= F, \\ G(F,T,F) &= T, \\ G(F,F,T) &= T, \\ G(F,F,F) &= F. \end{aligned} \tag{2.6}$$

Let us focus on the cases in (2.6) that produce the value $T$. Also, for each such case let us identify a corresponding wff $\gamma_i$ as was done in Example 2.3.2, namely,

$$\begin{aligned} G(T,T,T) &= T \quad \gamma_1 = (\mathbf{A}_1 \wedge \mathbf{A}_2 \wedge \mathbf{A}_3), \\ G(T,F,F) &= T \quad \gamma_2 = (\mathbf{A}_1 \wedge \neg\mathbf{A}_2 \wedge \neg\mathbf{A}_3), \\ G(F,T,F) &= T \quad \gamma_3 = (\neg\mathbf{A}_1 \wedge \mathbf{A}_2 \wedge \neg\mathbf{A}_3), \\ G(F,F,T) &= T \quad \gamma_4 = (\neg\mathbf{A}_1 \wedge \neg\mathbf{A}_2 \wedge \mathbf{A}_3). \end{aligned} \tag{2.7}$$

Now let $\alpha = \gamma_1 \vee \gamma_2 \vee \gamma_3 \vee \gamma_4$, that is, let $\alpha$ be the following wff:

$$(\mathbf{A}_1 \wedge \mathbf{A}_2 \wedge \mathbf{A}_3) \vee (\mathbf{A}_1 \wedge \neg\mathbf{A}_2 \wedge \neg\mathbf{A}_3) \vee (\neg\mathbf{A}_1 \wedge \mathbf{A}_2 \wedge \neg\mathbf{A}_3) \vee (\neg\mathbf{A}_1 \wedge \neg\mathbf{A}_2 \wedge \mathbf{A}_3).$$

Note that $\overline{v}(\alpha) = T$ if and only if $G(v(\mathbf{A}_1), v(\mathbf{A}_2), v(\mathbf{A}_3)) = T$. By applying (2.5), we see that $G(x_1, x_2, x_3) = H_\alpha(x_1, x_2, x_3) = \overline{v}(\alpha)$ for all $x_1, x_2, x_3 \in V$. So $G = H_\alpha$.

The above examples offer a foundation for the proof of the following theorem.

**Theorem 2.3.4.** *Let $V = \{T, F\}$ and let $G\colon V^n \to V$ be any truth function. Then there is a wff $\alpha$ such that $G = H_\alpha$, where $H_\alpha$ is defined by* (2.5).

*Proof.* If $G\colon V^n \to V$ is always false, then let $\alpha$ be the wff $(\mathbf{A}_1 \wedge \neg\mathbf{A}_1)$. Thus, $G = H_\alpha$ (see Example 2.3.1). Suppose now that $G$ has some output values being $T$, say, $k$ many such cases. Let us list all the cases that yield the value $T$. For each such case let us also identify a corresponding wff $\gamma_i$ as was done in Example 2.3.3 (see (2.7)), specifically,

$$\begin{aligned} G(x_{11}, x_{12}, \ldots, x_{1n}) &= T, \quad \gamma_1 = (\beta_{11} \wedge \beta_{12} \wedge \cdots \wedge \beta_{1n}), \\ G(x_{21}, x_{22}, \ldots, x_{2n}) &= T, \quad \gamma_2 = (\beta_{21} \wedge \beta_{22} \wedge \cdots \wedge \beta_{2n}), \\ &\vdots \qquad\qquad \vdots \\ G(x_{k1}, x_{k2}, \ldots, x_{kn}) &= T, \quad \gamma_k = (\beta_{k1} \wedge \beta_{k2} \wedge \cdots \wedge \beta_{kn}), \end{aligned}$$

where for $1 \leq i \leq k$ and $1 \leq j \leq n$, we have

$$\beta_{ij} = \begin{cases} \mathbf{A}_j, & \text{if } x_{ij} = T, \\ \neg\mathbf{A}_j, & \text{if } x_{ij} = F. \end{cases}$$

Note that $\overline{v}(\gamma_i) = T$ if and only if $v(\mathbf{A}_1) = x_{i1}, v(\mathbf{A}_2) = x_{i2}, \ldots, v(\mathbf{A}_n) = x_{in}$. Let $\alpha = \gamma_1 \vee \gamma_2 \vee \cdots \vee \gamma_k$. So $\overline{v}(\alpha) = T$ if and only if $G(v(\mathbf{A}_1), v(\mathbf{A}_2), \ldots, v(\mathbf{A}_n)) = T$. By applying (2.5), we see that $G(x_1, x_2, \ldots, x_n) = H_\alpha(x_1, x_2, \ldots, x_n) = \overline{v}(\alpha)$ for all $x_1, x_2, \ldots, x_n \in V$. So $G = H_\alpha$. □

Theorem 2.3.4 shows that every truth function is equal to a function of the form $H_\alpha$ for some wff $\alpha$. For this reason, the set $\{\wedge, \vee, \neg, \to, \leftrightarrow\}$ of logical connectives can be said to be *truth functionally complete*. In fact, the proof of Theorem 2.3.4 shows that $\{\wedge, \vee, \neg\}$

is also truth functionally complete. To clarify this assertion, let us say that a wff $\alpha$ is in *disjunctive normal form* if

$$\alpha = \gamma_1 \vee \gamma_2 \vee \cdots \vee \gamma_k,$$

where each $\gamma_i$ is a conjunction of the form

$$\gamma_i = (\beta_{i1} \wedge \beta_{i2} \wedge \cdots \wedge \beta_{in})$$

and each $\beta_{ij}$ is a sentence symbol or the negation of a sentence symbol. The proof of Theorem 2.3.4 shows that every truth function is equal to a function of the form $H_\alpha$, where $\alpha$ is in disjunctive normal form. Thus, $\{\wedge, \vee, \neg\}$ is truth functionally complete as well. This discussion now allows us to establish the following result.

**Theorem 2.3.5.** *Every wff $\varphi$ is tautologically equivalent to a wff $\alpha$ in disjunctive normal form.*

*Proof.* Let $\varphi$ be a wff and let $H_\varphi$ be the truth function defined by (2.5). The proof of Theorem 2.3.4 shows that $H_\varphi = H_\alpha$, where $\alpha$ is in disjunctive normal form. Exercise 3 below shows that $\varphi$ and $\alpha$ are tautologically equivalent. □

**Definition 2.3.6.** Let $\mathcal{C}$ be a subset of $\{\wedge, \vee, \neg, \to, \leftrightarrow\}$. Then $\mathcal{C}$ is *tautologically complete* if and only if every wff $\alpha$ is tautologically equivalent to a wff $\alpha'$ that contains only the connectives in $\mathcal{C}$.

The following result follows from Theorem 2.3.5.

**Theorem 2.3.7.** *The set $\{\wedge, \vee, \neg\}$ is tautologically complete.*

**Theorem 2.3.8.** *The sets $\{\neg, \wedge\}$ and $\{\neg, \vee\}$ are both tautologically complete.*

*Proof.* We will only prove that the set $\{\neg, \wedge\}$ is tautologically complete. A similar proof shows that $\{\neg, \vee\}$ is tautologically complete. By Theorem 2.3.7 we know that every wff is tautologically equivalent to a wff which contains only the connectives in $\{\wedge, \vee, \neg\}$. So it is sufficient to prove by induction that every wff $\alpha$ containing only connectives in $\{\wedge, \vee, \neg\}$ is tautologically equivalent to a wff $\alpha'$ with connectives only in $\{\neg, \wedge\}$.

*Base step:* Let $\alpha = \mathbf{A}_i$. As $\alpha$ has no connectives, $\alpha' = \alpha$ as required.

*Inductive step:* Let $\alpha$ and $\beta$ be arbitrary wffs that contain only the connectives in the set $\{\wedge, \vee, \neg\}$. Assume the induction hypothesis

$$\begin{aligned} &\alpha \vDash\!\!\dashv \alpha', \quad \text{where } \alpha' \text{ contains only the connectives in } \{\neg, \wedge\}, \\ &\beta \vDash\!\!\dashv \beta', \quad \text{where } \beta' \text{ contains only the connectives in } \{\neg, \wedge\}. \end{aligned} \tag{IH}$$

We prove that each of the wffs $(\neg\alpha), (\alpha \wedge \beta), (\alpha \vee \beta)$ is tautologically equivalent to a wff that contains only connectives in $\{\neg, \wedge\}$.

For $(\neg\alpha)$, from (IH) and Exercise 14(a) on page 38, it follows that $(\neg\alpha) \vDash\Dashv (\neg\alpha')$, where $(\neg\alpha')$ has connectives only in $\{\neg, \wedge\}$.

For $(\alpha \wedge \beta)$, from (IH) and Exercise 14(b) on page 38, we have $(\alpha \wedge \beta) \vDash\Dashv (\alpha' \wedge \beta')$, where $(\alpha' \wedge \beta')$ contains only connectives in $\{\neg, \wedge\}$.

For $(\alpha \vee \beta)$, from (IH) and Exercise 14(c) on page 38, we have

$$(\alpha \vee \beta) \vDash\Dashv \neg((\neg\alpha') \wedge (\neg\beta')),$$

where $\neg((\neg\alpha') \wedge (\neg\beta'))$ contains only connectives in $\{\neg, \wedge\}$. □

**Remark 2.3.9.** Let $\mathcal{F} = \{\mathcal{E}_\neg, \mathcal{E}_\wedge, \mathcal{E}_\vee\}$ and let $\mathcal{S} = \{\mathbf{A}_1, \mathbf{A}_2, \mathbf{A}_3, \dots\}$ be the set of all sentence symbols in the language $\mathcal{L}$. Let $\overline{\mathcal{S}}$ be the set generated from $\mathcal{S}$ by the functions in $\mathcal{F}$. Theorem 1.1.25 validates the proof by induction given in Theorem 2.3.8.

**Exercises 2.3.**

1. For each of the following wffs $\alpha$, using the tautologies in Section 2.2.1, find a wff that is tautologically equivalent to $\alpha$ which contains connectives only in $\mathcal{C}$:
   (a) $\alpha = (\mathbf{A} \to \mathbf{B}) \leftrightarrow (\mathbf{C} \to \mathbf{A})$ and $\mathcal{C} = \{\wedge, \vee, \neg\}$,
   (b) $\alpha = (\mathbf{A} \to \mathbf{B}) \leftrightarrow (\mathbf{C} \to \mathbf{A})$ and $\mathcal{C} = \{\neg, \wedge\}$.
2. Let $V = \{T, F\}$ and define the truth function $G\colon V^3 \to V$ by

$$G(x_1, x_2, x_3) = \begin{cases} T, & \text{if exactly two of } x_1, x_2, x_3 \text{ are } T, \\ F, & \text{otherwise.} \end{cases}$$

   Find a wff $\alpha$ in disjunctive normal form such that $G = H_\alpha$, where $H_\alpha$ is defined as in (2.5).
*3. Let $\alpha$ and $\beta$ be wffs with the same sentence symbols among $\mathbf{A}_1, \mathbf{A}_2, \dots, \mathbf{A}_n$. Let $H_\alpha$ and $H_\beta$ be defined as in (2.5). Prove that $H_\alpha = H_\beta$ if and only if $\alpha \vDash\Dashv \beta$.
4. Using the fact that $\{\neg, \wedge\}$ is tautologically complete, prove (by induction) that $\{\neg, \vee\}$ is tautologically complete.
*5. Given that $\{\neg, \wedge\}$ is tautologically complete, prove (by induction) that $\{\neg, \to\}$ is tautologically complete.
6. Let $\mathcal{F} = \{\mathcal{E}_\wedge, \mathcal{E}_\to\}$ and let $\mathcal{S} = \{\mathbf{A}_1, \mathbf{A}_2, \mathbf{A}_3, \dots\}$ be the set of all sentence symbols in the language $\mathcal{L}$. Let $\overline{\mathcal{S}}$ be the set generated from $\mathcal{S}$ by the functions in $\mathcal{F}$. Now let $v\colon \mathcal{S} \to \{F, T\}$ be defined by $v(\mathbf{A}_i) = T$ for all $i = 1, 2, 3, \dots$. Prove by induction on $\alpha$ that $\overline{v}(\alpha) = T$ for all $\alpha \in \overline{\mathcal{S}}$. Is $\{\wedge, \to\}$ tautologically complete?

**Exercise Notes:** For Exercises 4 and 5, read the first paragraph of the proof of Theorem 2.3.8. For Exercise 5, see Exercise 14(d) on page 38.

## 2.4 Compactness

Let $\Sigma$ be an infinite set of wffs. Suppose that every finite subset of $\Sigma$ is satisfiable. Thus, every finite subset of $\Sigma$ has a particular truth assignment that satisfies all of the wffs in this finite subset. Is there a truth assignment that will thus satisfy all of the infinite number of wffs in $\Sigma$? The compactness theorem addresses this interesting question. In this section, we shall give a proof of the compactness theorem, for propositional logic. We begin by formally introducing the following pertinent definition.

**Definition 2.4.1.** A set of wffs $\Sigma$ is *finitely satisfiable* if and only if every finite subset of $\Sigma$ is satisfiable.

The proof of the following theorem shows that a sophisticated mathematical argument can establish a powerful result about propositional logic. A similar argument will be applied in Section 4.2.

**Theorem 2.4.2** (Compactness theorem). *Let $\Sigma$ be a set of wffs. If $\Sigma$ is finitely satisfiable, then $\Sigma$ is satisfiable.*

*Proof.* The basic idea behind the proof is as follows: Given that $\Sigma$ is finitely satisfiable, we first construct a set of wffs $\Delta$ such that:
(1) $\Sigma \subseteq \Delta$;
(2) $\Delta$ is finitely satisfiable;
(3) for every wff $\alpha$, either $\alpha \in \Delta$ or $(\neg\alpha) \in \Delta$.

We shall then use $\Delta$ to define a truth assignment that satisfies $\Sigma$.

Note that Theorem 1.1.30 (on page 12) implies that the set of *all* wffs is a countable set, because the set of all wffs is a subset of the set of all finite sequences of symbols in the countable language $\mathcal{L}$. Thus, let

$$\alpha_1, \alpha_2, \alpha_3, \dots, \alpha_n, \dots \tag{2.8}$$

be a fixed enumeration (see Corollary 1.1.33) of all the wffs in the language $\mathcal{L}$. Define by recursion on $\mathbb{N}$ (see Theorem 1.1.23) the following sets:
(i) $\Delta_0 = \Sigma$,
(ii) $\Delta_{n+1} = \begin{cases} \Delta_n \cup \{\alpha_{n+1}\}, & \text{if } \Delta_n \cup \{\alpha_{n+1}\} \text{ is finitely satisfiable,} \\ \Delta_n \cup \{\neg\alpha_{n+1}\}, & \text{otherwise.} \end{cases}$

From the above recursive definition, we see that $\Delta_n \subseteq \Delta_{n+1}$ for all $n \in \mathbb{N}$. Thus,

$$\Sigma = \Delta_0 \subseteq \Delta_1 \subseteq \Delta_2 \subseteq \cdots.$$

We shall now establish four claims.

**Claim 1.** *For all $n \in \mathbb{N}$, $\Delta_n$ is finitely satisfiable.*

*Proof of Claim* 1. We shall use induction on $n$.

*Base step:* Let $n = 0$. Then $\Delta_0 = \Sigma$, which is finitely satisfiable by assumption.

*Inductive step:* Let $n \in \mathbb{N}$ be arbitrary. Assume the induction hypothesis

$$\Delta_n \text{ is finitely satisfiable.} \tag{IH}$$

We prove that $\Delta_{n+1}$ is finitely satisfiable. Since $\Delta_n$ is finitely satisfiable, Exercise 1 below implies that $\Delta_{n+1}$ is finitely satisfiable. (Claim 1) □

Let $\Delta = \bigcup_{n\in\mathbb{N}} \Delta_n$. Clearly, $\Sigma \subseteq \Delta$ as $\Delta_0 = \Sigma$. Recall that $\Delta_n \subseteq \Delta_{n+1}$ for all $n \in \mathbb{N}$.

**Claim 2.** *The set $\Delta$ is finitely satisfiable.*

*Proof of Claim* 2. We show that each finite subset of $\Delta$ is satisfiable. Let $\Pi$ be a finite subset of $\Delta$. Since $\Pi \subseteq \Delta = \bigcup_{n\in\mathbb{N}} \Delta_n$ and $\Pi$ is finite, it follows that $\Pi \subseteq \Delta_n$ for some $n$. By Claim 1, $\Delta_n$ is finitely satisfiable. Thus, $\Pi$ is satisfiable. (Claim 2) □

**Claim 3.** *For every wff $\alpha$, either $\alpha \in \Delta$ or $(\neg\alpha) \in \Delta$, but not both.*

*Proof of Claim* 3. Let $\alpha$ be any wff. Since (2.8) enumerates all of the wffs, there is an $n \in \mathbb{N}$ such that $\alpha = \alpha_n$. Either $\alpha_n \in \Delta_n$ or $(\neg\alpha_n) \in \Delta_n$ by (ii) of the above construction. Since $\Delta_n \subseteq \Delta$, it follows that $\alpha \in \Delta$ or $(\neg\alpha) \in \Delta$. If both $\alpha \in \Delta$ and $(\neg\alpha) \in \Delta$, then $\{\alpha, (\neg\alpha)\}$ would be a finite subset of $\Delta$. Since $\Delta$ is finitely satisfiable, we must conclude that $\{\alpha, (\neg\alpha)\}$ is satisfiable. This is a contradiction, as the set $\{\alpha, (\neg\alpha)\}$ is not satisfiable. Therefore, we cannot have both $\alpha \in \Delta$ and $(\neg\alpha) \in \Delta$. (Claim 3) □

We now continue with the proof of the theorem. Note that:
(1) $\Sigma \subseteq \Delta$;
(2) $\Delta$ is finitely satisfiable by Claim 2;
(3) for every wff $\alpha$, either $\alpha \in \Delta$ or $(\neg\alpha) \in \Delta$, but not both, by Claim 3.

We will now use $\Delta$ to define a truth assignment. Let $\mathcal{S}$ be the set of *all* the sentence symbols and define $v: \mathcal{S} \to \{F, T\}$ by

$$v(\mathbf{A}_i) = \begin{cases} T, & \text{if } \mathbf{A}_i \in \Delta, \\ F, & \text{if } \mathbf{A}_i \notin \Delta. \end{cases} \tag{2.9}$$

Let $\overline{v}: \overline{\mathcal{S}} \to \{F, T\}$ be the extension of $v$ as given by Theorem 2.2.4.

**Claim 4.** *For every wff $\varphi$, we have*

$$\overline{v}(\varphi) = T \quad \textit{if and only if} \quad \varphi \in \Delta. \tag{2.10}$$

*Proof of Claim* 4. We shall prove that (2.10) holds by induction on $\alpha$.

*Base step:* Let $\alpha = \mathbf{A}_i$ be any sentence symbol. Since $\overline{v}(\mathbf{A}_i) = v(\mathbf{A}_i)$, (2.9) implies that $\overline{v}(\mathbf{A}_i) = T$ iff $\mathbf{A}_i \in \Delta$.

*Inductive step:* Let $\alpha$ and $\beta$ be arbitrary wffs. Assume the induction hypothesis

$$\begin{aligned} \overline{v}(\alpha) = T \quad &\text{iff} \quad \alpha \in \Delta, \\ \overline{v}(\beta) = T \quad &\text{iff} \quad \beta \in \Delta. \end{aligned} \tag{IH}$$

We must prove that the same holds for each of the following wffs:

$$(\neg\alpha), (\alpha \wedge \beta), (\alpha \vee \beta), (\alpha \rightarrow \beta), (\alpha \leftrightarrow \beta).$$

Case $(\neg\alpha)$: We prove that $\overline{v}((\neg\alpha)) = T$ iff $(\neg\alpha) \in \Delta$.

($\Rightarrow$). Assume that $\overline{v}((\neg\alpha)) = T$. Since $\overline{v}((\neg\alpha)) = T$, it follows that $\overline{v}(\alpha) = F$. So, $\alpha \notin \Delta$ by (IH). Therefore, by (3), $(\neg\alpha) \in \Delta$.

($\Leftarrow$). Assume that $(\neg\alpha) \in \Delta$. Since $(\neg\alpha) \in \Delta$, it follows that $\alpha \notin \Delta$; because if $\alpha \in \Delta$, then by (2) $\{(\neg\alpha), \alpha\}$ is a satisfiable subset of $\Delta$, which is false. Thus, $\alpha \notin \Delta$ and the induction hypothesis (IH) implies that $\overline{v}(\alpha) = F$, and therefore $\overline{v}((\neg\alpha)) = T$.

Case $(\alpha \wedge \beta)$: We must prove that $\overline{v}((\alpha \wedge \beta)) = T$ iff $(\alpha \wedge \beta) \in \Delta$.

($\Rightarrow$). Assume that $\overline{v}((\alpha\wedge\beta)) = T$. We need to show that $(\alpha\wedge\beta) \in \Delta$. Since $\overline{v}((\alpha\wedge\beta)) = T$, it follows that $\overline{v}(\alpha) = T$ and $\overline{v}(\beta) = T$. So by (IH), $\alpha \in \Delta$ and $\beta \in \Delta$. Suppose, for a contradiction, that $(\alpha \wedge \beta) \notin \Delta$. Therefore, by (3), we conclude that $(\neg(\alpha \wedge \beta)) \in \Delta$. So $\{\alpha, \beta, \neg(\alpha \wedge \beta)\}$ is a finite subset of $\Delta$ and by (2) it is satisfiable, which is false. Therefore, $(\alpha \wedge \beta) \in \Delta$.

($\Leftarrow$). Assume that $(\alpha \wedge \beta) \in \Delta$. We will show that $\overline{v}((\alpha \wedge \beta)) = T$. As $(\alpha \wedge \beta) \in \Delta$, it must be the case that $\alpha \in \Delta$ and $\beta \in \Delta$. To see this, suppose that either $\alpha \notin \Delta$ or $\beta \notin \Delta$. If $\alpha \notin \Delta$, then by (3), we have $(\neg\alpha) \in \Delta$. So $\{(\alpha\wedge\beta), (\neg\alpha)\}$ is a finite subset of $\Delta$ and by (2) it is satisfiable, which is false. A similar argument shows that $\beta \in \Delta$. Hence, $\alpha \in \Delta$ and $\beta \in \Delta$ and by (IH), we have $\overline{v}(\alpha) = T$ and $\overline{v}(\beta) = T$. Therefore, $\overline{v}((\alpha\wedge\beta)) = T$ by Remark 2.2.4(2).

Cases $(\alpha \vee \beta)$, $(\alpha \rightarrow \beta)$, and $(\alpha \leftrightarrow \beta)$: See Exercise 4 below. (Claim 4) □

Claim 4 implies that $\overline{v}$ satisfies $\Delta$. Since $\Sigma \subseteq \Delta$, it follows that $\overline{v}$ satisfies $\Sigma$. Therefore, there is a truth assignment that satisfies $\Sigma$. (Theorem) □

**Remark 2.4.3.** If we were working in a language that had uncountably many sentence symbols, then Theorem 2.4.2 could be established by using Zorn's lemma (Lemma 1.1.36) to obtain a set of wffs $\Delta$ that satisfies items (1), (2), and (3) in the above proof. The remainder of the proof would then be as that after the proof of Claim 3.

The next useful corollary is just the contrapositive of the Compactness Theorem 2.4.2.

**Corollary 2.4.4.** *Let $\Sigma$ be a set of wffs. If $\Sigma$ is not satisfiable, then there is a finite subset $\Sigma'$ of $\Sigma$ that is not satisfiable.*

Let $\Sigma$ be a set of wffs and let $\tau$ be a wff. From Exercise 16 on page 38, we have

$$\Sigma \nvDash \tau \quad \text{if and only if} \quad \Sigma \cup \{\neg\tau\} \text{ is satisfiable.} \tag{2.11}$$

The above equivalence (2.11) will be used in the proof of the following corollary.

**Corollary 2.4.5.** *If $\Sigma \vDash \tau$, then there is a finite $\Sigma_0 \subseteq \Sigma$ such that $\Sigma_0 \vDash \tau$.*

*Proof.* Assume that $\Sigma \vDash \tau$. We shall prove that there is a finite $\Sigma_0 \subseteq \Sigma$ such that $\Sigma_0 \vDash \tau$. Suppose, for a contradiction, that for every finite $\Sigma_0 \subseteq \Sigma$, we have $\Sigma_0 \nvDash \tau$. The above (2.11) implies that

$$\text{for every finite } \Sigma_0 \subseteq \Sigma, \ \Sigma_0 \cup \{\neg\tau\} \text{ is satisfiable.} \tag{2.12}$$

Consider the set of wffs $\Pi = \Sigma \cup \{\neg\tau\}$. The above statement (2.12) implies that every finite subset of $\Pi$ is satisfiable (why?). The compactness theorem thus asserts that $\Pi = \Sigma \cup \{\neg\tau\}$ is satisfiable. Thus, by (2.11), we have that $\Sigma \nvDash \tau$, which contradicts our assumption. □

**Exercises 2.4.**

*1. Let $\Sigma$ be a set of wffs and let $\alpha$ be a wff. Assume that $\Sigma$ is finitely satisfiable. Prove that either $\Sigma \cup \{\alpha\}$ or $\Sigma \cup \{\neg\alpha\}$ is finitely satisfiable.

2. Let $\Sigma$ be a set of wffs. Show $\Sigma$ is not satisfiable if and only if $\Sigma \vDash (\mathbf{A} \wedge (\neg\mathbf{A}))$.

3. Using the above Exercise 2, show that Corollary 2.4.5 implies Theorem 2.4.2.

*4. In the proof of Theorem 2.4.2, complete the proof of Claim 4.

5. Let $\Sigma$ be the following 3-element set of wffs:

$$\Sigma = \{((\neg\mathbf{A} \vee \neg\mathbf{B}) \wedge \mathbf{C}), ((\neg\mathbf{A} \vee \neg\mathbf{C}) \wedge \mathbf{B}), ((\neg\mathbf{C} \vee \neg\mathbf{B}) \wedge \mathbf{A})\}.$$

   (a) Show that every subset of $\Sigma$ containing *less* than three elements is satisfiable.
   (b) Is $\Sigma$ is satisfiable?

6. Let $\Sigma$ and $\Pi$ be sets of wffs. Suppose that for every finite $\Pi_0 \subseteq \Pi$, the set $\Sigma \cup \Pi_0$ is satisfiable. Using the compactness theorem, prove that $\Sigma \cup \Pi$ is satisfiable.

7. Let $\Delta$ be a set of wffs such that:
   (1) $\Delta$ is finitely satisfiable and
   (2) for every wff $\alpha$, either $\alpha \in \Delta$ or $(\neg\alpha) \in \Delta$.

   Let $\gamma$ and $\beta$ be wffs.
   (a) Suppose that $\gamma \in \Delta$ and $\beta$ is tautologically equivalent to $\gamma$. Show that $\beta \in \Delta$.
   (b) Show that $\gamma \wedge \beta \in \Delta$ if and only if $\gamma \in \Delta$ and $\beta \in \Delta$.
   (c) Show that $\gamma \vee \beta \in \Delta$ if and only if either $\gamma \in \Delta$ or $\beta \in \Delta$.
   (d) Prove that $(\gamma \to \beta) \in \Delta$ if and only if $\gamma \notin \Delta$ or $\beta \in \Delta$.

**Exercise Notes:** For Exercise 1, if $\Sigma \cup \{\alpha\}$ and $\Sigma \cup \{\neg\alpha\}$ are not finitely satisfiable, then there are finite subsets $\Sigma_1$ and $\Sigma_2$ of $\Sigma$ such that $\Sigma_1 \cup \{\alpha\}$ and $\Sigma_2 \cup \{\neg\alpha\}$ are not satisfiable. Why? Now consider the finite set $\Sigma_1 \cup \Sigma_2$.

## 2.5 Deductions

Let $\Sigma$ be a set of wffs and let $\psi$ be a wff. We know that $\Sigma \models \psi$ means that every truth assignment that satisfies every formula in $\Sigma$ will also satisfy $\psi$. Rather than working with all such truth assignments, is it possible to "deduce" $\psi$ from $\Sigma$? If so, what methods of deduction would be required to demonstrate this fact? Whenever mathematicians compose a proof, they justify a final conclusion by means of a series of intermediate steps. Such steps typically appeal to some given assumptions and/or previously established theorems. A theorem in mathematics is a statement that is true. Therefore, mathematicians use assumptions and true statements to derive new results. In propositional logic, can we use assumptions and tautologies to derive new results? In this section we will positively address this question.

**Definition 2.5.1.** Let $\Sigma$ be a set of wffs and let $\psi$ be a wff. A *deduction of $\psi$ from $\Sigma$* is a finite sequence of wffs $\langle \alpha_1, \alpha_2, \ldots, \alpha_n \rangle$ such that $\alpha_n = \psi$ and for each $k \le n$,

1. $\alpha_k \in \Sigma$,
2. $\alpha_k$ is a tautology, or
3. $\alpha_k$ is obtained by modus ponens from some $\alpha_i = (\alpha_j \to \alpha_k)$ and $\alpha_j$, where $i, j < k$.

When there is a deduction of $\psi$ from $\Sigma$, we shall say that $\psi$ is *deducible* from $\Sigma$ or that $\psi$ is a *theorem* of $\Sigma$, and we shall write $\Sigma \vdash \psi$. Recall that the inference rule modus ponens is discussed on page 35.

A deduction can be viewed as a "formal proof," where each step is governed by rules (1)–(3) of Definition 2.5.1. We use the term deduction to avoid confusion with our own mathematical proofs.

We now present two examples of deductions for which each "deduction sequence" is written in a vertical form with a parallel justification for each step.

**Example 2.5.2.** Let $\Sigma = \{\alpha, \beta\}$. Show that $\Sigma \vdash (\alpha \wedge \beta)$.

**Solution.** The following (vertical) sequence satisfies the conditions in Definition 2.5.1:

| | |
|---|---|
| 1. $\alpha$ | in $\Sigma$, |
| 2. $\beta$ | in $\Sigma$, |
| 3. $\alpha \to \beta \to (\alpha \wedge \beta)$ | tautology (see Exercise 4 on page 37), |
| 4. $\beta \to (\alpha \wedge \beta)$ | from 1 and 3, by modus ponens, |
| 5. $(\alpha \wedge \beta)$ | from 2 and 4, by modus ponens. |

Therefore, $\Sigma \vdash (\alpha \wedge \beta)$.

**Example 2.5.3.** Let $\Sigma = \{\mathbf{H} \to \neg\mathbf{B}, \mathbf{D} \to \mathbf{B}, \mathbf{H}\}$. Show that $\Sigma \vdash (\neg\mathbf{D})$.

**Solution.** The (vertical) sequence below is a deduction:

| | |
|---|---|
| 1. $\mathbf{H} \to \neg\mathbf{B}$ | in $\Sigma$, |
| 2. $\mathbf{D} \to \mathbf{B}$ | in $\Sigma$, |
| 3. $\mathbf{H}$ | in $\Sigma$, |
| 4. $\neg\mathbf{B}$ | from 1 and 3, by modus ponens, |
| 5. $(\mathbf{D} \to \mathbf{B}) \to (\neg\mathbf{B} \to \neg\mathbf{D})$ | tautology (see contrapositive law),[1] |
| 6. $\neg\mathbf{B} \to \neg\mathbf{D}$ | from 2 and 5, by modus ponens, |
| 7. $\neg\mathbf{D}$ | from 4 and 6, by modus ponens. |

Hence, $\Sigma \vdash (\neg\mathbf{D})$.

The definition of $\Sigma \vDash \psi$ involves truth assignments and the definition of $\Sigma \vdash \psi$ involves a "deduction sequence." On the surface, these two concepts, $\Sigma \vDash \psi$ and $\Sigma \vdash \psi$, seem to be unrelated. However, they are in fact intimately related. Our next two theorems will confirm this purported intimacy. The first theorem shows that if the assumptions in $\Sigma$ are true and $\Sigma \vdash \psi$, then the deduction $\psi$ is also true, that is, our deduction system preserves truth and for this reason, it is said to be *sound*.

**Theorem 2.5.4** (Propositional soundness). *Let $\Sigma$ be a set of wffs and $\psi$ be a wff. If $\Sigma \vdash \psi$, then $\Sigma \vDash \psi$.*

*Proof.* Suppose that $\Sigma \vdash \psi$. So let $\langle \alpha_1, \alpha_2, \ldots, \alpha_n \rangle$, where $\alpha_n = \psi$, be a deduction of $\psi$ from $\Sigma$. For each natural number $k$, we shall prove that if $1 \le k \le n$, then $\Sigma \vDash \alpha_k$. We shall use strong induction on $k$. If $k = 1$, then by Definition 2.5.1, either $\alpha_1 \in \Sigma$ or $\alpha_1$ is a tautology. It thus follows that $\Sigma \vDash \alpha_1$ (see Exercise 2 on page 37). Now let $k$ be such that $1 < k \le n$ and assume the induction hypothesis

$$\Sigma \vDash \alpha_i \quad \text{for all } i < k. \tag{IH}$$

We must show that $\Sigma \vDash \alpha_k$. Again, if $\alpha_k \in \Sigma$ or $\alpha_k$ is a tautology, then we have $\Sigma \vDash \alpha_k$. On the other hand, suppose that $\alpha_k$ is obtained by modus ponens from some $\alpha_i = (\alpha_j \to \alpha_k)$ and $\alpha_j$, where $i, j < k$. By the induction hypothesis (IH), we have $\Sigma \vDash (\alpha_j \to \alpha_k)$ and $\Sigma \vDash \alpha_j$. Hence, $\Sigma \vDash \alpha_k$ (see page 35). This completes the induction, and thus, $\Sigma \vDash \psi$. □

Our next theorem shows that the converse of Theorem 2.5.4 holds, that is, for every wff $\psi$, if $\Sigma$ tautologically implies $\psi$, then there is a deduction of $\psi$ from $\Sigma$. Thus, our deduction system is said to be *complete*.

**Theorem 2.5.5** (Propositional completeness). *Let $\Sigma$ be a set of wffs and let $\psi$ be a wff. If $\Sigma \vDash \psi$, then $\Sigma \vdash \psi$.*

1 See Exercise 10 on page 38.

*Proof.* Assume that $\Sigma \models \psi$. Corollary 2.4.5 implies that there is a finite $\Sigma_0 \subseteq \Sigma$ such that $\Sigma_0 \models \tau$. Let $\Sigma_0 = \{\alpha_1, \alpha_2, \ldots \alpha_n\}$. Exercise 6 below implies that

$$\alpha_1 \to \alpha_2 \to \cdots \to \alpha_n \to \psi$$

is a tautology. By repeated use of modus ponens, it easily follows that $\Sigma_0 \vdash \psi$. As $\Sigma_0 \subseteq \Sigma$, we conclude that $\Sigma \vdash \psi$. □

Theorems 2.5.4 and 2.5.5 immediately establish the following result.

**Theorem 2.5.6.** *Let $\Sigma$ be a set of wffs and let $\psi$ be a wff. Then $\Sigma \models \psi$ if and only if $\Sigma \vdash \psi$.*

Our next theorem can sometimes be used to indirectly show that a deduction exists. Our proof of this result applies Theorem 2.5.6.

**Theorem 2.5.7** (Deduction theorem). *Let $\Sigma$ be a set of wffs and let $\alpha$ and $\beta$ be wffs. Then $\Sigma \vdash (\alpha \to \beta)$ if and only if $\Sigma \cup \{\alpha\} \vdash \beta$.*

*Proof.* Let $\Sigma$ be a set of wffs and let $\alpha$ and $\beta$ be wffs. We thus have

$$\begin{aligned} \Sigma \vdash (\alpha \to \beta) \quad &\text{iff} \quad \Sigma \models (\alpha \to \beta) && \text{by Theorem 2.5.6,} \\ &\text{iff} \quad \Sigma \cup \{\alpha\} \models \beta && \text{by Exercise 9 on page 38,} \\ &\text{iff} \quad \Sigma \cup \{\alpha\} \vdash \beta && \text{by Theorem 2.5.6.} \end{aligned}$$

□

Example 2.5.2 shows that $\{\gamma, \sigma\} \vdash (\gamma \wedge \sigma)$. So by Theorem 2.5.7, $\{\gamma\} \vdash (\sigma \to (\gamma \wedge \sigma))$, where $\Sigma = \{\gamma\}$, $\alpha$ is $\sigma$, and $\beta$ is $(\gamma \wedge \sigma)$.

In the conditional $\alpha \to \beta$, $\alpha$ is called the *hypothesis* and $\beta$ is called the *conclusion*. In a mathematical proof of a conditional "if _, then _", one typically assumes the hypothesis and then derives the conclusion. The deduction theorem affirms that this is a valid technique of proof in mathematics.

Our final theorem of this section verifies that "proof by contradiction" is a valid proof technique.

**Definition 2.5.8.** A set of wffs $\Sigma$ is said to be *inconsistent* if there exists a wff $\beta$ such that $\Sigma \vdash \beta$ and $\Sigma \vdash \neg\beta$. Moreover, $\Sigma$ is *consistent* if for no wff $\beta$ we have $\Sigma \vdash \beta$ and $\Sigma \vdash \neg\beta$.

**Theorem 2.5.9.** *Let $\Sigma$ be a set of wffs and let $\alpha$ be a wff. If $\Sigma \cup \{\neg\alpha\}$ is inconsistent, then $\Sigma \vdash \alpha$.*

*Proof.* Assume that $\Sigma \cup \{\neg\alpha\}$ is inconsistent. Hence, for some wff $\beta$ we have $\Sigma \cup \{\neg\alpha\} \vdash \beta$ and $\Sigma \cup \{\neg\alpha\} \vdash \neg\beta$. Thus, by Theorem 2.5.7 (the deduction theorem), we have $\Sigma \vdash (\neg\alpha \to \beta)$ and $\Sigma \vdash (\neg\alpha \to \neg\beta)$. By Exercise 5 on page 37,

$$(\neg\alpha \to \beta) \to (\neg\alpha \to \neg\beta) \to \alpha$$

is a tautology. Exercise 4 below now implies that $\Sigma \vdash \alpha$. □

There are alternative deduction systems for propositional logic that are equivalent to the one presented in this section. In particular, there are deduction systems that start with a finite set of tautologies and from this finite set one can then deduce all of the other tautologies. Consequently, the deductions in such systems can be much longer than those presented in this text. On the other hand, there are deduction systems that consist only of a finite number of inference rules and from these rules one can deduce all of the tautologies. Again, such deductions can be longer than the ones presented here. For each of these alternative systems, one can prove a corresponding analogue of Theorem 2.5.6, and thus these deduction systems are equivalent. In an introductory course in mathematical logic, we believe that one should present a deduction system that is succinct and produces the desired results.

**Exercises 2.5.**

1. Let $\Sigma$ be an infinite set of wffs and let $\alpha$ be a wff. Show that if $\Sigma \vdash \alpha$, then there is a finite $\Sigma_0 \subseteq \Sigma$ such that $\Sigma_0 \vdash \alpha$.
2. Let $\Sigma$ be a set of wffs and let $\alpha$ be a wff. Let $\Sigma_0 \subseteq \Sigma$. Show that if $\Sigma_0 \vdash \alpha$, then $\Sigma \vdash \alpha$.
3. Let $\Sigma$ be a set of wffs and let $\psi$ be a tautology. Show that $\Sigma \vdash \psi$.

*4. Let $\Sigma$ be a set of wffs and let $\theta, \psi$ be wffs. Suppose that $\Sigma \vdash \theta$ and $\Sigma \vdash (\theta \to \psi)$. Show that $\Sigma \vdash \psi$.

5. Let $\Sigma_1$ and $\Sigma_2$ be sets of wffs and let $\alpha$ and $\beta$ be wffs. Suppose that $\Sigma_1 \vdash \alpha$ and $\Sigma_2 \vdash \beta$. Show that $\Sigma_1 \cup \Sigma_2 \vdash (\alpha \wedge \beta)$.

*6. Let $n \geq 1$ be a natural number and let $\Sigma^n = \{\alpha_1, \alpha_2, \dots \alpha_n\}$ be a set of $n$ wffs. Prove that for all wffs $\theta$, if $\Sigma^n \vDash \theta$, then $\alpha_1 \to \alpha_2 \to \cdots \to \alpha_n \to \theta$ is a tautology.

7. Let $n \geq 1$ be a natural number and let $\Sigma^n = \{\alpha_1, \alpha_2, \dots \alpha_n\}$ be a set of $n$ wffs. Prove that for all wffs $\theta$, if $\alpha_1 \to \alpha_2 \to \cdots \to \alpha_n \to \theta$ is a tautology, then $\Sigma^n \vDash \theta$.
8. Give a deduction from the set $\Sigma = \{\neg \mathbf{A} \vee \mathbf{B}, \mathbf{B} \to \mathbf{C}, \mathbf{A}\}$ whose last component is $\mathbf{C}$.
9. Let $\Sigma$ be an infinite set of wffs. Show that if $\Sigma$ is inconsistent, then a finite subset of $\Sigma$ is inconsistent.

**Exercise Notes:** For Exercise 6, one can repeatedly apply Exercise 9 on page 38 (see Remark 2.2.13(b)). A more formal proof of this result can be obtained by induction on $n$. For Exercise 7, a proof of this result is obtained by induction on $n$ (see (2.4) on page 37).

# 3 First-order logic

Mathematical logic is a branch of mathematics that investigates the foundations of mathematics. In this chapter, we shall do the same. Specifically, in Section 3.1, we discuss first-order languages, together with some examples of first-order languages. First-order logic is rich enough to formalize virtually all of mathematics. In Section 3.2, we will investigate mathematical structures (models) and Tarski's definition of *truth* (satisfaction) in a structure. In Section 3.3, we shall examine the definition of *proof* (deduction) in a first-order language.

## 3.1 First-order languages

In this section, we will formally define the syntax of the language of first-order logic. First-order logic deals with formal statements that are expressed in terms of predicates, logical connectives, variables, and quantifiers. A preview of such a logic is given in Section 1.2.3. We will first identify the symbols of the language. An expression will then be any finite string of these symbols. Some expressions will be nonsensical, while others will be meaningful. Some of the meaningful expressions will denote terms which act as the nouns and pronouns of the language; the terms can be interpreted as naming an individual object. Once we have the terms of the language, we can define the atomic formulas of the language. Atomic formulas are analogous to the sentence symbols of propositional logic. We can then identify the correct rules of grammar and define the well-formed formulas (wffs) of the language. To specify the terms and wffs requires definition by recursion (see Section 1.1.5).

**Definition 3.1.1.** The alphabet of a first-order language $\mathcal{L}$ consists of the following distinct symbols:

A. Required symbols
   1. Parentheses: (, ).
   2. Logical connectives: $\to$, $\neg$.
   3. Variables: $v_1, v_2, v_3, v_4, \ldots, v_n, \ldots$.
   4. Predicate symbols: For each natural number $n > 0$, a (possibly empty) set of $n$-place predicate symbols.
   5. Quantifier symbol: $\forall$.

B. Optional symbols
   1. Equality symbol: $\doteq$, a 2-place relation.
   2. Constant symbols: A set of symbols, for example, $\{c_1, c_2, \ldots\}$.
   3. Function symbols: For each natural number $n > 0$, a (possibly empty) set of $n$-place function symbols.

https://doi.org/10.1515/9783110782073-003

A dot over a familiar symbol is used to emphasize that the symbol has to be interpreted. This is also done to make a distinction between the symbol itself and the object that it commonly represents.

A finite string of symbols from the language $\mathcal{L}$ will be called an *$\mathcal{L}$-expression*. For example, $v_3c_4c_1)v_3$ is an $\mathcal{L}$-expression that starts with the symbol $v_3$. Moreover, $v_3c_4$ is a proper initial segment of $v_3c_4c_1)v_3$. The $\mathcal{L}$-expression $v_3c_4c_1)v_3$ does not appear to be expressive. We will soon isolate the meaningful $\mathcal{L}$-expressions from those that are meaningless.

There exists a one-to-one correspondence between each $\mathcal{L}$-expression $\alpha$ and a finite sequence of symbols in $\mathcal{L}$ denoted by $\langle\alpha^*\rangle$, where $\alpha^*$ is the result of putting commas between all of the symbols in $\alpha$. So an $\mathcal{L}$-expression $\alpha$ is a *proper initial segment* of the $\mathcal{L}$-expression $\beta$ when $\langle\alpha^*\rangle$ is a proper initial segment of $\langle\beta^*\rangle$ (see Definition 1.1.6).

In the formal language $\mathcal{L}$, we have listed only the logical connectives $\to$ and $\neg$. Since these two connectives are tautologically complete (see Exercise 5 on page 43), there is no need to add more.

We also only identified the universal quantifier $\forall$. Since $\neg\forall x\neg Px$ is equivalent to $\exists xPx$ (see Quantifier Negation Laws 1.2.4(3)), the existential quantifier $\exists x$ can be viewed as an abbreviation for $\neg\forall x\neg$.

In the language $\mathcal{L}$, an $n$-place function symbol is intended to represent a function of $n$ relevant $\mathcal{L}$-expressions $\varepsilon_1, \varepsilon_2, \ldots, \varepsilon_n$. Let $f$ be a 3-place function with input $\varepsilon_1, \varepsilon_2, \varepsilon_3$. In mathematics $f(\varepsilon_1, \varepsilon_2, \varepsilon_3)$ would be the standard notation for the output value; however, if $f$ is a function symbol in the language $\mathcal{L}$, we shall represent this output value by using the Polish notation $f\varepsilon_1\varepsilon_2\varepsilon_3$, where there are no parentheses or commas.

An $n$-place predicate symbol is intended to represent a property or relationship of $n$ relevant $\mathcal{L}$-expressions; for example, if $P$ is a 4-place predicate symbol in the language $\mathcal{L}$, then $P\varepsilon_1\varepsilon_2\varepsilon_3\varepsilon_4$ can be viewed as an assertion about $\varepsilon_1, \varepsilon_2, \varepsilon_3, \varepsilon_4$. Predicate symbols are sometimes also called *relation symbols*.

The predicate, constant, and function symbols can be viewed as the parameters of the language. To specify a language we must identify the particular predicate, constant, and function symbols that we wish to use. Suppose our language $\mathcal{L}$ contains the equality symbol, the predicate symbols $P_1, P_2, \ldots$, constant symbols $c_1, c_2, \ldots$, and function symbols $f_1, f_2, \ldots$. In the future we shall describe a language $\mathcal{L}$ by expressing it as a set of these selected parameters, that is, we shall say that $\mathcal{L}$ is the set

$$\mathcal{L} = \{P_1, P_2, \ldots, c_1, c_2, \ldots, f_1, f_2, \ldots, \dot{=}\}.$$

**Example 3.1.2** (Language of groups). When working with groups, one employs the language $\mathcal{L} = \{e, *, \dot{=}\}$, which has a 2-place function symbol $*$ for the group operation and a constant symbol $e$ for the identity element. We can write $v_1 * v_2$ and $v_1 \dot{=} v_2$ to represent the Polish notation $\dot{=}v_1v_2$ and $*v_1v_2$, respectively.

**Example 3.1.3** (Language of set theory). Set theory uses the language $\mathcal{L} = \{\dot{\in}, \dot{=}\}$, where $\dot{\in}$ is a 2-place predicate symbol. We shall write $v_1 \mathbin{\dot{\in}} v_2$ to denote $\dot{\in} v_1 v_2$.

**Example 3.1.4** (Language of elementary number theory). In number theory one can use the language $\mathcal{L} = \{\dot{<}, \dot{0}, \dot{S}, \dot{+}, \dot{\times}, \dot{E}, \dot{=}\}$, which has a 2-place relation symbol $\dot{<}$, a constant symbol $\dot{0}$, a 1-place function symbol $\dot{S}$ (successor), and 2-place function symbols $\dot{+}$ (addition), $\dot{\times}$ (multiplication), and $\dot{E}$ (exponentiation). Using Polish notation with these function symbols,

$$
\begin{aligned}
\dot{S}x &\text{ denotes “}x \mathbin{\dot{+}} 1,\text{”}\\
\dot{+}xy &\text{ denotes “}x \mathbin{\dot{+}} y,\text{”}\\
\dot{\times}xy &\text{ denotes “}x \mathbin{\dot{\times}} y,\text{”}\\
\dot{E}xy &\text{ denotes “}x^y.\text{”}
\end{aligned}
$$

We will write $x \mathbin{\dot{<}} y$ to represent $\dot{<}xy$. The Polish notation $\dot{S}\dot{S}\dot{0}$ can be translated to $\dot{S}(\dot{S}(\dot{0}))$. Continuing in this vein, how does one translate $\dot{+}\dot{S}\dot{S}\dot{0}\dot{S}\dot{S}\dot{S}\dot{0}$? Since $\dot{S}\dot{S}\dot{S}\dot{0}$ translates to $\dot{S}(\dot{S}(\dot{S}(\dot{0})))$ and $\dot{S}\dot{S}\dot{0}$ translates to $\dot{S}(\dot{S}(\dot{0}))$, we see that $\dot{+}SS\dot{0}\dot{S}\dot{S}\dot{0}$ translates to $\dot{S}(\dot{S}(\dot{0})) \mathbin{\dot{+}} \dot{S}(\dot{S}(\dot{S}(\dot{0})))$. For any natural number $n$, we will write

$$\dot{S}^n\dot{0} = \overbrace{\dot{S}\dot{S}\cdots\dot{S}}^{n \text{ times}}\,\dot{0}.$$

**Example 3.1.5** (A language for real analysis). In real analysis, one could use the language $\mathcal{L} = \{\dot{<}, \dot{-}, \dot{0}, c, |\,|, f, \dot{=}\}$ which has a 2-place relation symbol $\dot{<}$ (less than), constant symbols $\dot{0}$ (zero) and $c$, a 2-place function symbol $\dot{-}$ (subtraction), and 1-place function symbols $|\,|$ (absolute value) and $f$.

### 3.1.1 Terms and atomic formulas

We will now describe how to generate the terms of a language $\mathcal{L}$. The method we shall use to construct the terms is not new; it is just an application of Theorem 1.1.24. Let $U$ be the set of all $\mathcal{L}$-expressions. For each $n$-place function symbol $f$ in the language $\mathcal{L}$, define the (mathematical) function $\mathcal{E}_f : U^n \to U$ by

$$\mathcal{E}_f(\varepsilon_1, \varepsilon_2, \ldots, \varepsilon_n) = f\varepsilon_1\varepsilon_2\cdots\varepsilon_n. \tag{3.1}$$

**Definition 3.1.6.** Let $\mathcal{F} = \{\mathcal{E}_f : f \text{ is a function symbol in } \mathcal{L}\}$ and let $\mathcal{T}$ be the set of all the variables and constants in the language $\mathcal{L}$. Let $\overline{\mathcal{T}}$ be the set generated from $\mathcal{T}$ by the functions in $\mathcal{F}$. An $\mathcal{L}$-expression $\tau$ is an *$\mathcal{L}$-term* if and only if $\tau \in \overline{\mathcal{T}}$.

So the set of $\mathcal{L}$-terms $\overline{\mathcal{T}}$ is the set of $\mathcal{L}$-expressions that can be constructed by starting with the variables and constants and by repeatedly applying the functions in $\mathcal{F}$. This may all seem a bit abstract. Let us try to make Definition 3.1.6 a little more concrete by revisiting Theorem 1.1.24 in the current context.

Let $C_0 = \mathcal{T}$ be the set of all the variables and constants in the language $\mathcal{L}$. Constants and variables are $\mathcal{L}$-terms, and we can use these to build more $\mathcal{L}$-terms. Let $h$ be a 3-place function symbol in $\mathcal{L}$. Thus,

$$\mathcal{E}_h(\varepsilon_1, \varepsilon_2, \varepsilon_3) = h\varepsilon_1\varepsilon_2\varepsilon_3,$$

by (3.1). Now, using $\mathcal{E}_h$, the following set produces more $\mathcal{L}$-terms:

$$\mathcal{E}_h[C_0] = \{h\varepsilon_1\varepsilon_2\varepsilon_3 : \varepsilon_1, \varepsilon_2, \varepsilon_3 \in C_0\}.$$

Thus, $t \in \mathcal{E}_h[C_0]$ if and only if $t = h\varepsilon_1\varepsilon_2\varepsilon_3$, where $\varepsilon_1, \varepsilon_2, \varepsilon_3$ are constants or variables. For example, $t_1 = hv_7c_5v_3$ and $t_2 = hc_4c_1v_2$ are $\mathcal{L}$-terms in $\mathcal{E}_h[C_0]$. This is only a small sample of all the $\mathcal{L}$-terms that can be constructed. For any other function symbol $g$ in $\mathcal{L}$, there are new $\mathcal{L}$-terms in $\mathcal{E}_g[C_0]$ as well. To put all of these $\mathcal{L}$-terms together in one set, we define $C_1 = C_0 \cup \bigcup\{\mathcal{E}_f[C_0] : \mathcal{E}_f \in \mathcal{F}\}$, where $\mathcal{F}$ is as in Definition 3.1.6. So $t_1, t_2 \in C_1$. Using all of the $\mathcal{L}$-terms in $C_1$, we can build more $\mathcal{L}$-terms by letting $C_2 = C_1 \cup \bigcup\{\mathcal{E}_f[C_1] : \mathcal{E}_f \in \mathcal{F}\}$. So, for example, if $g$ is a 4-place function symbol in $\mathcal{L}$, then $gv_4t_1c_1t_2 \in C_2$. Note that

$$gv_4t_1c_1t_2 = gv_4hv_7c_5v_3c_1hc_4c_1v_2.$$

By repeating this process, we get the set of $\mathcal{L}$-terms

$$C_{n+1} = C_n \cup \bigcup\{\mathcal{E}_f[C_n] : \mathcal{E}_f \in \mathcal{F}\}$$

for each $n$. Then the set of all the $\mathcal{L}$-terms is $\overline{\mathcal{T}} = \bigcup_{n\in\mathbb{N}} C_n$, that is, the set $\overline{\mathcal{T}}$ is generated from $\mathcal{T}$ by the functions in $\mathcal{F}$. Definition 3.1.6 and the previous discussion justify the following useful remark.

**Remark 3.1.7.** An $\mathcal{L}$-expression $t$ is an $\mathcal{L}$-term of a language $\mathcal{L}$ if and only if either
1. $t$ is a variable, or
2. $t$ is a constant symbol, or
3. $t$ is $ft_1t_2\cdots t_n$, where $f$ is an $n$-place function symbol of $\mathcal{L}$ and each $t_i$ is an $\mathcal{L}$-term.

If the language $\mathcal{L}$ has no function symbols, then the terms are just the constants and the variables.

### 3.1.2 Induction on terms principle

Since there is a procedure for building all of the $\mathcal{L}$-terms by starting with the variables and constants and then repeatedly applying the function symbols of $\mathcal{L}$, Theorem 1.1.25 yields the following *term induction principle* and associated proof strategy.

**Term Induction Principle.** Let $\mathbb{S}(t)$ be a statement about $\mathcal{L}$-terms $t$. If

1. $\mathbb{S}(v)$ and $\mathbb{S}(c)$ hold for all the variables $v$ and constants $c$ and
2. for all $n$-place function symbols $f$ in $\mathcal{L}$ and all $\mathcal{L}$-terms $t_1, t_2, \ldots, t_n$, if $S(t_i)$ holds for each $1 \leq i \leq n$, then $\mathbb{S}(ft_1t_2 \cdots t_n)$,

then $\mathbb{S}(t)$ is true for *all* $\mathcal{L}$-terms $t$.

**Proof Strategy.** In order to prove an assertion "for all $\mathcal{L}$-terms $t$, $\mathbb{S}(t)$" by induction on $\mathcal{L}$-terms, use the following proof diagram:

*Base step:* Prove $\mathbb{S}(v)$ and $\mathbb{S}(c)$ for all the variables $v$ and constants $c$.
*Inductive step:* Let $f$ be an $n$-place function symbol and let $t_1, t_2, \ldots, t_n$ be $\mathcal{L}$-terms.
Assume $\mathbb{S}(t_i)$ for each $1 \leq i \leq n$.
Prove $\mathbb{S}(ft_1t_2 \cdots t_n)$.

#### Applications of the term induction principle

**Theorem 3.1.8.** *Let $\mathcal{L}$ be a language. For all $\mathcal{L}$-terms $t$ and $\tau$, neither $t$ nor $\tau$ is a proper initial segment of the other.*

*Proof.* For all $\mathcal{L}$-expressions $\tau$ and $t$, we shall write $\tau \prec t$ to mean that $\tau$ is a proper initial segment of $t$ and write $\tau \nprec t$ to mean that $\tau$ is *not* a proper initial segment of $t$. Consider the following statement about $\mathcal{L}$-terms $t$:

$$\text{For all } \mathcal{L}\text{-terms } \tau, \tau \nprec t \text{ and } t \nprec \tau.$$

We prove that the above statement holds for all $\mathcal{L}$-terms $t$ by induction.

*Base step:* Let $t$ be a variable or a constant symbol and let $\tau$ be any $\mathcal{L}$-term. As $t$ has no proper initial segments, we see that $\tau \nprec t$. Since $t$ is either a variable or a constant symbol, Exercise 1 below implies that $t \nprec \tau$.

*Inductive step:* Let $f$ be an $n$-place function symbol in $\mathcal{L}$ and also let $t_1, t_2, \ldots, t_n$ be $\mathcal{L}$-terms. Assume the induction hypothesis:

$$\text{For all } \mathcal{L}\text{-terms } \tau, \tau \nprec t_i \text{ and } t_i \nprec \tau, \text{ whenever } 1 \leq i \leq n. \tag{IH}$$

Let $\tau$ be any $\mathcal{L}$-term. We must prove that

$$\tau \nprec ft_1t_2 \cdots t_n \quad \text{and} \quad ft_1t_2 \cdots t_n \nprec \tau.$$

First we show that $\tau \nprec ft_1t_2\cdots t_n$. Assume, to the contrary, that (▲) $\tau \prec ft_1t_2\cdots t_n$. Hence, $\tau$ must start with the symbol $f$. Since $f$ is an $n$-place function symbol and $\tau$ is a term, $\tau$ must have the form $\tau = f\tau_1\tau_2\cdots\tau_n$, where $\tau_1, \tau_2, \ldots, \tau_n$ are $\mathcal{L}$-terms. From (▲), we obtain $f\tau_1\tau_2\cdots\tau_n \prec ft_1t_2\cdots t_n$. By dropping the common starting symbol $f$, we conclude that

$$\tau_1\tau_2\cdots\tau_n \prec t_1t_2\cdots t_n. \tag{3.2}$$

As $\tau_1$ and $t_1$ are $\mathcal{L}$-terms, the induction hypothesis (IH) implies that $\tau_1$ and $t_1$ cannot be proper initial segments of one another. Thus, (3.2) implies that $\tau_1 = t_1$. Hence,

$$\tau_2\cdots\tau_n \prec t_2\cdots t_n. \tag{3.3}$$

By similar reasoning, (3.3) implies that $\tau_2 = t_2$. Continuing in this manner, we infer that $\tau_1 = t_1, \tau_2 = t_2, \ldots, \tau_{n-1} = t_{n-1}$. From (3.2), we now conclude that $t_n \prec \tau_n$, which contradicts (IH). Hence, $\tau \nprec ft_1t_2\cdots t_n$. A very similar argument shows that $ft_1t_2\cdots t_n \nprec \tau$. □

**Theorem 3.1.9** (Unique readability of terms). *Let $\mathcal{L}$ be a language and let $\mathcal{T}$ be the set of all the variables and constant symbols. Moreover, let $\overline{\mathcal{T}}$ be the set of terms generated from $\mathcal{T}$ by the functions in $\mathcal{F} = \{\mathcal{E}_f : f$ is a function symbol in $\mathcal{L}\}$. When the functions in $\mathcal{F}$ are restricted to the set of $\mathcal{L}$-terms $\overline{\mathcal{T}}$, we have the following:*
(a) *The range of each function in $\mathcal{F}$ is disjoint from $\mathcal{T}$.*
(b) *Any two distinct functions in $\mathcal{F}$ have disjoint ranges.*
(c) *Every function in $\mathcal{F}$ is one-to-one.*

*That is, the set of all terms is freely generated from the set $\mathcal{T}$ by the functions in $\mathcal{F}$.*

*Proof.* Let $\mathcal{L}$, $\mathcal{T}$, $\mathcal{F}$, and $\overline{\mathcal{T}}$ be as stated in the theorem and let all the functions in $\mathcal{F}$ be restricted to $\overline{\mathcal{T}}$.
(a) Let $\mathcal{E}_f \in \mathcal{F}$. Since every term in the range of $\mathcal{E}_f$ starts with the function symbol $f$, we see that the range of $\mathcal{E}_f$ is disjoint from $\mathcal{T}$.
(b) Let $f$ and $g$ be two distinct function symbols in $\mathcal{L}$. Since $f \neq g$, we see that the ranges of $\mathcal{E}_f$ and $\mathcal{E}_f$ are disjoint.
(c) Let $f$ be an $n$-place function symbol in $\mathcal{L}$. We must show that $\mathcal{E}_f$ is one-to-one. Let $\tau_1, \tau_2, \ldots, \tau_n$ and $t_1, t_2, \ldots, t_n$ be $\mathcal{L}$-terms. Assume that

$$\mathcal{E}_f(\tau_1, \tau_2, \ldots, \tau_n) = \mathcal{E}_f(t_1, t_2, \ldots, t_n).$$

Thus, $f\tau_1\tau_2\cdots\tau_n = ft_1t_2\cdots t_n$. Theorem 3.1.8, together with its proof, allows us to conclude that $\tau_1 = t_1, \tau_2 = t_2, \ldots, \tau_n = t_n$. □

The above unique readability theorem (Theorem 3.1.9) will now allow us, via Theorem 1.1.27, to recursively define a function on the $\mathcal{L}$-terms using a function that is only defined on the variables and constants of a language $\mathcal{L}$.

We are now able to define the atomic formulas.

**Definition 3.1.10.** An *atomic formula* of a language $\mathcal{L}$ is an expression of the form $Pt_1t_2 \cdots t_n$, where $P$ is an $n$-place predicate symbol of $\mathcal{L}$ and each $t_i$ is an $\mathcal{L}$-term.

For example, if the language $\mathcal{L}$ has the 2-place equality symbol, then the expression $\doteq v_i v_j$ is an atomic formula as the variables are $\mathcal{L}$-terms. In the future, we shall let $\doteq v_i v_j$ be denoted by $v_i \doteq v_j$. In the language of set theory, $\dot{\in} v_i v_j$ is an atomic formula, which we will denote by $v_i \dot{\in} v_j$. As we will see, the atomic formulas will play a role similar to that of the sentence symbols of propositional logic.

Theorem 3.1.8 directly implies an analogous result for atomic formulas.

**Theorem 3.1.11.** *Let $\mathcal{L}$ be a language. No proper initial segment of an atomic formula is itself an atomic formula.*

### 3.1.3 Well-formed formulas

We will now formally define the concept of a well-formed formula (wff or formula) in first-order logic. Informally speaking, wffs are the atomic formulas and those expressions that can be built up from the atomic formulas using the logical connectives and the quantifier symbol.

Before defining wffs, we need the following *formula building functions*. Let $\alpha$ and $\beta$ be expressions. Then

$$\begin{aligned} \mathcal{E}_{\neg}(\alpha) &= (\neg\alpha), \\ \mathcal{E}_{\to}(\alpha, \beta) &= (\alpha \to \beta), \\ \mathcal{E}_{Q_i}(\alpha) &= \forall v_i \alpha, \end{aligned} \tag{3.4}$$

where $i = 1, 2, 3, \ldots$. The following definition is a special case of Theorem 1.1.24.

**Definition 3.1.12.** Let $\mathcal{L}$ be a language, let $\mathcal{F} = \{\mathcal{E}_{\neg}, \mathcal{E}_{\to}, \mathcal{E}_{Q_1}, \mathcal{E}_{Q_2}, \ldots\}$, and let $\mathcal{S}$ be a set of all of the atomic formulas in the language $\mathcal{L}$. Let $\overline{\mathcal{S}}$ be the set generated from $\mathcal{S}$ by the functions in $\mathcal{F}$. An $\mathcal{L}$-expression $\alpha$ is a *wff* if and only if $\alpha \in \overline{\mathcal{S}}$.

We will also say that $\alpha$ is an *$\mathcal{L}$-formula*, or an *$\mathcal{L}$-wff*, to mean that $\alpha$ is a wff in the language $\mathcal{L}$. The following remark is justified by Definition 3.1.12.

**Remark 3.1.13.** An $\mathcal{L}$-expression $\psi$ is a wff if and only if either
1. $\psi$ is an atomic formula, or
2. $\psi$ has the form $(\neg\alpha)$, where $\alpha$ is a wff, or
3. $\psi$ has the form $(\alpha \to \beta)$, where $\alpha$ and $\beta$ are wffs, or
4. $\psi$ has the form $\forall v_i \alpha$, where $\alpha$ is a wff and $i \geq 1$.

**Example 3.1.14.** Let $\mathcal{L} = \{P, f, c, \doteq\}$, where $P$ is a 2-place predicate symbol, $f$ is a 1-place function symbol, and $c$ is a constant symbol. The following are $\mathcal{L}$-formulas:

1. $\forall v_1(Pv_1c \to v_1 \doteq fc)$, where $Pv_1c$ and $v_1 \doteq fc$ are atomic formulas,
2. $\forall v_1(fv_2 \doteq fc \to (\neg Pfv_1c))$, where $fv_2 \doteq fc$ and $Pfv_1c$ are atomic formulas,
3. $\forall v_1\forall v_2(fv_1 \doteq fv_2 \to v_1 \doteq v_2)$, where $fv_1 \doteq fv_2$ and $v_1 \doteq v_2$ are atomic formulas.

### 3.1.4 Induction on wffs principle

Since Definition 3.1.12 ensures that there is a procedure for building each wff by first starting with atomic formulas and then applying the connective symbols $\neg, \to$ and the quantifier symbol $\forall$, Theorem 1.1.25 validates the following *induction on wffs principle*.

**Wff Induction Principle.** Let $\mathbb{S}(\alpha)$ be a statement about an arbitrary wff $\alpha$. If
1. $\mathbb{S}(\phi)$ is true for all atomic formulas $\phi$ and
2. for all wffs $\alpha$ and $\beta$, $\mathbb{S}(\alpha)$ and $\mathbb{S}(\beta)$ imply that $\mathbb{S}((\neg\alpha))$, $\mathbb{S}((\alpha \to \beta))$, and $\mathbb{S}(\forall v_i\alpha)$,

then $\mathbb{S}(\alpha)$ is true for *all* wffs $\alpha$.

**Proof Strategy.** In order to prove an assertion "for all wffs $\alpha$, $\mathbb{S}(\alpha)$" by induction, use the following diagram:

*Base step:* Prove $\mathbb{S}(\phi)$ for all atomic formulas $\phi$.
*Inductive step:* Let $\alpha$ and $\beta$ be arbitrary wffs.
Assume $\mathbb{S}(\alpha)$ and $\mathbb{S}(\beta)$.
Prove $\mathbb{S}((\neg\alpha))$, $\mathbb{S}((\alpha \to \beta))$, and $\mathbb{S}(\forall v_i\alpha)$.

**Applications of wff induction principle**

**Theorem 3.1.15.** *Let $\mathcal{L}$ be a language. For all wffs $\alpha$ and $\beta$, neither $\alpha$ nor $\beta$ is a proper initial segment of the other.*

*Proof.* For all $\mathcal{L}$-expressions $\psi$ and $\varphi$, we shall write $\psi \prec \varphi$ to mean that $\psi$ is a proper initial segment of $\varphi$ and write $\psi \not\prec \varphi$ to mean that $\psi$ is *not* a proper initial segment of $\varphi$. Consider the following statement about wffs $\alpha$:

$$\text{For all wffs } \gamma, \gamma \not\prec \alpha \text{ and } \alpha \not\prec \gamma.$$

We prove that the above statement holds for all wffs $\alpha$ by induction.

*Base step:* Let $\alpha$ be an atomic formula and let $\gamma$ be an arbitrary wff. To show that $\gamma \not\prec \alpha$, assume, to the contrary, that $\gamma \prec \alpha$. Since $\alpha$ is an atomic formula, there exist an $n$-place predicate symbol $P$ and terms $t_1, t_2, \ldots, t_n$ such that $\alpha = Pt_1t_2\cdots t_n$. Hence, (▲) $\gamma \prec Pt_1t_2\cdots t_n$. So $\gamma$ starts with the predicate symbol $P$. As $\gamma$ is a wff, it follows that $\gamma$ must also be an atomic formula. Thus, (▲) contradicts Theorem 3.1.11. A similar argument shows that $\alpha \not\prec \gamma$.

*Inductive step:* Let $\alpha$ and $\beta$ be wffs. Assume the induction hypothesis:

$$\text{For all wffs } \gamma \text{, we have } \gamma \nprec \alpha, \alpha \nprec \gamma \text{ and } \gamma \nprec \beta, \beta \nprec \gamma. \tag{IH}$$

Let $\gamma$ be any wff. We must prove that $\gamma \nprec \psi$ and $\psi \nprec \gamma$ whenever $\psi$ has the form
(1) $(\neg\alpha)$,
(2) $(\alpha \to \beta)$, or
(3) $\forall v_i \alpha$.

Let us first consider case (2). To show that $\gamma \nprec (\alpha \to \beta)$, assume (for a contradiction) that (▼) $\gamma \prec (\alpha \to \beta)$. So $\gamma$ starts with the symbol (. Thus, as $\gamma$ is a wff, $\gamma$ must have the form $(\neg\vartheta)$ or $(\zeta \to \varphi)$, where $\vartheta$, $\zeta$, and $\varphi$ are wffs (see Remark 3.1.13). If $\gamma$ had the form $(\neg\vartheta)$, then (▼) would imply that the wff $\alpha$ starts with the symbol $\neg$, which is impossible. Hence, $\gamma = (\zeta \to \varphi)$, and thus $(\zeta \to \varphi) \prec (\alpha \to \beta)$. Therefore, by dropping the left parenthesis, the induction hypothesis (IH) implies that $\zeta = \alpha$. It now follows that $\varphi \prec \beta$ (why?), which contradicts (IH). An analogous argument shows that $(\alpha \to \beta) \nprec \gamma$. The proofs of cases (1) and (3) also follow by a similar argument. □

It is now straightforward to establish the following important theorem (see the proof of Theorem 2.2.1).

**Theorem 3.1.16** (Unique readability of wffs). *Let $\mathcal{L}$ be a language and let $\mathcal{S}$ be the set of all the atomic formulas. Moreover, let $\mathcal{F} = \{\mathcal{E}_\neg, \mathcal{E}_\to, \mathcal{E}_{Q_1}, \mathcal{E}_{Q_2}, \dots\}$. Let $\overline{\mathcal{S}}$ be the set generated from $\mathcal{S}$ by the functions in $\mathcal{F}$. When the functions in $\mathcal{F}$ are restricted to the set of wffs $\overline{\mathcal{S}}$, we have the following:*
(a) *The range of each function in $\mathcal{F}$ is disjoint from $\mathcal{S}$.*
(b) *Any two distinct functions in $\mathcal{F}$ have disjoint ranges.*
(c) *Every function in $\mathcal{F}$ is one-to-one.*

*That is, the set of all wffs is freely generated from the set $\mathcal{S}$ by the functions in $\mathcal{F}$.*

Theorem 3.1.16 will now allow us, via Theorem 1.1.27, to recursively define a function on the wffs of a language $\mathcal{L}$ using a function that is only defined on the atomic formulas of the language.

### 3.1.5 Free variables

A variable $v$ is free in a wff if it occurs at least once in the formula without being introduced by the quantified expression $\forall v$. In set theory one uses the language $\mathcal{L} = \{\dot{\in}, \dot{=}\}$, which has the 2-place predicate symbol $\dot{\in}$. Here are two wffs from the language of set theory and their translations into English.

1. $\forall v_1(v_1 \mathrel{\dot\in} v_1)$ English: "Every set is an element of itself."
2. $\forall v_3(v_3 \mathrel{\dot\in} v_1 \to v_3 \mathrel{\dot\in} v_2)$ English: "Every element in $v_1$ is also in $v_2$."

There is a critical difference between these two formulas. The first formula translates to a complete English sentence, whereas the second formula translates to an English expression containing the two variables $v_1$ and $v_2$. In the second formula, we shall say that $v_1$ and $v_2$ are "free variables." Note the variable in the first formula is attached to a quantifier. In this case we say that the variable is *bound by a quantifier*. For another example, let $\mathcal{L} = \{P, c, \doteq\}$, where $P$ is a 2-place predicate symbol and $c$ is a constant symbol. Then the variable $v_1$ is free in the wff $(\forall v_1 P v_1 v_1 \to (\neg P v_1 c))$ because the appearance of $v_1$ after $\to$ is not attached to the quantifier.

The above descriptions of "free" and "not free" variables may seem a bit vague. We shall now give a precise mathematical definition for the concept of a variable occurring free in a wff. The following definition by recursion is an application of Theorems 1.1.27 and 3.1.16, as will be demonstrated. It is this definition that one should use to determine whether or not a variable appears free in a wff.

**Definition 3.1.17.** Let $\mathcal{L}$ be a language with variable $v$. The concept that $v$ *occurs free* in a wff of $\mathcal{L}$ is defined recursively as follows:
1. $v$ *occurs free* in $\phi$ if $\phi$ is an atomic formula *and* $v$ appears as a symbol in $\phi$;
2. $v$ *occurs free* in $(\neg\alpha)$ if and only if $v$ occurs free in $\alpha$;
3. $v$ *occurs free* in $(\alpha \to \beta)$ if and only if $v$ occurs free in $\alpha$ *or* $v$ occurs free in $\beta$;
4. $v$ *occurs free* in $\forall v_i \alpha$ if and only if $v$ occurs free in $\alpha$ *and* $v \neq v_i$.

We will now formally justify the validity of Definition 3.1.17. Before continuing, it is recommended that one revisit Theorem 1.1.27. Let $\mathcal{L}$ be a language and let $V$ be the set of all finite sets of the variables of $\mathcal{L}$. Let $\mathcal{F} = \{\mathcal{E}_\neg, \mathcal{E}_\to, \mathcal{E}_{Q_1}, \mathcal{E}_{Q_2}, \dots\}$, let $S$ be the set of atomic formulas, and let $\overline{S}$ be the set generated from $S$ by the functions in $\mathcal{F}$. Of course, $\overline{S}$ is the set of all the wffs. For each function in $\mathcal{F}$ we define the associated functions $F_\neg\colon V \to V$, $F_\to\colon V^2 \to V$, $F_{Q_i}\colon V \to V$ (for each $i \geq 1$) as follows:

$$
\begin{aligned}
F_\neg(a) &= a,\\
F_\to(a, b) &= a \cup b,\\
F_{Q_i}(a) &= a \setminus \{v_i\}.
\end{aligned}
$$

Now let $h\colon S \to V$ be defined by

$$h(\phi) = \text{set of variables, if any, that occur in the atomic formula } \phi.$$

Theorems 1.1.27 and 3.1.16 imply that there is a unique $\overline{h}\colon \overline{S} \to V$ such that:
(1) $\overline{h}(\alpha) = h(\alpha)$ if $\alpha$ is an atomic formula,
(2) $\overline{h}((\neg\alpha)) = \overline{h}(\alpha)$,

(3) $\overline{h}((\alpha \to \beta)) = \overline{h}(\alpha) \cup \overline{h}(\beta)$,
(4) $\overline{h}(\forall v_i \alpha) = \overline{h}(\alpha) \setminus \{v_i\}$.

For every wff $\alpha$, it follows that $\overline{h}(\alpha)$ is the set of all the *free variables* in $\alpha$. Thus, if $\overline{h}(\alpha) = \varnothing$, then $\alpha$ has no free variables.

**Definition 3.1.18.** Let $\alpha$ be a wff in a language $\mathcal{L}$. If no variable occurs free in $\alpha$, then $\alpha$ is called a *sentence*, or an $\mathcal{L}$-*sentence*.

Let $\alpha$ be a sentence. When one translates $\alpha$ into English, one will obtain a complete English sentence. On the other hand, if $\beta$ is a wff in which variables occur free, then an English translation of $\beta$ will lead to an English expression containing variables.

### 3.1.6 Notational abbreviations

The limitations that our first-order languages have imposed upon us should be clear. For example, we cannot use the logical connectives $\wedge$ and $\vee$, and we cannot use the existential quantifier $\exists v$. These restrictions will now be removed by using the method of "abbreviations," which will translate our wffs into a more readable form. This method does not change our formal definition of a wff; it will only enhance the readability of our wffs. However, whenever we define or prove new results about a first-order language, we will use Definition 3.1.12 as our definition of a wff.

Since the set of logical connectives $\{\neg, \to\}$ is tautologically complete, the usage of the logical connectives $\wedge$ and $\vee$ can be expressed in terms of the connectives $\neg$ and $\to$. Moreover, the existential quantifier $\exists$ can be expressed in terms of $\neg$ and $\forall$ (recall the Quantifier Negation Law 1.2.4(3)). Hence, we will be using the following abbreviations and conventions. The word "abbreviated" is intended to mean "easier to read."

1. The expression $(\alpha \wedge \beta)$ is the abbreviated form of $(\neg(\alpha \to (\neg\beta)))$.
2. The expression $(\alpha \vee \beta)$ is the abbreviated form of $((\neg\alpha) \to \beta)$.
3. The expression $(\alpha \leftrightarrow \beta)$ is equivalent to $((\alpha \to \beta) \wedge (\beta \to \alpha))$ and thus the abbreviated form of
$$(\neg((\alpha \to \beta) \to (\neg(\beta \to \alpha)))).$$
4. The expression $\exists v\alpha$ is the abbreviated form of $(\neg\forall v(\neg\alpha))$.
5. $x \dot{=} y$ is the abbreviated form of $\dot{=} xy$. Similar abbreviations will apply to several other 2-place predicate and function symbols; namely, $x \dot{<} y$ is the abbreviated form of $\dot{<} xy$ and $x \dot{+} y$ is the abbreviated form of $\dot{+} xy$.
6. $x \not\dot{=} y$ is the abbreviated form of $\neg \dot{=} xy$. Similar abbreviations apply to the negation of a few other 2-place predicate symbols. For example, $x \not\dot{<} y$ is the abbreviated form of $\neg \dot{<} xy$.

7. The outermost parentheses need not be explicitly written. So we can write $\forall x\alpha \to \beta$ rather than $(\forall x\alpha \to \beta)$ and $\neg\alpha$ rather than $(\neg\alpha)$.
8. $\neg$, $\forall$, and $\exists$ apply to as little as possible. For example,
   (a) $\neg\alpha \wedge \beta$ denotes $(\neg\alpha) \wedge \beta$, and not $\neg(\alpha \wedge \beta)$;
   (b) $\forall x\alpha \to \beta$ denotes $(\forall x\alpha \to \beta)$, and not $\forall x(\alpha \to \beta)$;
   (c) $\exists x\alpha \wedge \beta$ denotes $(\exists x\alpha \wedge \beta)$, and not $\exists x(\alpha \wedge \beta)$.
9. $\wedge$ and $\vee$ will apply to as little as possible, given that convention 8 is observed. For example, $\alpha \wedge \beta \to \neg\gamma \vee \delta$ denotes $((\alpha \wedge \beta) \to ((\neg\gamma) \vee \delta))$.
10. When one connective is used repeatedly, grouping is to the right. For example, we will write $\alpha \wedge \beta \wedge \gamma$ to denote $\alpha \wedge (\beta \wedge \gamma)$ and $\alpha \to \beta \to \gamma$ to denote $\alpha \to (\beta \to \gamma)$.
11. We will add parentheses when necessary to ensure readability.

Given an abbreviated wff, one can eliminate all of the abbreviations and obtain the unabbreviated version. For example, consider the abbreviated wff $\exists x(\alpha \wedge \beta)$. We can begin to "expand" it to the original wff as follows:

$$\begin{aligned} \exists x(\alpha \wedge \beta) &\Leftrightarrow (\neg\forall x(\neg(\alpha \wedge \beta))) && \text{by item 4 above,} \\ &\Leftrightarrow (\neg\forall x(\neg(\neg(\alpha \to (\neg\beta))))) && \text{by item 1 above.} \end{aligned}$$

Thus, $\exists x(\alpha \wedge \beta)$ expands to $(\neg\forall x(\neg(\neg(\alpha \to (\neg\beta)))))$. Now, if required, expand $\alpha$ and $\beta$, and do a substitution. For another example, let us rewrite $\exists x\alpha \to \beta$. We obtain

$$\begin{aligned} \exists x\alpha \to \beta &\Leftrightarrow (\exists x\alpha \to \beta) && \text{adding parentheses,} \\ &\Leftrightarrow ((\neg\forall x(\neg\alpha)) \to \beta) && \text{by item 4 above.} \end{aligned}$$

Throughout the text, we will attempt to use the following conventions:
- Predicate symbols: Upper-case symbols. Also, $\dot{\in}$, $\dot{<}$, and $\dot{=}$.
- Variables: $v_i, u, v, x, y, z$.
- Function symbols: $f, g, h$. Also, $S, \dot{+}, \dot{-}, \dot{\times}$, etc.
- Constant symbols: $c_1, c_2, \ldots, a, b, c, \ldots$. Also, $\dot{0}$.
- Terms: $t, \tau$.
- Wffs: Lower-case Greek letters.
- Sets of wffs: Upper-case Greek letters.

### 3.1.7 Examples of languages

**Example 3.1.19** (Language of groups). The language $\mathcal{L} = \{e, *, \dot{=}\}$ is used in group theory. The language $\mathcal{L}$ has a 2-place function symbol $*$ for the group operation and a constant symbol $e$ for the identity element. The quantifier $\forall$ is intended to mean "for all elements in the group." Using $\mathcal{L}$ and writing $*xy$ as $x * y$, we can express the following group axioms:

1. $\forall v_1 \forall v_2 \forall v_3 (v_1 * (v_2 * v_3) \doteq (v_1 * v_2) * v_3)$, (associativity)
2. $\forall v_1 (v_1 * e \doteq v_1)$, (identity element)
3. $\forall v_1 \exists v_2 (v_1 * v_2 \doteq e)$. (inverses exists)

**Example 3.1.20** (Language of set theory). In set theory one employs the language $\mathcal{L} = \{\dot{\in}, \doteq\}$, which has a 2-place predicate symbol $\dot{\in}$. It is intended that $\forall$ should mean "for all sets." Using $\mathcal{L}$ and writing $\dot{\in}xy$ as $x \mathrel{\dot{\in}} y$, one can express the following:

1. $\forall v_1 \exists v_2 (v_1 \mathrel{\dot{\in}} v_2)$: "every set is an element of some set,"
2. $\forall v_3 (v_3 \mathrel{\dot{\in}} v_1 \to v_3 \mathrel{\dot{\in}} v_2)$: "every element in $v_1$ is also in $v_2$,"
3. $\forall v_2 \forall v_3 ((v_3 \mathrel{\dot{\in}} v_1 \wedge v_2 \mathrel{\dot{\in}} v_1) \to v_2 \doteq v_3)$: "$v_1$ has at most one element."

**Example 3.1.21** (A language for real analysis). If we are working in real analysis, then we could use a language like $\mathcal{L} = \{\dot{<}, \dot{-}, \dot{0}, c, \ell, |\,|, f\}$, which has a 2-place relation symbol $\dot{<}$ (less than); constant symbols $\dot{0}$ (zero), $c$, and $\ell$; a 2-place function symbol $\dot{-}$ (subtraction); the 1-place function symbol $|\,|$ (absolute value); and the 1-place function symbol $f$. It is intended that $\forall$ should mean "for all real numbers." The following wff in the language $\mathcal{L}$ asserts that $\lim_{x \to c} f(x) = \ell$:

$$\forall v_1 \exists v_2 (\dot{0} \mathrel{\dot{<}} v_1 \to \forall v_3 ((\dot{0} \mathrel{\dot{<}} |v_3 - c| \wedge |v_3 - c| \mathrel{\dot{<}} v_2) \to |fv_3 - \ell| \mathrel{\dot{<}} v_1)).$$

**Example 3.1.22** (Language of elementary number theory). In elementary number theory one can use the language $\mathcal{L} = \{\dot{<}, \dot{0}, \dot{S}, \dot{+}, \dot{\times}, \dot{E}, \doteq\}$, which has a 2-place relation symbol $\dot{<}$ (less than), a constant symbol $\dot{0}$, a 1-place function symbol $\dot{S}$ (successor; $\dot{S}\dot{0}$ denotes 1, $\dot{S}\dot{S}\dot{0}$ denotes 2, etc.), and three 2-place function symbols $\dot{+}$ (addition), $\dot{\times}$ (multiplication), and $\dot{E}$ (exponentiation; $\dot{E}xy$ usually denotes $x^y$). The universal quantifier $\forall$ is intended to mean "for all natural numbers." The wff

$$\forall v_1 ((\dot{S}\dot{S}\dot{S}\dot{0} \mathrel{\dot{<}} v_1 \vee v_1 \doteq \dot{S}\dot{S}\dot{S}\dot{0}) \to \forall v_2 \forall v_3 \forall v_4 (\dot{E}v_2v_1 \mathbin{\dot{+}} \dot{E}v_3v_1 \not\doteq \dot{E}v_4v_1))$$

in the language $\mathcal{L}$ asserts Fermat's last theorem: "For all $v_1 \geq 3$, the equation $v_2^{v_1} + v_3^{v_1} = v_4^{v_1}$ has no solutions."

**Exercises 3.1.**

1. Prove that every $\mathcal{L}$-term cannot have a variable or constant symbol as a proper initial segment.
2. Prove Theorem 3.1.11.
3. Let $\mathcal{L}$ be a language. Prove that every wff has an even number of parentheses.
4. Let $t_1, t_2, \ldots, t_n$ be $\mathcal{L}$-terms, where $n > 1$. Show that $t_1 t_2 \cdots t_n$ is not an $\mathcal{L}$-term.
5. Let $\mathcal{L}$ be a language and let $\alpha_1, \alpha_2, \ldots, \alpha_n$ be wffs, where $n > 1$. Show that $\alpha_1 \alpha_2 \cdots \alpha_n$ is not a wff.
6. Let $\mathcal{L}$ be a language and let $\alpha_1, \alpha_2, \ldots, \alpha_n, \beta_1, \beta_2, \ldots, \beta_n$ be wffs, where $n \geq 1$. Show that if $\alpha_1 \alpha_2 \cdots \alpha_n = \beta_1 \beta_2 \cdots \beta_n$, then $\alpha_i = \beta_i$ for all $i \leq n$.

7. Let $\mathcal{L}$ be a language and let $\alpha_1, \alpha_2, \ldots, \alpha_k, \beta_1, \beta_2, \ldots, \beta_n$ be wffs, where $1 \le k < n$. Show that $\alpha_1\alpha_2 \cdots \alpha_k \ne \beta_1\beta_2 \cdots \beta_n$.
8. Let $\mathcal{L}$ be a language and let $\alpha_1, \alpha_2, \ldots, \alpha_n, \beta_1, \beta_2, \ldots, \beta_m$ be wffs, where $n, m \ge 1$. Show that if $\alpha_1\alpha_2 \cdots \alpha_n = \beta_1\beta_2 \cdots \beta_m$, then $m = n$ and $\alpha_i = \beta_i$ for all $i \le n$.
9. In the proof of Theorem 3.1.15, complete the proof of the inductive step by establishing cases (1) and (3).
10. Let $\mathcal{L} = \{\dot{<}, \dot{0}, \dot{S}, \dot{+}, \dot{\times}, \dot{E}, \dot{=}\}$ be as in Example 3.1.22.
    (a) Construct one term using all of the function symbols $\dot{S}$, $\dot{\times}$, and $\dot{+}$.
    (b) Construct one term using all of the function symbols $\dot{S}$, $\dot{+}$, $\dot{\times}$, and $\dot{E}$.
    (c) Using only your terms in (a) and (b), give an example of an atomic formula.

*11. Let $\mathcal{L}$ be a language and let $\mathcal{T}$ be the set of all the variables and constant symbols. Let $\overline{\mathcal{T}}$ be the set of the terms of $\mathcal{L}$. Let $x \in \mathcal{T}$ be a variable and let $t \in \overline{\mathcal{T}}$. Define $h: \mathcal{T} \to \overline{\mathcal{T}}$ by

$$h(v) = \begin{cases} t, & \text{if } v = x, \\ v, & \text{if } v \ne x. \end{cases} \tag{3.5}$$

By Theorems 3.1.9 and 1.1.27, there is a unique function $\overline{h}: \overline{\mathcal{T}} \to \overline{\mathcal{T}}$ such that:
(1) $\overline{h}(v) = h(v)$ for each $v \in \mathcal{T}$;
(2) $\overline{h}(ft_1t_2 \cdots t_n) = f\overline{h}(t_1)\overline{h}(t_2) \cdots \overline{h}(t_n)$ for each $n$-place function symbol $f$ and terms $t_1, t_2, \ldots, t_n$.
For all $\tau \in \overline{\mathcal{T}}$, let $\tau_t^x = \overline{h}(\tau)$. Prove by induction on terms that for all terms $\tau$, $\tau_t^x$ is the term obtained by replacing all occurrences of $x$ in $\tau$ with $t$.

12. Let $\mathcal{L} = \{\dot{<}, \dot{0}, \dot{S}, \dot{+}, \dot{\times}, \dot{E}, \dot{=}\}$ be as in Example 3.1.22. For each of the following wffs, find the free variables, if any:
    (a) $x \dot{<} \dot{S}\dot{S}\dot{0}$,
    (b) $\dot{0} \dot{<} \dot{S}\dot{S}\dot{0}$,
    (c) $x \dot{<} \dot{S}\dot{S}y$,
    (d) $x \dot{=} \dot{S}\dot{0} \vee y \dot{=} \dot{0}$,
    (e) $\forall x(x \dot{<} \dot{S}\dot{S}\dot{0}) \to (x \dot{=} \dot{S}\dot{0} \vee x \dot{=} \dot{0})$,
    (f) $\forall x(x \dot{<} \dot{S}\dot{S}\dot{0} \to (x \dot{=} \dot{S}\dot{0} \vee x \dot{=} \dot{0}))$.
    Which of these wffs are sentences?
13. Eliminate all of the abbreviations and obtain the unabbreviated version of the following wffs, where $P, H, C, D$ are 1-place predicate symbols:
    (a) $\exists v_1 P v_1 \vee P v_1$,
    (b) $\forall v_1 P v_1 \wedge H v_1 \to \exists v_2 \neg C v_2 \vee D v_2$.
14. Prove Theorem 3.1.16.
15. Let $\mathcal{L} = \{\dot{<}, \dot{0}, \dot{S}, \dot{+}, \dot{\times}, \dot{E}, \dot{=}\}$ be as in Example 3.1.22. Write wffs that express each of the following:
    (a) "$v_5$ is even,"
    (b) "$v_5$ is odd,"

(c) "$v_5$ is a prime number,"
(d) "there is no largest even number,"
(e) "$v_1$ is a perfect square,"
(f) "every natural number is the sum of four perfect squares."

16. Let $\mathcal{L} = \{P, c\}$, where $P$ is a 3-place predicate symbol and $c$ is a constant symbol. Using the induction on wffs principle, prove that for every wff $\alpha$, the number of symbols $n$ (counting repetitions and parentheses) in $\alpha$ can be written as a linear combination of 2 and 3, that is, $n = 2a + 3b$, where $a$ and $b$ are integers. (Note: abbreviations are not allowed and a variable $v_i$ is counted as one symbol.)

**Exercise Notes:** For Exercise 1, use induction on terms. For Exercise 2, no induction is required. Read the inductive step of the proof of Theorem 3.1.15. For Exercise 7, use proof by contradiction and then conclude that $\alpha_k = \beta_k\beta_{k+1}\cdots\beta_n$.

## 3.2 Truth and structures

In Section 3.1, we investigated the syntax of first-order languages. This syntax involves certain rules of grammar that dictate the correct formation of a wff. Semantics, on the other hand, involves giving meaning to these logical formulas. In this section, we will pursue the semantics of these languages and attach meaning to their wffs. This involves the definition of a structure which interprets the parameters of the language. To define the concept of a wff being "true" in such a structure requires a precise mathematical definition. This formidable definition is due to Alfred Tarski and formalizes the intuitive meanings of the logical connectives and the quantifiers. Because of its mathematical precision, Tarski's semantic conception of truth is often said to be the best formulation of truth in a structure.

### 3.2.1 Structures for first-order languages

In order to determine the truth value of a wff that contains quantified variables, we must investigate *structures* in which one can deal with the possible values that the variables may possess. Structures will also address the following questions:

- What are the objects that the universal quantifier $\forall$ refers to?
- What objects do the constant, function, and predicate symbols represent?

Given a language of the form

$$\mathcal{L} = \{P_1, P_2, \ldots, c_1, c_2, \ldots, f_1, f_2, \ldots, \doteq\},$$

a *structure* $\mathfrak{A}$ for the language $\mathcal{L}$, or an $\mathcal{L}$-*structure*, is a sequence of the form

$$\mathfrak{A} = \langle A; P_1^{\mathfrak{A}}, P_2^{\mathfrak{A}}, \dots, c_1^{\mathfrak{A}}, c_2^{\mathfrak{A}}, \dots, f_1^{\mathfrak{A}}, f_2^{\mathfrak{A}}, \dots \rangle$$

such that:

1. The set $A$ is nonempty and is called the domain of $\mathfrak{A}$. The set $A$ is sometimes denoted by $|\mathfrak{A}|$.
2. $\mathfrak{A}$ assigns to each $n$-place predicate symbol $P$ an $n$-place relation $P^{\mathfrak{A}} \subseteq A^n$.
3. $\mathfrak{A}$ assigns to each constant symbol $c$ a member $c^{\mathfrak{A}}$ of the universe $A$.
4. $\mathfrak{A}$ assigns to each $n$-place function symbol $f$ an $n$-place function $f^{\mathfrak{A}} : A^n \to A$.
5. The equality symbol $\doteq$ will always be interpreted as "equality."

The idea is that $\mathfrak{A}$ assigns meaning to each of the parameters of the language $\mathcal{L}$. The quantifier $\forall$ is to mean "for every element in $A$." The symbol $c$ is the name of an element $c^{\mathfrak{A}}$ in $A$. Each *atomic* formula $Pt_1t_2 \cdots t_n$ is to be interpreted as asserting that the $n$-tuple of elements in $A$, named by $t_1, \dots, t_n$, is in the relation $P^{\mathfrak{A}}$. Each term of the form $ft_1t_2 \cdots t_n$ can be interpreted as being the value of the function $f^{\mathfrak{A}} : A^n \to A$ when applied to the elements in $A$ which are named by $t_1, \dots, t_n$.

**Example 3.2.1** (Groups). Consider the language $\mathcal{L} = \{e, *, \doteq\}$ of groups discussed in Example 3.1.19. A structure for this language is $\mathfrak{A} = \langle \mathbb{Q}^+; e^{\mathfrak{A}}, *^{\mathfrak{A}} \rangle$, where $\mathbb{Q}^+$ is the set of all positive rational numbers, $e^{\mathfrak{A}} = 1$, and $*^{\mathfrak{A}}$ is the usual multiplication of rational numbers. This structure is a group as it satisfies the following group axioms:

1. $\forall v_1 \forall v_2 \forall v_3 (v_1 * (v_2 * v_3) \doteq (v_1 * v_2) * v_3)$, (associativity)
2. $\forall v_1 (v_1 * e \doteq v_1)$, (identity element)
3. $\forall v_1 \exists v_2 (v_1 * v_2 \doteq e)$. (inverses exists)

**Example 3.2.2** (Set theory). Let $\mathcal{L} = \{\dot{\in}, \doteq\}$ be the language of set theory presented in Example 3.1.20. A structure for this language is $\mathfrak{A} = \langle \mathbb{N}; \in^{\mathfrak{A}} \rangle$, where

$$\dot{\in}^{\mathfrak{A}} = \{\langle m, n \rangle : m < n \text{ and } m, n \in \mathbb{N}\}.$$

The structure $\mathfrak{A}$ satisfies the sentence $\forall v_1 \exists v_2 (v_1 \dot{\in} v_2)$, as for every $v_1 \in \mathbb{N}$ there is a $v_2 \in \mathbb{N}$ such that $v_1 < v_2$.

**Example 3.2.3.** Consider the language $\mathcal{L} = \{L, f, c\}$, where $L$ is a 2-place predicate symbol, $f$ is a 1-place function symbol, and $c$ is a constant symbol. Now let $\mathfrak{A}$ be the structure $\mathfrak{A} = \langle \mathbb{N}; L^{\mathfrak{A}}, f^{\mathfrak{A}}, c^{\mathfrak{A}} \rangle$, where:

(a) $A = \mathbb{N}$,
(b) $L^{\mathfrak{A}}$ is the set of pairs $\langle m, n \rangle$ such that $m < n$,
(c) $f^{\mathfrak{A}} = S$ is the successor function $S(n) = n + 1$,
(d) $c^{\mathfrak{A}} = 0$ is the natural number zero.

The sentence $\forall x Lxfx$ is true in the structure $\mathfrak{A}$, because $n < n + 1$ for all $n \in \mathbb{N}$.

In the above examples, we have used the ambiguous notions that "$\mathfrak{A}$ satisfies $\varphi$" or that "$\varphi$ is true in $\mathfrak{A}$," where $\varphi$ is a sentence of the language. Is the concept of being true in a structure so vague that one cannot hope to give an accurate formalization of this concept? That is, can one give a precise mathematical definition for the concept of a formula being "true in a structure"?

In 1933, the mathematician Alfred Tarski published a paper in which he discussed the conditions that a definition for a "true sentence" should satisfy. In 1956, he and his colleague Robert Vaught at UC Berkeley published a revised version of this paper to serve as a definition of truth in a structure for first-order languages. In the next section, we shall present Tarski's definition of truth.

### 3.2.2 Satisfaction (Tarski's definition)

We will define a *satisfaction relation* between a structure and wffs. Let $\mathfrak{A}$ be a structure for a language $\mathcal{L}$ with domain $A$. The *satisfaction relation* between $\mathfrak{A}$ and a wff will be defined by means of the following ordered steps:

(a) We first assign each variable in $\mathcal{L}$ to an element in $A$.
(b) Using the assignment in (a), we assign each term in $\mathcal{L}$ to an element in $A$.
(c) Using the assignment in (b), we define the satisfaction relation on the wffs.

**Definition 3.2.4.** Let $\mathfrak{A}$ be an $\mathcal{L}$-structure with domain $A$. Let $V$ be the set of all the variables of $\mathcal{L}$. A function $\nu\colon V \to A$ is called a *variable assignment*. Now let $\mathcal{T}$ be the set of all the variables and constant symbols in $\mathcal{L}$. Given a variable assignment $\nu$, we shall call the function $s\colon \mathcal{T} \to A$ defined by

$$s(v) = \begin{cases} \nu(v_i), & \text{if } v = v_i, \text{ a variable,} \\ c^{\mathfrak{A}}, & \text{if } v = c, \text{ a constant symbol,} \end{cases} \tag{3.6}$$

an *assignment*.

Let $\mathfrak{A}$ be a structure for a language $\mathcal{L}$, with domain $A$. Let $s\colon \mathcal{T} \to A$ be an assignment as defined in (3.6). Let $\mathcal{F} = \{\mathcal{E}_f : \text{ is a function symbol in } \mathcal{L}\}$ (see (3.1)). Theorem 3.1.9 implies that $\overline{\mathcal{T}}$ is the set of all the terms and is freely generated from $\mathcal{T}$ by the functions in $\mathcal{F}$. For each function $\mathcal{E}_f$ in $\mathcal{F}$, we also have the corresponding function $f^{\mathfrak{A}}$ assigned by the structure $\mathfrak{A}$. Theorem 1.1.27 now implies the following result.

**Theorem 3.2.5.** *Let $\mathfrak{A}$ be an $\mathcal{L}$-structure with domain $A$ and let $s\colon \mathcal{T} \to A$ be an assignment. Then there is a unique function $\overline{s}\colon \overline{\mathcal{T}} \to A$ satisfying the following:*

(1) $\overline{s}(v) = s(v)$ *for each variable $v$;*
(2) $\overline{s}(c) = c^{\mathfrak{A}}$ *for each constant symbol $c$;*
(3) $\overline{s}(ft_1t_2\cdots t_n) = f^{\mathfrak{A}}(\overline{s}(t_1), \overline{s}(t_2), \ldots, \overline{s}(t_n))$ *for each $n$-place function symbol $f$ and terms $t_1, t_2, \ldots, t_n$.*

Theorem 3.2.5 shows that for any given assignment $s$, the extension $\overline{s}$ assigns all of the terms in $\mathcal{L}$ to elements in $A$. So each term $t$ in $\mathcal{L}$ can be viewed as a name for the element $\overline{s}(t)$ in $A$.

As we will see, a wff is satisfiable in a structure if it holds under some assignment of its variables. In the definition below, the satisfaction relation is first defined on the atomic formulas. The atomic formulas are the building blocks for constructing all of the wffs. Then we define the satisfaction relation by recursion on the more complicated formulas which are built from the building blocks using $\neg$, $\to$, and $\forall$. The validity of this recursive definition follows from Theorems 1.1.27 and 3.1.16, as will be shown.

**Definition 3.2.6** (Tarski's definition). Let $\mathcal{L}$ be a language and let $\mathfrak{A}$ be an $\mathcal{L}$-structure. We define the relation $\mathfrak{A} \models \varphi[s]$, for all assignments $s$ and all formulas $\varphi$, by recursion as follows:
(1) $\mathfrak{A} \models \doteq t_1 t_2[s]$ iff $\overline{s}(t_1) = \overline{s}(t_2)$, for terms $t_1$ and $t_2$;
(2) $\mathfrak{A} \models Pt_1 \cdots t_n[s]$ iff $\langle \overline{s}(t_1), \ldots, \overline{s}(t_n) \rangle \in P^{\mathfrak{A}}$, for atomic formulas $Pt_1 \cdots t_n$;
(3) $\mathfrak{A} \models (\neg\varphi)[s]$ iff $\mathfrak{A} \not\models \varphi[s]$;
(4) $\mathfrak{A} \models (\varphi \to \psi)[s]$ iff (if $\mathfrak{A} \models \varphi[s]$, then $\mathfrak{A} \models \psi[s]$);
(5) $\mathfrak{A} \models \forall v\varphi[s]$ iff for all $d \in A$, $\mathfrak{A} \models \varphi[s_{v|d}]$.

In (5), the assignment $s_{v|d}$ is exactly like $s$ except at the variable $v$, where $s_{v|d}(v) = d$, that is,

$$s_{v|d}(v') = \begin{cases} s(v'), & \text{if } v' \neq v, \\ d, & \text{if } v' = v. \end{cases} \tag{3.7}$$

**Example 3.2.7.** Let $\mathcal{L} = \{L, f, c\}$ and let $\mathfrak{A} = \langle \mathbb{N}; L^{\mathfrak{A}}, f^{\mathfrak{A}}, c^{\mathfrak{A}} \rangle$ be the structure for this language as defined in Example 3.2.3, that is,
(a) $A = \mathbb{N}$,
(b) $L^{\mathfrak{A}}$ is the set of pairs $\langle m, n \rangle$ such that $m < n$,
(c) $f^{\mathfrak{A}} = S$ is the successor function $S(n) = n + 1$,
(d) $c^{\mathfrak{A}} = 0$ is the natural number zero.

Let $\mathcal{T}$ be the set of all the variables and constant symbols and let $s\colon \mathcal{T} \to \mathbb{N}$ be the assignment satisfying (▲) $s(v_i) = i - 1$ for $i = 1, 2, \ldots$. So $s(v_1) = 0$, $s(v_2) = 1$. Thus,
(1) $\overline{s}(c) = 0$, by Theorem 3.2.5(2);
(1) $\overline{s}(ffc) = f^{\mathfrak{A}}(\overline{s}(fc)) = f^{\mathfrak{A}}(f^{\mathfrak{A}}(\overline{s}(c))) = S(S(0)) = 2$, by Theorem 3.2.5(3) and (c);
(2) $\overline{s}(ffv_3) = S(S(2)) = 4$ and $\overline{s}(fv_6) = S(5) = 6$, by Theorem 3.2.5(3) and (c);
(3) $\mathfrak{A} \models Lcfv_6[s]$, because $\langle \overline{s}(c), \overline{s}(fv_6) \rangle = \langle 0, 6 \rangle \in L^{\mathfrak{A}}$, by Definition 3.2.6(2);
(4) $\mathfrak{A} \models \forall v_2 Lcfv_2[s]$, because

| | | | |
|---|---|---|---|
| $\mathfrak{A} \models \forall v_2 Lcfv_2[s]$ | iff | for all $n \in \mathbb{N}, \mathfrak{A} \models Lcfv_2[s_{v_2|n}]$, | Definition 3.2.6(5) |
| | iff | for all $n \in \mathbb{N}, \langle \overline{s}_{v_2|n}(c), \overline{s}_{v_2|n}(fv_2)\rangle \in L^{\mathfrak{A}}$, | Definition 3.2.6(2) |
| | iff | for all $n \in \mathbb{N}, \langle 0, f^{\mathfrak{A}}(\overline{s}_{v_2|n}(v_2))\rangle \in L^{\mathfrak{A}}$, | Theorem 3.2.5(2)(3) |
| | iff | for all $n \in \mathbb{N}, \langle 0, f^{\mathfrak{A}}(n)\rangle \in L^{\mathfrak{A}}$, | by (3.7) |
| | iff | for all $n \in \mathbb{N}, \langle 0, S(n)\rangle \in L^{\mathfrak{A}}$, | since $f^{\mathfrak{A}} = S$ |
| | iff | for all $n \in \mathbb{N}, \langle 0, n+1\rangle \in L^{\mathfrak{A}}$, | $S(n) = n+1$ |
| | iff | for all $n \in \mathbb{N}, 0 < n+1$; | by (b) |

thus, $\mathfrak{A} \models \forall v_2 Lcfv_2[s]$ because it is true that for all $n \in \mathbb{N}$, $0 < n+1$;

(5) $\mathfrak{A} \not\models \forall v_2 Lv_3 fv_2[s]$; otherwise,

| | | | |
|---|---|---|---|
| $\mathfrak{A} \models \forall v_2 Lv_3fv_2[s]$ | iff | for all $n \in \mathbb{N}, \mathfrak{A} \models Lv_3fv_2[s_{v_2|n}]$, | Definition 3.2.6(5) |
| | iff | for all $n \in \mathbb{N}, \langle \overline{s}_{v_2|n}(v_3), \overline{s}_{v_2|n}(fv_2)\rangle \in L^{\mathfrak{A}}$, | Definition 3.2.6(2) |
| | iff | for all $n \in \mathbb{N}, \langle 2, \overline{s}_{v_2|n}(fv_2)\rangle \in L^{\mathfrak{A}}$, | by (3.7) and (▲) |
| | iff | for all $n \in \mathbb{N}, \langle 2, f^{\mathfrak{A}}(\overline{s}_{v_2|n}(v_2))\rangle \in L^{\mathfrak{A}}$, | Theorem 3.2.5(3) |
| | iff | for all $n \in \mathbb{N}, \langle 2, f^{\mathfrak{A}}(n)\rangle \in L^{\mathfrak{A}}$, | by (3.7) |
| | iff | for all $n \in \mathbb{N}, \langle 2, S(n)\rangle \in L^{\mathfrak{A}}$, | as $f^{\mathfrak{A}} = S$ |
| | iff | for all $n \in \mathbb{N}, \langle 2, n+1\rangle \in L^{\mathfrak{A}}$, | $S(n) = n+1$ |
| | iff | for all $n \in \mathbb{N}, 2 < n+1$; | a falsehood |

thus, $\mathfrak{A} \not\models \forall v_2 Lv_3 fv_2[s]$, as $2 \nless 0+1$ and $0 \in \mathbb{N}$.

**Remark 3.2.8** (Extended definition of satisfaction). The abbreviations presented in Section 3.1.6 allow us to extend Definition 3.2.6. Let $\mathfrak{A}$ be an $\mathcal{L}$-structure. Then for all assignments $s$ and all formulas $\alpha$ and $\beta$, one can establish the following extension of Definition 3.2.6:

(1) $\mathfrak{A} \models \doteq t_1 t_2[s]$ iff $\overline{s}(t_1) = \overline{s}(t_2)$, for terms $t_1$ and $t_2$;
(2) $\mathfrak{A} \models Pt_1 \cdots t_n[s]$ iff $\langle \overline{s}(t_1), \ldots, \overline{s}(t_n)\rangle \in P^{\mathfrak{A}}$, for atomic formulas $Pt_1 \cdots t_n$;
(3) $\mathfrak{A} \models (\neg\alpha)[s]$ iff $\mathfrak{A} \not\models \alpha[s]$;
(4) $\mathfrak{A} \models (\alpha \wedge \beta)[s]$ iff ($\mathfrak{A} \models \alpha[s]$ and $\mathfrak{A} \models \beta[s]$);
(5) $\mathfrak{A} \models (\alpha \vee \beta)[s]$ iff ($\mathfrak{A} \models \alpha[s]$ or $\mathfrak{A} \models \beta[s]$);
(6) $\mathfrak{A} \models (\alpha \rightarrow \beta)[s]$ iff (if $\mathfrak{A} \models \alpha[s]$, then $\mathfrak{A} \models \beta[s]$);
(7) $\mathfrak{A} \models (\alpha \leftrightarrow \beta)[s]$ iff ($\mathfrak{A} \models \alpha[s]$ iff $\mathfrak{A} \models \beta[s]$);
(8) $\mathfrak{A} \models \forall v\alpha[s]$ iff for all $d \in A$, $\mathfrak{A} \models \alpha[s_{v|d}]$;
(9) $\mathfrak{A} \models \exists v\alpha[s]$ iff for some $d \in A$, $\mathfrak{A} \models \alpha[s_{v|d}]$.

When applying Remark 3.2.8 on a wff with multiple quantifiers, note the next remark.

**Remark 3.2.9.** Let $\mathfrak{A}$ be an $\mathcal{L}$-structure and let $s: \mathcal{T} \to A$ be an assignment, where $\mathcal{T}$ is the set of the variables and constant symbols in $\mathcal{L}$. For a variable $v$ and $d \in A$, recall that the function $s_{v|d}$ is exactly like $s$, except at $v$, where $s_{v|d}(v) = d$, that is,

$$s_{v|d}(v') = \begin{cases} s(v'), & \text{if } v' \neq v, \\ d, & \text{if } v' = v. \end{cases}$$

Let $x$ be a variable where $x \neq v$ and let $e \in A$. Then, the function $(s_{v|d})_{x|e}$ is exactly like $s$, except at the variables $v$ and $x$, where

$$(s_{v|d})_{x|e}(v) = d \quad \text{and} \quad (s_{v|d})_{x|e}(x) = e.$$

Observe that

$$(s_{v|d})_{x|e} = (s_{x|e})_{v|d}. \tag{3.8}$$

On the other hand, one can show that $(s_{v|d})_{v|e} = s_{v|e}$ and $(s_{x|e})_{x|d} = s_{x|d}$.

We end this section by showing, as promised, that Definition 3.2.6 is an application of Theorems 1.1.27 and 3.1.16. Let $\mathfrak{A}$ be an $\mathcal{L}$-structure with domain $A$. Let $\mathcal{S}$ be the set of all the atomic formulas of $\mathcal{L}$, let $\mathcal{F} = \{\mathcal{E}_\neg, \mathcal{E}_\to, \mathcal{E}_{Q_1}, \mathcal{E}_{Q_2}, \dots\}$ (see (3.4)), and let $\overline{\mathcal{S}}$ be the set generated from $\mathcal{S}$ by the functions in $\mathcal{F}$. By Theorem 3.1.16, we know that $\overline{\mathcal{S}}$ is the set of all the wffs and that $\overline{\mathcal{S}}$ is freely generated from the set $\mathcal{S}$ by the functions in $\mathcal{F}$. Let $U$ be the set of all assignments and let $\mathcal{U}$ be the set of all subsets of $U$. For each function in $\mathcal{F}$ we define the associated functions $F_\neg: \mathcal{U} \to \mathcal{U}$, $F_\to: \mathcal{U}^2 \to \mathcal{U}$, $F_{Q_i}: \mathcal{U} \to \mathcal{U}$ (for each $i \geq 1$) as follows:

$$\begin{aligned} F_\neg(a) &= U \setminus a, \\ F_\to(a, b) &= (U \setminus a) \cup b, \\ F_{Q_i}(a) &= \{s \in U : \text{ for all } d \in A, s_{v_i|d} \in a\}. \end{aligned}$$

Now define $h: \mathcal{S} \to \mathcal{U}$ by

$$\begin{aligned} h(\doteq t_1 t_2) &= \{s \in U : \bar{s}(t_1) = \bar{s}(t_2)\}, \\ h(P t_1 t_2 \cdots t_n) &= \{s \in U : \langle \bar{s}(t_1), \dots, \bar{s}(t_n) \rangle \in P^{\mathfrak{A}}\}, \end{aligned}$$

for each atomic formula $\doteq t_1 t_2$ and $P t_1 t_2 \cdots t_n$. Theorems 1.1.27 and 3.1.16 now imply that there is a unique function $\bar{h}: \overline{\mathcal{S}} \to \mathcal{U}$ such that:

(1) $\bar{h}(\alpha) = h(\alpha)$ if $\alpha$ is an atomic formula,
(2) $\bar{h}((\neg\alpha)) = U \setminus \bar{h}(\alpha)$,
(3) $\bar{h}((\alpha \to \beta)) = (U \setminus \bar{h}(\alpha)) \cup \bar{h}(\beta)$,
(4) $\bar{h}(\forall v_i \alpha) = \{s \in U : \text{ for all } d \in A, s_{v_i|d} \in \bar{h}(\alpha)\}$.

Define the relation $\mathfrak{A} \models \varphi[s]$ between $\varphi$ and $s$ by

$$\mathfrak{A} \models \varphi[s] \quad \text{if and only if} \quad s \in \overline{h}(\varphi) \tag{3.9}$$

for all wffs $\varphi$ and all assignments $s$. Using (3.9) and conditions (1)–(4), one can show that the relation $\mathfrak{A} \models \varphi[s]$ satisfies Definition 3.2.6.

**Satisfaction relation for sentences**

A sentence in a first-order language has no free variables. So given a structure for this language, one may suspect that if a sentence is true in the structure, then its truth should be independent of any assignment to the variables in the language. Our focus in this section is on addressing this suspicion. First we must show that if two assignments agree on all of the variables in a term, then the two assignments will assign the term to the same element in the domain of the structure.

**Lemma 3.2.10.** *Let $\mathfrak{A}$ be a structure for a language $\mathcal{L}$. Suppose that $s$ and $s'$ are assignments that agree on all of the variables in a term $t$. Then $\overline{s}(t) = \overline{s}'(t)$, where $\overline{s}' = \overline{s'}$.*

*Proof.* We prove the following statement by induction on terms: *Whenever assignments $s$ and $s'$ agree on all of the variables in a term $t$, then $\overline{s}(t) = \overline{s}'(t)$.*

*Base step:* Let $c$ and $v$ be a constant and a variable, respectively, of the language $\mathcal{L}$. Clearly, $\overline{s}(c) = \overline{s}'(c)$ by the definition of $\overline{s}$ and $\overline{s}'$, for any two assignments $s$ and $s'$. If $s$ and $s'$ agree on the variables in $v$, then $s(v) = s'(v)$, and thus $\overline{s}(v) = \overline{s}'(v)$.

*Inductive step:* Let $f$ be an arbitrary $n$-place function symbol in $\mathcal{L}$ and let $t_1, t_2, \ldots, t_n$ be arbitrary terms. Assume the induction hypothesis

$$\text{For each } i \leq n, \text{ if } s \text{ and } s' \text{ agree on the variables in } t_i, \text{ then } \overline{s}(t_i) = \overline{s}'(t_i). \tag{IH}$$

We must prove that the same holds for the term $ft_1t_2\cdots t_n$. Let $s$ and $s'$ be assignments that agree on the variables in $ft_1t_2\cdots t_n$. Then for each $i \leq n$ $s$ and $s'$ agree on the variables in $t_i$. Hence

$$\begin{aligned}\overline{s}(ft_1t_2\cdots t_n) &= f^{\mathfrak{A}}(\overline{s}(t_1), \overline{s}(t_2), \ldots, \overline{s}(t_n)) && \text{by Theorem 3.2.5(3),}\\ &= f^{\mathfrak{A}}(\overline{s}'(t_1), \overline{s}'(t_2), \ldots, \overline{s}'(t_n)) && \text{by (IH),}\\ &= \overline{s}'(ft_1t_2\cdots t_n) && \text{by Theorem 3.2.5(3).}\end{aligned}$$

□

We can now extend Lemma 3.2.10 to formulas as well.

**Theorem 3.2.11.** *Let $\mathfrak{A}$ be an $\mathcal{L}$-structure. For all assignments $s$ and $s'$ that agree on all of the free variables in the wff $\varphi$, we have*

$$\mathfrak{A} \models \varphi[s] \quad \textit{if and only if} \quad \mathfrak{A} \models \varphi[s'].$$

*Proof.* We prove the following statement by induction on wffs: *If assignments s and s′ agree on all of the free variables in* $\varphi$, *then* $\mathfrak{A} \models \varphi[s]$ *if and only if* $\mathfrak{A} \models \varphi[s']$.

*Base step:* Let $\phi = Pt_1t_2 \cdots t_n$ be an atomic formula and let $s$ and $s'$ be assignments that agree on all of the free variables in $Pt_1 \cdots t_n$. Hence, $s$ and $s'$ agree on all of the free variables in $t_1, \ldots, t_n$. Thus,

$$\begin{aligned} \mathfrak{A} \models Pt_1t_2 \cdots t_n[s] \quad &\text{iff} \quad \langle \bar{s}(t_1), \ldots, \bar{s}(t_n) \rangle \in P^{\mathfrak{A}} && \text{by Definition 3.2.6(2)},\\ &\text{iff} \quad \langle \bar{s}'(t_1), \ldots, \bar{s}'(t_n) \rangle \in P^{\mathfrak{A}} && \text{by Lemma 3.2.10},\\ &\text{iff} \quad \mathfrak{A} \models Pt_1t_2 \cdots t_n[s'] && \text{by Definition 3.2.6(2)}. \end{aligned}$$

Therefore, the proof of the base step is complete.

*Inductive step:* Let $\alpha$ and $\beta$ be arbitrary wffs. Assume the induction hypothesis

$$\begin{aligned} \mathfrak{A} \models \alpha[s] \quad &\text{iff} \quad \mathfrak{A} \models \alpha[s'],\\ \mathfrak{A} \models \beta[s] \quad &\text{iff} \quad \mathfrak{A} \models \beta[s'], \end{aligned} \tag{IH}$$

for all assignments $s$ and $s'$ that agree on the free variables in the formulas $\alpha$ and $\beta$, respectively. We must prove that the same holds for each of the following:

$$(\neg\alpha),\ (\alpha \to \beta),\ \forall v\alpha.$$

CASE $(\neg\alpha)$: Let $s$ and $s'$ be assignments that agree on all of the free variables in $(\neg\alpha)$. It follows that $s$ and $s'$ agree on all of the free variables in $\alpha$. Therefore, the first part of the induction hypothesis (IH) holds. Hence

$$\begin{aligned} \mathfrak{A} \models (\neg\alpha)[s] \quad &\text{iff} \quad \mathfrak{A} \not\models \alpha[s] && \text{by Definition 3.2.6(3)},\\ &\text{iff} \quad \mathfrak{A} \not\models \alpha[s'] && \text{by (IH)},\\ &\text{iff} \quad \mathfrak{A} \models (\neg\alpha)[s'] && \text{by Definition 3.2.6(3)}. \end{aligned}$$

CASE $(\alpha \to \beta)$: Let $s$ and $s'$ be assignments that agree on all of the free variables in the wff $(\alpha \to \beta)$. It follows that $s$ and $s'$ agree on all of the free variables in both $\alpha$ and $\beta$. Therefore, the induction hypothesis (IH) holds. Hence

$$\begin{aligned} \mathfrak{A} \models (\alpha \to \beta)[s] \quad &\text{iff} \quad \mathfrak{A} \models \alpha[s] \text{ implies } \mathfrak{A} \models \beta[s] && \text{by Definition 3.2.6(4)},\\ &\text{iff} \quad \mathfrak{A} \models \alpha[s'] \text{ implies } \mathfrak{A} \models \beta[s'] && \text{by (IH)},\\ &\text{iff} \quad \mathfrak{A} \models (\alpha \to \beta)[s'] && \text{by Definition 3.2.6(4)}. \end{aligned}$$

CASE $\forall v\alpha$: Let $s$ and $s'$ be assignments that agree on all of the free variables in $\forall v\alpha$. Since $v$ is not free in $\forall v\alpha$, it does not follow that $s$ and $s'$ agree on the variable $v$. However, for any $d \in A$, it does follow that $s_{v|d}$ and $s'_{v|d}$ agree on $v$ and hence on all the variables in

$\alpha$. Thus, the induction hypothesis (IH) implies that $\mathfrak{A} \models \alpha[s_{v|d}]$ if and only if $\mathfrak{A} \models \alpha[s'_{v|d}]$, for any $d \in A$. Therefore,

$$\begin{aligned} \mathfrak{A} \models \forall v\alpha[s] \quad &\text{iff} \quad \text{for every } d \in A, \mathfrak{A} \models \varphi[s_{v|d}] && \text{by Definition 3.2.6(5),} \\ &\text{iff} \quad \text{for every } d \in A, \mathfrak{A} \models \varphi[s'_{v|d}] && \text{by (IH),} \\ &\text{iff} \quad \mathfrak{A} \models \forall v\alpha[s'] && \text{by Definition 3.2.6(5).} \end{aligned}$$

□

The following corollary shows that for a *sentence* $\varphi$, the truth or falsity of $\mathfrak{A} \models \varphi[s]$ is independent of the assignment $s$. Thus, we can write $\mathfrak{A} \models \varphi$ if for some (hence every) assignment $s$, we have $\mathfrak{A} \models \varphi[s]$.

**Corollary 3.2.12.** *Let $\mathfrak{A}$ be a structure for a language $\mathcal{L}$. Let $\varphi$ be a sentence. Then*

$$\mathfrak{A} \models \varphi[s] \text{ for every assignment } s \text{ if and only if } \mathfrak{A} \models \varphi[s'] \text{ for some assignment } s'.$$

*Proof.* Let $\mathfrak{A}$ be a structure for a language $\mathcal{L}$ with domain $A$. Let $\varphi$ be any sentence.

($\Rightarrow$). Assume that $\mathfrak{A} \models \varphi[s]$ for every assignment $s$. Then (since $A$ is nonempty) it follows that $\mathfrak{A} \models \varphi[s']$ for some assignment $s'$.

($\Leftarrow$). Assume that $\mathfrak{A} \models \varphi[s']$ for some assignment $s'$. We shall show that $\mathfrak{A} \models \varphi[s]$ for every assignment $s$. Let $s$ be an arbitrary assignment. Since $\varphi$ is a sentence, it has no free variables. Thus, $s$ and $s'$ agree on all the free variables in $\varphi$. Theorem 3.2.11 now implies that $\mathfrak{A} \models \varphi[s]$. □

Corollary 3.2.12 supports our next definition.

**Definition 3.2.13.** Let $\mathfrak{A}$ be an $\mathcal{L}$-structure. Let $\varphi$ be any sentence. We shall say that $\varphi$ is *true* in $\mathfrak{A}$ or that $\mathfrak{A}$ is a *model* of $\varphi$, denoted by $\mathfrak{A} \models \varphi$, if for some (or every) assignment $s$, we have $\mathfrak{A} \models \varphi[s]$. In addition, let $\Sigma$ be a set of sentences. We shall say that $\mathfrak{A}$ is a *model* of $\Sigma$ if $\mathfrak{A} \models \varphi$ for all $\varphi$ in $\Sigma$.

**Example 3.2.14.** Let $\mathcal{L} = \{\dot{0}, \dot{1}, \dot{+}, \dot{\times}, \dot{=}\}$ be the language having equality, two 2-place function symbols $\dot{+}$ and $\dot{\times}$, and two constant symbols $\dot{0}, \dot{1}$. Now consider the two structures $\mathfrak{R} = \langle \mathbb{R}; 0, 1, +, \times \rangle$ and $\mathfrak{Q} = \langle \mathbb{Q}; 0, 1, +, \times \rangle$, where $+$ and $\times$ are the standard addition and multiplication operations. Find a sentence $\varphi$ in the language $\mathcal{L}$ that is true in one of these structures but false in the other.

**Solution.** Let $\varphi$ be the sentence $\exists v(v \dot{\times} v = \dot{1} \dot{+} \dot{1})$. Then $\mathfrak{R} \models \varphi$ since $\sqrt{2} \in \mathbb{R}$. However, $\mathfrak{Q} \not\models \varphi$ because $\sqrt{2} \notin \mathbb{Q}$.

### 3.2.3 Logical implication

Logical implication is a truth preserving relation between a given set of premises and a conclusion; namely, whenever the premises are all true, the conclusion is true. The definition of logical implication in first-order logic is very similar to the definition of tautological implication in propositional logic (see Definition 2.2.10). However, the following definition of logical implication is more complicated than that of tautological implication, in part because Tarski's definition of satisfaction is complex.

For the duration of this section, let $\mathcal{L}$ be a given language and let $\mathcal{T}$ be the set of all the variables and constant symbols in $\mathcal{L}$.

**Definition 3.2.15.** Let $\Gamma$ be a set of wffs and let $\varphi$ be a wff. Then $\Gamma$ *logically implies* $\varphi$, denoted by $\Gamma \vDash \varphi$, if and only if for *every* structure $\mathfrak{A}$ and *every* assignment $s\colon \mathcal{T} \to A$, if $\mathfrak{A} \vDash \alpha[s]$ for every $\alpha$ in $\Gamma$, then $\mathfrak{A} \vDash \varphi[s]$.

**Remark 3.2.16.** Some special cases concerning Definition 3.2.15 deserve mention.
(a) If $\Gamma$ is the empty set $\varnothing$, then every structure models $\Gamma$.
(b) It follows from (a) that $\varnothing \vDash \varphi$ if and only if $\mathfrak{A} \vDash \varphi[s]$ for every structure $\mathfrak{A}$ and every assignment $s$.
(c) If there is no structure and assignment that will satisfy all of the wffs in $\Gamma$, then it is vacuously true that $\Gamma \vDash \varphi$, for any $\varphi$.
(d) If $\Gamma$ is a singleton $\{\gamma\}$, then we write $\gamma \vDash \varphi$ in place of $\{\gamma\} \vDash \varphi$.

**Definition 3.2.17.** Let $\varphi$ be a wff. Then $\varphi$ is *logically valid* (written as $\vDash \varphi$) if and only if for *every* structure $\mathfrak{A}$ and *every* assignment $s\colon \mathcal{T} \to A$, we have $\mathfrak{A} \vDash \varphi[s]$.

**Definition 3.2.18.** Two wffs $\varphi$ and $\psi$ are *logically equivalent* (denoted by $\varphi \vDash\!\!\!\!\dashv \psi$) if $\varphi \vDash \psi$ and $\psi \vDash \varphi$.

Corollary 3.2.12 implies that for sentences, Definition 3.2.15 does not depend on the assignments. So logical implication can be stated more concisely for sentences.

**Corollary 3.2.19.** *Let $\Sigma$ be a set of sentences and let $\psi$ be a sentence. Then:*
1. *$\Sigma \vDash \psi$ if and only if every model of $\Sigma$ is also a model of $\psi$;*
2. *$\psi$ is logically valid if and only if $\psi$ is true in every structure.*

**Example 3.2.20.** Let $\mathcal{L} = \{Q, P\}$, where $Q$ is a 1-place predicate symbol and $P$ is a 2-place predicate symbol. Show that the following hold:
1. $\forall v_1 Q v_1 \vDash Q v_2$,
2. $Q v_1 \nvDash \forall v_1 Q v_1$,
3. $\vDash \neg\neg\alpha \to \alpha$,
4. $\forall v_1 Q v_1 \vDash \exists v_2 Q v_2$,
5. $\exists x \forall y Pxy \vDash \forall y \exists x Pxy$,
6. $\forall y \exists x Pxy \nvDash \exists x \forall y Pxy$,
7. $\vDash \exists x(Qx \to \forall y Qy)$.

**Solution.** We will show why items 1–7 hold.
1. To show $\forall v_1 Q v_1 \vDash Q v_2$, let $\mathfrak{A}$ be an $\mathcal{L}$-structure with assignment $s\colon \mathcal{T} \to A$ such that $\mathfrak{A} \vDash \forall v_1 Q v_1[s]$. We must show that $\mathfrak{A} \vDash Q v_2[s]$, that is, we must show that $s(v_2) \in Q^{\mathfrak{A}}$.

Let $d = s(v_2)$. Since $\mathfrak{A} \models \forall v_1 Q v_1[s]$, it follows that $\mathfrak{A} \models Qv_1[s_{v_1|d}]$. Thus, $s_{v_1|d}(v_1) \in Q^{\mathfrak{A}}$, so $d \in Q^{\mathfrak{A}}$. Therefore, $s(v_2) \in Q^{\mathfrak{A}}$.

2. To show $Qv_1 \not\models \forall v_1 Q v_1$, we must find a structure $\mathfrak{A}$ and an assignment $s: \mathcal{T} \to A$ such that $\mathfrak{A} \models Qv_1[s]$ and $\mathfrak{A} \not\models \forall v_1 Q v_1[s]$. Let $A = \{1, 2\}$, $Q^{\mathfrak{A}} = \{2\}$, $P^{\mathfrak{A}} = \varnothing$ and let $\mathfrak{A} = \langle A; Q^{\mathfrak{A}}, P^{\mathfrak{A}} \rangle$. For an assignment $s$ such that $s(v_1) = 2$, one can now show that $\mathfrak{A} \models Qv_1[s]$ and $\mathfrak{A} \not\models \forall v_1 Q v_1[s]$.
3. To show $\models \neg\neg\alpha \to \alpha$, let $\mathfrak{A}$ be an $\mathcal{L}$-structure with assignment $s: \mathcal{T} \to A$. Assume that $\mathfrak{A} \models \neg\neg\alpha[s]$. We must show that $\mathfrak{A} \models \alpha[s]$. Since $\mathfrak{A} \models \neg\neg\alpha[s]$, it follows that $\mathfrak{A} \not\models \neg\alpha[s]$ by Remark 3.2.8(3). Since $\mathfrak{A} \not\models \neg\alpha[s]$, it follows that $\mathfrak{A} \models \alpha[s]$, again by Remark 3.2.8(3).
4. To show $\forall v_1 Q v_1 \models \exists v_2 Q v_2$, let $\mathfrak{A}$ be an $\mathcal{L}$-structure with assignment $s: \mathcal{T} \to A$. Assume that $\mathfrak{A} \models \forall v_1 Q v_1[s]$. We must show that $\mathfrak{A} \models \exists v_2 Q v_2[s]$, that is, we must show that for some $d \in A$ we have $\mathfrak{A} \models Qv_2[s_{v_2|d}]$. Since $\mathfrak{A} \models \forall v_1 Q v_1[s]$, it follows that for all $d \in A$, $\mathfrak{A} \models Qv_1[s_{v_1|d}]$. Thus, because the universe of $\mathfrak{A}$ is nonempty, there is a $d \in A$ such that $\mathfrak{A} \models Qv_1[s_{v_1|d}]$. Hence, $s_{v_1|d}(v_1) = d \in Q^{\mathfrak{A}}$, so $s_{v_2|d}(v_2) = d \in Q^{\mathfrak{A}}$. We conclude that $\mathfrak{A} \models \exists v_2 Q v_2[s]$.
5. To show $\exists x \forall y Pxy \models \forall y \exists x Pxy$, let $\mathfrak{A}$ be an $\mathcal{L}$-structure and let $s: \mathcal{T} \to A$ be such that $\mathfrak{A} \models \exists x \forall y Pxy[s]$. By Remark 3.2.8(8)(9), there exists a $d \in A$ such that for all $e \in A$, we have $\mathfrak{A} \models Pxy[(s_{x|d})_{y|e}]$ (see Remark 3.2.9). So, for all $e \in A$, there is a $d \in A$ such that $\mathfrak{A} \models Pxy[(s_{y|e})_{x|d}]$ (see equation (3.8)). Therefore, by Remark 3.2.8(8)(9), $\mathfrak{A} \models \forall y \exists x Pxy[s]$.
6. To show $\forall y \exists x Pxy \not\models \exists x \forall y Pxy$, we must find a structure $\mathfrak{A}$ and $s: \mathcal{T} \to A$ such that $\mathfrak{A} \models \forall y \exists x Pxy[s]$ and $\mathfrak{A} \not\models \exists x \forall y Pxy[s]$. Let $A = \mathbb{Z}$, $P^{\mathfrak{A}} = <$ (the standard less than relation), $Q^{\mathfrak{A}} = \varnothing$ and let $\mathfrak{A} = \langle A; Q^{\mathfrak{A}}, P^{\mathfrak{A}} \rangle$. For any assignment $s$, one can now show that $\mathfrak{A} \models \forall y \exists x Pxy[s]$ (because there is no largest integer) and $\mathfrak{A} \not\models \exists x \forall y Pxy[s]$ (because there is no smallest integer).
7. To show that $\models \exists x(Qx \to \forall y Qy)$, let $\mathfrak{A}$ be an $\mathcal{L}$-structure and let $s: \mathcal{T} \to A$. We must show that $\mathfrak{A} \models \exists x(Qx \to \forall y Qy)[s]$. There are two cases to consider.

   Case (i): $Q^{\mathfrak{A}} = A$. Let $d \in A$. As $Q^{\mathfrak{A}} = A$, we see that $d \in Q^{\mathfrak{A}}$. So $\mathfrak{A} \models Qx[s_{x|d}]$ and $\mathfrak{A} \models \forall y Qy[s_{x|d}]$. It thus follows that $\mathfrak{A} \models (Qx \to \forall y Qy)[s_{x|d}]$, by Remark 3.2.8(6). Therefore, $\mathfrak{A} \models \exists x(Qx \to \forall y Qy)[s]$, by Remark 3.2.8(9).

   Case (ii): $Q^{\mathfrak{A}} \neq A$. Let $d \in A$ be such that $d \notin Q^{\mathfrak{A}}$. We see that $\mathfrak{A} \not\models Qx[s_{x|d}]$. It thus follows (vacuously) that if $\mathfrak{A} \models Qx[s_{x|d}]$, then $\mathfrak{A} \models \forall y Qy[s]$. Hence, by Remark 3.2.8(6), $\mathfrak{A} \models (Qx \to \forall y Qy)[s_{x|d}]$. Therefore, $\mathfrak{A} \models \exists x(Qx \to \forall y Qy)[s]$, by Remark 3.2.8(9).

### 3.2.4 Definability over a structure

Let $\mathfrak{A}$ be an $\mathcal{L}$-structure with universe $A$. Some subsets of $A$ and relations on $A$ can be singled out by using a wff and the satisfaction relation. In this case, we can say that the subset or relation is definable over $\mathfrak{A}$. This is an important concept that we will pursue in this section. Theorem 3.2.11 justifies the following definition.

**Definition 3.2.21.** Let $\mathfrak{A}$ be an $\mathcal{L}$-structure and let $\varphi$ be a wff having all of its free variables in the list $v_1, v_2, \ldots, v_k$. For all $a_1, a_2, \ldots, a_k$ in $A$, the notation

$$\mathfrak{A} \models \varphi[\![a_1, a_2, \ldots, a_k]\!]$$

means that for some (hence for any) assignment $s\colon \mathcal{T} \to A$ such that $s(v_i) = a_i$ for each $i = 1, 2, \ldots, k$, we have $\mathfrak{A} \models \varphi[s]$.

**Example 3.2.22.** Let $\mathcal{L} = \{L, f, c\}$ and let $\mathfrak{A}$ be as in Example 3.2.7. Let $\varphi$ be the wff $\exists v_2 L f v_2 v_1$. Then $\mathfrak{A} \models \varphi[\![2]\!]$ and $\mathfrak{A} \not\models \varphi[\![1]\!]$.

Let $\mathcal{L} = \{\dot{0}, \dot{1}, \dot{+}, \dot{\times}, \doteq\}$ be the language having equality, two 2-place function symbols $\dot{+}$ and $\dot{\times}$, and two constant symbols $\dot{0}, \dot{1}$. Let $\mathfrak{R} = \langle \mathbb{R}; 0, 1, +, \times\rangle$ be the structure, where $+$ and $\times$ are the standard operations of addition and multiplication. The structure $\mathfrak{R}$ is called the *real field*. Note that for any $a \in \mathbb{R}$, it follows that $a \geq 0$ if and only if $a = x^2$ for some $x \in \mathbb{R}$. This fact implies that there is a wff $\varphi$ with a free variable such that

$$\mathfrak{R} \models \varphi[\![a]\!] \quad \text{iff} \quad a \geq 0.$$

Let $\varphi$ be the wff $\exists x(v_1 \doteq x \dot{\times} x)$. Then

$$\mathfrak{R} \models \exists x(v_1 \doteq x \dot{\times} x)[\![a]\!] \quad \text{iff} \quad a \geq 0.$$

For this reason, we shall say that the interval $[0, \infty)$ is *definable* over $\mathfrak{R}$ and that the formula $\exists x(v_1 \doteq x \dot{\times} x)$ *defines* $[0, \infty)$ in $\mathfrak{R}$.

Moreover, for any $a, b \in \mathbb{R}$, $a \leq b$ if and only if $b = a + x^2$ for some $x \in \mathbb{R}$. Thus, the ordering relation "less than or equal to" is also definable over the structure $\mathfrak{R}$, that is, there is a wff $\psi$ with two free variables such that

$$\mathfrak{R} \models \psi[\![a, b]\!] \quad \text{iff} \quad a \leq b.$$

Let $\psi$ be the wff $\exists x(v_2 \doteq v_1 \dot{+} x \dot{\times} x)$. Then

$$\mathfrak{R} \models \exists x(v_2 \doteq v_1 \dot{+} x \dot{\times} x)[\![a, b]\!] \quad \text{iff} \quad a \leq b.$$

Thus, we can say that the relation $\{\langle a, b\rangle \in \mathbb{R} \times \mathbb{R} \mid a \leq b\}$ is *definable* over $\mathfrak{R}$ and that the formula $\exists x(v_2 \doteq v_1 \dot{+} x \dot{\times} x)$ *defines* this relation in $\mathfrak{R}$.

We now give a precise description of the concept of definability over a structure.

**Definition 3.2.23.** Let $\mathfrak{A}$ be an $\mathcal{L}$-structure with domain $A$. Let $\varphi$ be a wff having all of its free variables in the list $v_1, v_2, \ldots, v_k$. Then the $k$-ary relation on $A$

$$\{\langle a_1, a_2, \ldots, a_k\rangle \mid \mathfrak{A} \models \varphi[\![a_1, a_2, \ldots, a_k]\!]\}$$

is *definable over* $\mathfrak{A}$ and the formula $\varphi$ *defines* this $k$-ary relation in $\mathfrak{A}$.

**Example 3.2.24** (Sublanguage of elementary number theory). Recall the language of number theory $\mathcal{L} = \{\dot{<}, \dot{0}, \dot{S}, \dot{+}, \dot{\times}, \dot{E}, \doteq\}$ in Example 3.1.4. Consider the sublanguage $\mathcal{L}' = \{\dot{0}, \dot{S}, \dot{+}, \dot{\times}, \doteq\}$ of $\mathcal{L}$ and also the $\mathcal{L}'$-structure $\mathcal{N} = \langle \mathbb{N}; 0, S, +, \times\rangle$, where $\dot{0}^{\mathcal{N}} = 0$, $\dot{S}^{\mathcal{N}} = S$ (successor function), and $\dot{+}^{\mathcal{N}} = +$, $\dot{\times}^{\mathcal{N}} = \times$ are the standard operations of addition and multiplication, respectively. We identify some subsets of $\mathbb{N}$ and relations on $\mathbb{N}$ that are definable over $\mathcal{N}$:

1. Let $m, n \in \mathbb{N}$. Clearly, $m < n$ if and only if $n = m + k$ for some $k \in \mathbb{N}$, where $k \geq 1$. Moreover, $k \geq 1$ when $k = i + 1$ for an $i \in \mathbb{N}$. This allows us to now show that the relation $\{\langle m, n\rangle : m < n\}$ is definable over $\mathcal{N}$. The formula
$$\exists v_3(v_2 \doteq v_1 \dot{+} \dot{S}v_3)$$
is such that
$$m < n \quad \text{iff} \quad \mathcal{N} \vDash \exists v_3(v_2 \doteq v_1 \dot{+} \dot{S}v_3)[\![m, n]\!].$$
2. For each $n \in \mathbb{N}$, it follows that $\{n\}$ is definable. For example, for the wff $v_1 \doteq \dot{S}\dot{S}\dot{S}\dot{0}$, we see that $\{3\} = \{n : \mathcal{N} \vDash v_1 \doteq \dot{S}\dot{S}\dot{S}\dot{0}[\![n]\!]\}$.
3. The set of prime numbers is also definable over $\mathcal{N}$. Observe that $p \in \mathbb{N}$ is a prime if and only if $1 < p$ and for all $m, n \in \mathbb{N}$, if $m \cdot n = p$, then $m = 1$ or $n = 1$. Let us first try the following formula, where 1 is represented by $\dot{S}\dot{0}$, $p$ is represented by the variable $v_1$, and $v_2$ and $v_3$ represent $m$ and $n$, respectively. Thus, we obtain
$$\dot{S}\dot{0} < v_1 \wedge \forall v_2 \forall v_3(v_2 \dot{\times} v_3 \doteq v_1 \to (v_2 \doteq \dot{S}\dot{0} \vee v_3 \doteq \dot{S}\dot{0})).$$
As $<$ is not part of the language $\mathcal{L}'$, we must replace $\dot{S}\dot{0} < v_1$ with an appropriate $\mathcal{L}'$-wff. By item 1, we have $1 < p$ if and only if $\mathcal{N} \vDash \exists v_3(v_1 \doteq \dot{S}\dot{0} \dot{+} \dot{S}v_3)[\![p]\!]$. Thus, the set of primes is definable over $\mathcal{N}$ by the $\mathcal{L}'$-wff
$$\exists v_3(v_1 \doteq \dot{S}\dot{0} \dot{+} \dot{S}v_3) \wedge \forall v_2 \forall v_3(v_2 \dot{\times} v_3 \doteq v_1 \to (v_2 \doteq \dot{S}\dot{0} \vee v_3 \doteq \dot{S}\dot{0})).$$

Some relations on a structure are definable over the structure and some are not. The concept of a homomorphism (see Section 3.2.6) can sometimes be used to show that a relation is not definable over a given structure.

### 3.2.5 Classes of structures

A structure consists of a set along with functions and relations that are defined on the set. In mathematics, one often studies a particular collection of structures because they each satisfy a specific set of axioms. For example, groups, rings, fields, and vector spaces are four types of structures that each satisfy four different sets of axioms. In this section, we want to pursue this theme in terms of structures of a particular language that satisfy all of the sentences in a specific set.

**Definition 3.2.25.** Let $\Sigma$ be a set of sentences in a given language $\mathcal{L}$. Then $\text{Mod}(\Sigma)$ denotes the class (collection) of all $\mathcal{L}$-structures in which every sentence in $\Sigma$ is true, that is, $\text{Mod}(\Sigma)$ is the collection of all $\mathcal{L}$-structures $\mathfrak{A}$ such that $\mathfrak{A} \models \varphi$ for all $\varphi \in \Sigma$.

For a single sentence $\psi$ we shall write $\text{Mod}(\psi)$ rather than $\text{Mod}(\{\psi\})$.

**Example 3.2.26.** Consider the language $\mathcal{L} = \{e, *, \doteq\}$ of groups as discussed in Example 3.2.1. Let $\Sigma$ be the set consisting of the following three group axioms:
1. $\forall v_1 \forall v_2 \forall v_3 (v_1 * (v_2 * v_3) \doteq (v_1 * v_2) * v_3)$,
2. $\forall v_1 (v_1 * e \doteq v_1)$,
3. $\forall v_1 \exists v_2 (v_1 * v_2 \doteq e)$.

Then $\text{Mod}(\Sigma)$ is the collection of all groups.

Let $\mathcal{K}$ be a collection of structures for a language. Suppose that every structure in $\mathcal{K}$ satisfies one particular sentence and any structure that satisfies this sentence is also in $\mathcal{K}$. When this is the case, $\mathcal{K}$ is called an *elementary class* (EC).

**Definition 3.2.27.** Let $\mathcal{K}$ be a class of structures for a given language $\mathcal{L}$. Then $\mathcal{K}$ is said to be an *EC* if $\mathcal{K} = \text{Mod}(\psi)$ for some $\mathcal{L}$-sentence $\psi$.

The term "elementary" is a synonym for "first-order" and the term "class" is a synonym for the word "collection." Our next definition is just an extension of Definition 3.2.27.

**Definition 3.2.28.** A class $\mathcal{K}$ of $\mathcal{L}$-structures is said to be an *EC in the wider sense* ($\text{EC}_\Delta$) if $\mathcal{K} = \text{Mod}(\Sigma)$ for some set of $\mathcal{L}$-sentences $\Sigma$.

Two structures for a language may be different, but they may be alike with respect to satisfying the exact same sentences in the language.

**Definition 3.2.29.** Let $\mathfrak{A}$ and $\mathfrak{B}$ be $\mathcal{L}$-structures. Then $\mathfrak{A}$ and $\mathfrak{B}$ are *elementarily equivalent*, denoted by $\mathfrak{A} \equiv \mathfrak{B}$, if and only if for every sentence $\varphi$

$$\mathfrak{A} \models \varphi \quad \text{iff} \quad \mathfrak{B} \models \varphi.$$

That is, two structures are elementarily equivalent if they satisfy the same sentences. Different worlds can sometimes share the same truths.

**Definition 3.2.30.** Let $\mathcal{K}$ be a class of $\mathcal{L}$-structures. Then $\mathcal{K}$ is *elementarily closed* if for all $\mathcal{L}$-structures $\mathfrak{A}$ and $\mathfrak{B}$, if $\mathfrak{A} \in \mathcal{K}$ and $\mathfrak{A} \equiv \mathfrak{B}$, then $\mathfrak{B} \in \mathcal{K}$.

Let $\Sigma$ be a set of sentences in a language $\mathcal{L}$. Then $\mathcal{K} = \text{Mod}(\Sigma)$ is elementarily closed, because if $\mathfrak{A} \in \mathcal{K}$ and $\mathfrak{A} \equiv \mathfrak{B}$, then $\mathfrak{B} \models \varphi$ for all $\varphi \in \Sigma$, so $\mathfrak{B} \in \mathcal{K}$.

**Definition 3.2.31.** Let $\mathfrak{A}$ be an $\mathcal{L}$-structure. The theory of $\mathfrak{A}$, denoted by $\mathrm{Th}(\mathfrak{A})$, is the set of all $\mathcal{L}$-sentences true in $\mathfrak{A}$, that is,

$$\mathrm{Th}(\mathfrak{A}) = \{\varphi : \varphi \text{ is a sentence and } \mathfrak{A} \models \varphi\}.$$

Let $\mathfrak{A}$ be an $\mathcal{L}$-structure. Then $\mathfrak{A} \in \mathrm{Mod}(\mathrm{Th}(\mathfrak{A}))$ and, as noted above, $\mathrm{Mod}(\mathrm{Th}(\mathfrak{A}))$ is elementarily closed. So, in particular, given any $\mathcal{L}$-structure $\mathfrak{A}$ there is always a set of sentences $\Sigma$ such that $\mathfrak{A} \in \mathrm{Mod}(\Sigma)$.

### 3.2.6 Homomorphisms

In linear algebra there is an interest in functions from one vector space $V$ into another vector space $W$ that preserve vector addition and scalar multiplication.

**Definition.** If $T\colon V \to W$ is a function from a vector space $(V, +, \cdot)$ to the vector space $(W, \oplus, *)$, then $T$ is called a *linear transformation* if for all vectors $\mathbf{x}$ and $\mathbf{y}$ in $V$ and for all scalars $c$, the following hold:
(1) $T(\mathbf{x} + \mathbf{y}) = T(\mathbf{x}) \oplus T(\mathbf{y})$,
(2) $T(c \cdot \mathbf{x}) = c * T(\mathbf{x})$.

In group theory one defines what it means for a function from one group $G$ to another group $G'$ to preserve the algebraic structure of the group $G$.

**Definition.** Let $(G, *)$ and $(G', \circledast)$ be two groups. A function $\varphi : G \to G'$ is called a *homomorphism* if for all $a, b \in G$, $\varphi(a * b) = \varphi(a) \circledast \varphi(b)$.

In this section, we will generalize these fundamental concepts to structures.

#### Functions that preserve operations and relations of structures

In mathematics, one uses a function to relate one set to another set. In mathematical logic, a homomorphism relates one structure with another structure. More specifically, a homomorphism is a structure preserving function between two structures of the same language. The word homomorphism is derived from ancient Greek, where "*homos*" means "same" and "*morphe*" means "form."

**Definition 3.2.32.** Let $\mathfrak{A} = \langle A; \dots \rangle$ and $\mathfrak{B} = \langle B; \dots \rangle$ be $\mathcal{L}$-structures. A function $h\colon A \to B$ is called a *homomorphism* if $h$ has the following properties:
(1) For each $n$-place predicate symbol $P$ and for all $a_1, a_2, \dots, a_n \in A$, we have

$$\langle a_1, a_2, \dots, a_n \rangle \in P^{\mathfrak{A}} \quad \text{iff} \quad \langle h(a_1), h(a_2), \dots, h(a_n) \rangle \in P^{\mathfrak{B}}.$$

(2) For each $n$-place function symbol $f$ and for all $a_1, a_2, \dots, a_n \in A$, we have

$$h(f^{\mathfrak{A}}(a_1, a_2, \ldots, a_n)) = f^{\mathfrak{B}}(h(a_1), h(a_2), \ldots, h(a_n)).$$

(3) For each constant symbol $c$, we have $h(c^{\mathfrak{A}}) = c^{\mathfrak{B}}$.

Conditions (1)–(3) are often expressed, respectively, as: "*h preserves the relations, the functions, and the constants.*"

**Example 3.2.33.** Consider the language $\mathcal{L} = \{\dot{+}, \dot{\times}\}$ and let $\mathfrak{A} = \langle \mathbb{N}; \dot{+}^{\mathfrak{A}}, \dot{\times}^{\mathfrak{A}} \rangle$, where $\dot{+}^{\mathfrak{A}}$ and $\dot{\times}^{\mathfrak{A}}$ are the standard addition and multiplication operations, respectively, on the natural numbers. Define a new structure, whose domain has just two elements, by $\mathfrak{B} = \langle \{o, e\}; \dot{+}^{\mathfrak{B}}, \dot{\times}^{\mathfrak{B}} \rangle$, where $\dot{+}^{\mathfrak{B}}$ and $\dot{\times}^{\mathfrak{B}}$ are given by the following addition and multiplication tables:

| $\dot{+}^{\mathfrak{B}}$ | $e$ | $o$ |
|---|---|---|
| $e$ | $e$ | $o$ |
| $o$ | $o$ | $e$ |

| $\dot{\times}^{\mathfrak{B}}$ | $e$ | $o$ |
|---|---|---|
| $e$ | $e$ | $e$ |
| $o$ | $e$ | $o$ |

The addition table can be viewed as saying that "even plus even is even," "even plus odd is odd," and "odd plus odd is even," and similarly for the multiplication table.

Now define $h: \mathbb{N} \to \{o, e\}$ by

$$h(n) = \begin{cases} e, & \text{if } n \text{ is even,} \\ o, & \text{if } n \text{ is odd.} \end{cases}$$

Then $h$ is a homomorphism, as clause (2) of Definition 3.2.32 is satisfied as follows:

$$h(n \dot{+}^{\mathfrak{A}} m) = h(n) \dot{+}^{\mathfrak{B}} h(m),$$
$$h(n \dot{\times}^{\mathfrak{A}} m) = h(n) \dot{\times}^{\mathfrak{B}} h(m).$$

For example, if $m$ and $n$ are both odd, then $n \dot{\times}^{\mathfrak{A}} m$ is odd. Thus, $h(n \dot{\times}^{\mathfrak{A}} m) = o$ and $h(n) \dot{\times}^{\mathfrak{B}} h(m) = o \dot{\times}^{\mathfrak{B}} o = o$. Hence, $h(n \dot{\times}^{\mathfrak{A}} m) = h(n) \dot{\times}^{\mathfrak{B}} h(m)$.

**Definition 3.2.34.** Let $\mathfrak{A} = \langle A; \ldots \rangle$ and $\mathfrak{B} = \langle B; \ldots \rangle$ be structures for the language $\mathcal{L}$. Let $h: A \to B$ be a homomorphism.

- We shall say $h$ is a homomorphism of $\mathfrak{A}$ *into* $\mathfrak{B}$.
- We shall say that $h: A \to B$ is an *isomorphism* or an *isomorphic embedding* if $h$ is one-to-one. In this case, we shall say that $h$ is an isomorphism of $\mathfrak{A}$ *into* $\mathfrak{B}$.
- When $h: A \to B$ is onto $B$, we shall say that $h$ is a homomorphism of $\mathfrak{A}$ *onto* $\mathfrak{B}$.
- If $h$ is both one-to-one and onto $B$, then $\mathfrak{A}$ and $\mathfrak{B}$ are *isomorphic*, denoted by $\mathfrak{A} \cong \mathfrak{B}$. In this case, we shall say that $h$ is an isomorphism of $\mathfrak{A}$ *onto* $\mathfrak{B}$.

**Example 3.2.35.** Let $\mathcal{L} = \{\dot{<}\}$ and let $\mathcal{P} = \langle \mathbb{P}; \dot{<}^{\mathcal{P}} \rangle$, where $\mathbb{P} = \{1, 2, 3, \ldots\}$ and $\dot{<}^{\mathcal{P}}$ is the standard "less than" relation on $\mathbb{P}$. Let $\mathcal{N} = \langle \mathbb{N}; \dot{<}^{\mathcal{N}} \rangle$, where $\mathbb{N} = \{0, 1, 2, 3, \ldots\}$ and $\dot{<}^{\mathcal{N}}$

is the standard "less than" relation on $\mathbb{N}$. Define $h: \mathbb{P} \to \mathbb{N}$ by $h(n) = n - 1$. Then $h$ is a homomorphism, as clause (1) of Definition 3.2.32 is satisfied as follows:

$$n \mathrel{\dot{<}^{\mathcal{P}}} m \quad \text{iff} \quad h(n) \mathrel{\dot{<}^{\mathcal{N}}} h(m).$$

Since $h$ is one-to-one, we conclude that $h$ is an isomorphic embedding. In addition, because $h$ is onto $\mathbb{N}$, we see that the structures $\mathcal{P}$ and $\mathcal{N}$ are isomorphic.

**Definition 3.2.36.** Let $\mathfrak{A} = \langle A; \dots \rangle$ and $\mathfrak{B} = \langle B; \dots \rangle$ be $\mathcal{L}$-structures. We shall say that $\mathfrak{A}$ is a *substructure* of $\mathfrak{B}$ if $A \subseteq B$ and the following conditions hold:

(a) For each $n$-place predicate symbol $P$ and for all $a_1, a_2, \dots, a_n \in A$, we have

$$\langle a_1, a_2, \dots, a_n \rangle \in P^{\mathfrak{A}} \quad \text{iff} \quad \langle a_1, a_2, \dots, a_n \rangle \in P^{\mathfrak{B}}.$$

(b) For each $n$-place function symbol $f$ and for all $a_1, a_2, \dots, a_n \in A$, we have

$$f^{\mathfrak{A}}(a_1, a_2, \dots, a_n) = f^{\mathfrak{B}}(a_1, a_2, \dots, a_n).$$

(c) For each constant symbol $c$, we have $c^{\mathfrak{A}} = c^{\mathfrak{B}}$.

**Example 3.2.37.** Consider the language $\mathcal{L} = \{\dot{+}\}$ and let $\mathcal{Q} = \langle \mathbb{Q}; \dot{+}^{\mathcal{Q}} \rangle$, where $\mathbb{Q}$ is the set of rational numbers and $\dot{+}^{\mathcal{Q}}$ is the standard addition operation on $\mathbb{Q}$. Let $\mathcal{R} = \langle \mathbb{R}; \dot{+}^{\mathcal{R}} \rangle$, where $\mathbb{R}$ is the set of real numbers and $\dot{+}^{\mathcal{R}}$ is the standard addition operation on $\mathbb{R}$. Then $\mathcal{Q}$ is a substructure of $\mathcal{R}$ because clause (b) of Definition 3.2.36 is satisfied, that is, the operations $\dot{+}^{\mathcal{Q}}$ and $\dot{+}^{\mathcal{R}}$ agree on the rational numbers.

Let $\mathfrak{A}$ and $\mathfrak{B}$ be structures for the language $\mathcal{L}$. Then $\mathfrak{A}$ is a *substructure* of $\mathfrak{B}$ if and only if $A \subseteq B$ and the identity function $i: A \to B$ is a homomorphism.

### The homomorphism theorem

We will soon state and prove our primary theorem about homomorphisms. The last part of this theorem will provide us with a technique for showing that some relations are not definable over a structure (see Theorem 3.2.41).

We begin by making some relevant remarks. Let $\mathfrak{A} = (A; \dots)$ and $\mathfrak{B} = (B; \dots)$ be $\mathcal{L}$-structures and let $\mathcal{T}$ be the set of all the variables and constants of $\mathcal{L}$. Let $s: \mathcal{T} \to A$ be an assignment. Thus, by Theorem 3.2.5, there is a unique extension $\overline{s}: \overline{\mathcal{T}} \to A$, where $\overline{\mathcal{T}}$ is the set of all the terms of $\mathcal{L}$. Suppose that $h$ is a homomorphism of $\mathfrak{A}$ into $\mathfrak{B}$. Then $h \circ s: \mathcal{T} \to B$ is also an assignment, where $(h \circ s)(v) = h(s(v))$ for all $v \in \mathcal{T}$. Thus, by Theorem 3.2.5, there is a unique extension $\overline{h \circ s}: \overline{\mathcal{T}} \to B$. We also note that a *quantifier-free* wff is one in which no quantifier appears in the formula.

**Theorem 3.2.38** (Homomorphism theorem). *Let $\mathfrak{A}$ and $\mathfrak{B}$ be $\mathcal{L}$-structures, let $h$ be a homomorphism of $\mathfrak{A}$ into $\mathfrak{B}$, and let $s: \mathcal{T} \to A$ be an assignment, where $\mathcal{T}$ is the set of all the variables and constant symbols of $\mathcal{L}$ and $A$ is the domain of $\mathfrak{A}$.*

(a) *For every term t of the language,* $h(\overline{s}(t)) = \overline{h \circ s}(t)$.
(b) *For every quantifier-free wff* $\varphi$ *that does not contain the equality symbol,*

$$\mathfrak{A} \models \varphi[s] \quad \textit{iff} \quad \mathfrak{B} \models \varphi[h \circ s].$$

(c) *If h is one-to-one, then for every quantifier-free wff* $\varphi$ *that can contain the equality symbol,*

$$\mathfrak{A} \models \varphi[s] \quad \textit{iff} \quad \mathfrak{B} \models \varphi[h \circ s].$$

(d) *If h is onto B, then for every wff* $\varphi$ *that does not contain the equality symbol,*

$$\mathfrak{A} \models \varphi[s] \quad \textit{iff} \quad \mathfrak{B} \models \varphi[h \circ s].$$

(e) *If h is one-to-one and onto B, then for* every *wff* $\varphi$,

$$\mathfrak{A} \models \varphi[s] \quad \textit{iff} \quad \mathfrak{B} \models \varphi[h \circ s].$$

*Proof.* We shall prove (a)–(e) below. Let $h$ be a homomorphism of $\mathfrak{A}$ into $\mathfrak{B}$. In the proofs of (a)–(c), let $s\colon \mathcal{T} \to A$ be an arbitrary assignment.

(a) We must first prove that for every term $t$ of the language, $h(\overline{s}(t)) = \overline{h \circ s}(t)$. This will be accomplished by induction on terms (see Section 3.1.2 on page 56).

*Base step:* For a constant symbol $c$, we have $h(\overline{s}(c)) = h(c^{\mathfrak{A}}) = c^{\mathfrak{B}}$ by Theorem 3.2.5(2) and Definition 3.2.32(3). Also, $\overline{h \circ s}(c) = c^{\mathfrak{B}}$ by Theorem 3.2.5(2) applied to $\overline{h \circ s}$. Therefore, $h(\overline{s}(c)) = \overline{h \circ s}(c)$. In the case where $v$ is a variable,

$$h(\overline{s}(v)) = h(s(v)) = (h \circ s)(v) = \overline{h \circ s}(v),$$

by Theorem 3.2.5(1) applied to $\overline{s}$ and $\overline{h \circ s}$.

*Inductive step:* Let $f$ be an $n$-place function symbol in $\mathcal{L}$ and let $t_1, t_2, \ldots, t_n$ be terms. Assume the induction hypothesis

$$\text{For each } i \leq n, \text{ we have } h(\overline{s}(t_i)) = \overline{h \circ s}(t_i). \tag{IH}$$

We prove that the same holds for the term $ft_1t_2 \cdots t_n$, that is, we prove that

$$h(\overline{s}(ft_1t_2 \cdots t_n)) = \overline{h \circ s}(ft_1t_2 \cdots t_n).$$

We do this as follows:

$$\begin{aligned} h(\overline{s}(ft_1t_2 \cdots t_n)) &= h(f^{\mathfrak{A}}(\overline{s}(t_1), \overline{s}(t_2), \ldots, \overline{s}(t_n))) && \text{by Theorem 3.2.5(3),} \\ &= f^{\mathfrak{B}}(h(\overline{s}(t_1)), h(\overline{s}(t_2)), \ldots, h(\overline{s}(t_n))) && \text{by Definition 3.2.32(2),} \end{aligned}$$

$$= f^{\mathfrak{B}}(\overline{h \circ s}(t_1), \overline{h \circ s}(t_2), \ldots, \overline{h \circ s}(t_n)) \quad \text{by (IH)},$$
$$= \overline{h \circ s}(ft_1t_2 \cdots t_n) \quad \text{by Theorem 3.2.5(3)}.$$

(b) For every $\varphi$ that is a quantifier-free wff *not* containing the symbol $\doteq$, we prove that

$$\mathfrak{A} \models \varphi[s] \quad \text{iff} \quad \mathfrak{B} \models \varphi[h \circ s]. \tag{▲}$$

We use induction on wffs.

*Base step:* We show that (▲) holds for all atomic formulas. So, let $P$ be an $n$-place predicate symbol and let $t_1, \ldots, t_n$ be terms. We show that

$$\mathfrak{A} \models Pt_1t_2 \cdots t_n[s] \quad \text{iff} \quad \mathfrak{B} \models Pt_1t_2 \cdots t_n[h \circ s]$$

as follows:

$$\begin{aligned}
\mathfrak{A} \models Pt_1t_2 \cdots t_n[s] \quad &\text{iff} \quad \langle \overline{s}(t_1), \ldots, \overline{s}(t_n) \rangle \in P^{\mathfrak{A}} && \text{by Definition 3.2.6(2)},\\
&\text{iff} \quad \langle h(\overline{s}(t_1)), \ldots, h(\overline{s}(t_n)) \rangle \in P^{\mathfrak{B}} && \text{by Definition 3.2.32(1)},\\
&\text{iff} \quad \langle \overline{h \circ s}(t_1), \ldots, \overline{h \circ s}(t_n) \rangle \in P^{\mathfrak{B}} && \text{by part (a) above},\\
&\text{iff} \quad \mathfrak{B} \models Pt_1t_2 \cdots t_n[h \circ s] && \text{by Definition 3.2.6(2)}.
\end{aligned}$$

*Inductive step:* Let $\alpha$ and $\beta$ be quantifier-free formulas that do not contain the equality symbol. Assume the induction hypothesis

$$\begin{aligned}
\mathfrak{A} \models \alpha[s] \quad &\text{iff} \quad \mathfrak{B} \models \alpha[h \circ s],\\
\mathfrak{A} \models \beta[s] \quad &\text{iff} \quad \mathfrak{B} \models \beta[h \circ s].
\end{aligned} \tag{IH}$$

One must now prove that $(\neg\alpha)$ and $(\alpha \to \beta)$ both satisfy condition (▲). We first prove that $(\neg\alpha)$ satisfies condition (▲) with the following argument:

$$\begin{aligned}
\mathfrak{A} \models \neg\alpha[s] \quad &\text{iff} \quad \mathfrak{A} \not\models \alpha[s] && \text{by Definition 3.2.6(3)},\\
&\text{iff} \quad \mathfrak{B} \not\models \alpha[h \circ s] && \text{as } \alpha \text{ satisfies (IH)},\\
&\text{iff} \quad \mathfrak{B} \models \neg\alpha[h \circ s] && \text{by Definition 3.2.6(3)}.
\end{aligned}$$

We now prove that $(\alpha \to \beta)$ satisfies condition (▲) as follows:

$$\begin{aligned}
\mathfrak{A} \models (\alpha \to \beta)[s] \quad &\text{iff} \quad \mathfrak{A} \models \alpha[s] \text{ implies } \mathfrak{A} \models \beta[s] && \text{by Definition 3.2.6(4)},\\
&\text{iff} \quad \mathfrak{B} \models \alpha[h \circ s] \text{ implies } \mathfrak{B} \models \beta[h \circ s] && \text{as } \alpha, \beta \text{ satisfy (IH)},\\
&\text{iff} \quad \mathfrak{B} \models (\alpha \to \beta)[h \circ s] && \text{by Definition 3.2.6(4)}.
\end{aligned}$$

(c) We must show that if $h$ is one-to-one, then for every quantifier-free formula $\varphi$,

$$\mathfrak{A} \models \varphi[s] \quad \text{iff} \quad \mathfrak{B} \models \varphi[h \circ s].$$

The argument is by induction, just as in the above proof of part (b). However, in the base step of (b), we need to add the following proof showing that the atomic formula $\doteq t_1 t_2$ satisfies (▲). Let $t_1$ and $t_2$ be terms. We then have the following:

$$\begin{array}{llll}\mathfrak{A} \models \doteq t_1 t_2[s] & \text{iff} & \bar{s}(t_1) = \bar{s}(t_2) & \text{by Definition 3.2.6(1),}\\ & \text{iff} & h(\bar{s}(t_1)) = h(\bar{s}(t_2)) & \text{because } h \text{ is one-to-one,}\\ & \text{iff} & \overline{h \circ s}(t_1) = \overline{h \circ s}(t_2) & \text{by part (a) above,}\\ & \text{iff} & \mathfrak{B} \models \doteq t_1 t_2[h \circ s] & \text{by Definition 3.2.6(1).}\end{array}$$

The rest of the argument is exactly like the one given for (b).

(d) Let $h$ be onto $B$. For every formula $\varphi$ *not* containing the symbol $\doteq$, we prove that

$$\mathfrak{A} \models \varphi[s] \quad \text{iff} \quad \mathfrak{B} \models \varphi[h \circ s], \quad \textit{for all assignments } s. \tag{3.10}$$

The argument is by induction. For atomic formulas $\varphi$, the proof of (3.10) is as in the above proof of part (b). However, in the inductive step, we need to include the quantifier symbol $\forall$. The proof of this case follows. Let $\alpha$ be a formula that does not contain the equality symbol. Assume the induction hypothesis

$$\mathfrak{A} \models \alpha[s] \quad \text{iff} \quad \mathfrak{B} \models \alpha[h \circ s], \quad \textit{for all assignments } s. \tag{IH}$$

We must show that

$$\mathfrak{A} \models \forall v\alpha[s] \quad \text{iff} \quad \mathfrak{B} \models \forall v\alpha[h \circ s], \quad \text{for all assignments } s.$$

Let $s\colon \mathcal{T} \to A$ be an arbitrary assignment and let $B$ be the domain of $\mathfrak{B}$. Thus,

$$\begin{array}{llll}\mathfrak{A} \models \forall v\alpha[s] & \text{iff} & \text{for all } a \in A,\ \mathfrak{A} \models \alpha[s_{v|a}] & \text{by Definition 3.2.6(5),}\\ & \text{iff} & \text{for all } a \in A,\ \mathfrak{B} \models \alpha[h \circ s_{v|a}] & \text{by (IH); } s_{v|a} \text{ is an assignment,}\\ & \text{iff} & \text{for all } a \in A,\ \mathfrak{B} \models \alpha[(h \circ s)_{v|h(a)}] & \text{as } h \circ s_{v|a} = (h \circ s)_{v|h(a)},\\ & \text{iff} & \text{for all } b \in B,\ \mathfrak{B} \models \alpha[(h \circ s)_{v|b}] & \text{as } h\colon A \to B \text{ is onto } B,\\ & \text{iff} & \mathfrak{B} \models \forall v\alpha[(h \circ s)] & \text{by Definition 3.2.6(5).}\end{array}$$

(e) Suppose that $h$ is one-to-one and onto $B$. We must show that for *every formula* $\varphi$,

$$\mathfrak{A} \models \varphi[s] \quad \text{iff} \quad \mathfrak{B} \models \varphi[h \circ s], \quad \text{for all assignments } s.$$

The argument is by induction on wffs. For the base step one uses the arguments given for the base steps in the above proofs of (b) and (c). For the inductive step one uses the arguments given for the inductive steps in the above proofs of (b) and (d).

This completes the proof of the homomorphism theorem. □

### Applications of the homomorphism theorem

Before we state and prove the next theorem, recall that $h$ is said to be an isomorphism of $\mathfrak{A}$ *onto* $\mathfrak{B}$ when $h: A \to B$ is both a one-to-one and onto homomorphism. In this case, $\mathfrak{A}$ and $\mathfrak{B}$ are said to be *isomorphic*, denoted by $\mathfrak{A} \cong \mathfrak{B}$. Also, recall Definition 3.2.29.

**Theorem 3.2.39.** *Let $\mathfrak{A}$ and $\mathfrak{B}$ be structures for a language $\mathcal{L}$. Suppose that $\mathfrak{A}$ and $\mathfrak{B}$ are isomorphic. Then $\mathfrak{A} \equiv \mathfrak{B}$.*

*Proof.* Assume that $\mathfrak{A} \cong \mathfrak{B}$. Let $\varphi$ be a sentence. We will show that $\mathfrak{A} \vDash \varphi$ iff $\mathfrak{B} \vDash \varphi$. Let $h$ be an isomorphism of $\mathfrak{A}$ *onto* $\mathfrak{B}$ and let $s: \mathcal{T} \to A$ be an assignment, where $\mathcal{T}$ is the set of all the variables and constant symbols of $\mathcal{L}$. Then

$$\begin{aligned} \mathfrak{A} \vDash \varphi \quad &\text{iff} \quad \mathfrak{A} \vDash \varphi[s] && \text{by Definition 3.2.13,} \\ &\text{iff} \quad \mathfrak{B} \vDash \varphi[h \circ s] && \text{by Theorem 3.2.38(e),} \\ &\text{iff} \quad \mathfrak{B} \vDash \varphi && \text{by Definition 3.2.13.} \end{aligned}$$

□

**Application 1.** Let $\mathcal{L} = \{\dot{<}, \dot{=}\}$ and let $\mathcal{P} = \langle \mathbb{P}; \dot{<}^{\mathcal{P}} \rangle$, where $\mathbb{P} = \{1, 2, 3, \dots\}$ and $\dot{<}^{\mathcal{P}}$ is the standard "less than" relation on $\mathbb{P}$. Let $\mathcal{N} = \langle \mathbb{N}; \dot{<}^{\mathcal{N}} \rangle$, where $\mathbb{N} = \{0, 1, 2, 3, \dots\}$ and $\dot{<}^{\mathcal{N}}$ is the standard "less than" relation on $\mathbb{N}$. Now define $h: \mathbb{P} \to \mathbb{N}$ by

$$h(n) = n - 1.$$

As discussed in Example 3.2.35, $h$ is a one-to-one and onto homomorphism, and thus $\mathcal{P}$ and $\mathcal{N}$ are isomorphic. So by Theorem 3.2.39, for any $\mathcal{L}$-sentence $\varphi$, we have $\mathcal{P} \vDash \varphi$ iff $\mathcal{N} \vDash \varphi$, that is, $\mathcal{P}$ and $\mathcal{N}$ are elementarily equivalent.

On the other hand, let us define $h': \mathbb{P} \to \mathbb{N}$ to be

$$h'(n) = n.$$

Then $h'$ is also a homomorphism and is one-to-one, but it is not onto $\mathbb{N}$. Let $s: T \to \mathbb{P}$ be an assignment. Hence, by Theorem 3.2.38(c), if $\varphi$ is quantifier-free, then

$$\mathcal{P} \vDash \varphi[s] \quad \text{iff} \quad \mathcal{N} \vDash \varphi[h' \circ s]. \tag{3.11}$$

The equivalence (3.11) may fail for a formula $\varphi$ that contains quantifiers. To illustrate this, let $s$ be an assignment such that $s(v_1) = 1$. Since $h'$ is the identity function, it follows that $h' \circ s = s$. Let $\varphi$ be the quantified statement

$$\forall v_2 (v_1 \dot{\neq} v_2 \to v_1 \dot{<} v_2).$$

Since $s(v_1) = 1$ and $(h' \circ s)(v_1) = 1$, it follows that $\mathcal{P} \vDash \varphi[s]$ and $\mathcal{N} \nvDash \varphi[h' \circ s]$. So, because $h'$ is one-to-one and not onto $\mathbb{N}$, (3.11) can fail when $\varphi$ contains quantifiers.

**Definition 3.2.40.** Let $\mathfrak{A} = \langle A; \dots \rangle$ be an $\mathcal{L}$-structure. A function $h: A \to A$ is called an *automorphism* of the structure $\mathfrak{A}$ if $h$ is an isomorphism of $\mathfrak{A}$ onto $\mathfrak{A}$.

Let $\mathfrak{A} = \langle A; \dots \rangle$ be a structure for a language $\mathcal{L}$. Recall that a $k$-ary relation $R$ on $A$ is definable over $\mathfrak{A}$ if there is a wff $\varphi$ with free variables $v_1, \dots, v_k$ such that

$$\langle a_1, a_2, \dots, a_k \rangle \in R \quad \text{iff} \quad \mathfrak{A} \models \varphi[\![a_1, a_2, \dots, a_k]\!]$$

for all $a_1, a_2, \dots, a_k \in A$.

**Theorem 3.2.41.** *Let $\mathfrak{A}$ be a structure for a language $\mathcal{L}$ and let $h$ be an automorphism of the structure $\mathfrak{A}$. Let $R$ be a $k$-ary relation that is definable over $\mathfrak{A}$. Then*

$$\langle a_1, a_2, \dots, a_k \rangle \in R \quad \textit{iff} \quad \langle h(a_1), h(a_2), \dots, h(a_k) \rangle \in R$$

*for all $a_1, a_2, \dots, a_k \in A$.*

*Proof.* Let $\varphi$ be a wff that defines $R$. For all $a_1, a_2, \dots, a_k \in A$, we have

$$\begin{aligned} \langle a_1, a_2, \dots, a_k \rangle \in R \quad &\text{iff} \quad \mathfrak{A} \models \varphi[\![a_1, a_2, \dots, a_k]\!] && \text{because } \varphi \text{ defines } R, \\ &\text{iff} \quad \mathfrak{A} \models \varphi\big[\![h(a_1), h(a_2), \dots, h(a_k)]\!\big] && \text{by Theorem 3.2.38(e)}, \\ &\text{iff} \quad \langle h(a_1), h(a_2), \dots, h(a_k) \rangle \in R && \text{because } \varphi \text{ defines } R. \end{aligned}$$ □

**Application 2.** Let $\mathcal{L} = \{\dot{<}, \dot{=}\}$ and let $\mathfrak{R} = \langle \mathbb{R}; < \rangle$, where $\mathbb{R}$ is the set of real numbers and $<$ is the usual "less than" relation on $\mathbb{R}$. Define $h\colon \mathbb{R} \to \mathbb{R}$ by

$$h(x) = x^3.$$

Then $h$ is a homomorphism, because clause (1) of Definition 3.2.32 holds as follows:

$$x < y \quad \text{iff} \quad h(x) < h(y).$$

Since $h$ is one-to-one and onto $\mathbb{R}$, we conclude that $h$ is an automorphism of the structure $\mathfrak{R}$. Note that $\mathbb{N} \subseteq \mathbb{R}$. We can now show that $\mathbb{N}$ is not definable over $\mathfrak{R}$. Suppose, for a contradiction, that there is a wff $\varphi$ such that

$$a \in \mathbb{N} \quad \text{iff} \quad \mathfrak{R} \models \varphi[\![a]\!]$$

for all $a \in \mathbb{R}$. Theorem 3.2.41 then implies that for all $a \in \mathbb{R}$,

$$a \in \mathbb{N} \quad \text{iff} \quad h(a) \in \mathbb{N}.$$

Let $a = \sqrt[3]{2}$. Since $h(a) = 2 \in \mathbb{N}$, the above equivalence implies that $\sqrt[3]{2} \in \mathbb{N}$. This contradiction shows that $\mathbb{N}$ is not definable over $\mathfrak{R}$.

**Exercises 3.2.**

1. Verify item (9) of Remark 3.2.8.
2. Let $\mathfrak{A}$ be an $\mathcal{L}$-structure and let $s$ be an assignment, as in Definition 3.2.4. Let $\psi$ be a wff. Show that either $\mathfrak{A} \models \psi[s]$ or $\mathfrak{A} \models (\neg\psi)[s]$, but not both.

3. Let $\mathcal{L} = \{L, f, c\}$ and let $\mathfrak{A}$ be as in Example 3.2.7. Let $s\colon \mathcal{T} \to \mathbb{N}$ be an assignment satisfying $s(v_i) = i + 1$. So, $s(v_1) = 2$ and $s(v_2) = 3$. Show that:
   (a) $\mathfrak{A} \models \exists v_2 L v_2 fc[s]$,
   (b) $\mathfrak{A} \models \exists v_2 L v_2 v_1[s]$,
   (c) $\mathfrak{A} \not\models \exists v_2 L v_2 c[s]$.
4. Let $\mathcal{L} = \{L, f, c\}$ and let $\mathfrak{A}$ be as in Example 3.2.7. Show that:
   (a) $\mathfrak{A} \models \forall v_1 \exists v_2 L f v_1 v_2$,
   (b) $\mathfrak{A} \not\models \exists v_2 \forall v_1 L f v_1 v_2$.
5. Let $\mathcal{L} = \{\dot{+}, \dot{0}, \dot{1}, \dot{2}, \doteq\}$, where $\dot{+}$ is a 2-place function symbol and $\dot{0}, \dot{1}, \dot{2}$ are constant symbols.
   (a) Find a structure for the language $\mathcal{L}$ in which the two sentences $\dot{1} \dot{+} \dot{1} = \dot{2}$ and $\forall v(v \dot{+} \dot{0} \doteq v)$ are *true*.
   (b) Find a structure for the language $\mathcal{L}$ in which the two sentences $\dot{1} \dot{+} \dot{1} = \dot{2}$ and $\forall v(v \dot{+} \dot{0} \doteq v)$ are *false*.
6. Let $\mathcal{L} = \{f, \doteq\}$ be a language where $f$ is a 1-place function symbol.
   (a) Find a sentence $\varphi$ so that $\mathfrak{A} \models \varphi$ if and only if $f^{\mathfrak{A}}\colon A \to A$ is one-to-one, for any structure $\mathfrak{A} = \langle A; f^{\mathfrak{A}} \rangle$.
   (b) Find a sentence $\varphi$ so that $\mathfrak{A} \models \varphi$ if and only if $f^{\mathfrak{A}}\colon A \to A$ is onto $A$, for any structure $\mathfrak{A} = \langle A; f^{\mathfrak{A}} \rangle$.
7. Let $\mathcal{L} = \{\dot{<}\}$ be the language having just the 2-place relation symbol $\dot{<}$. Consider the structures $\mathfrak{A} = \langle \mathbb{N}; \dot{<}^{\mathfrak{A}} \rangle$ and $\mathfrak{B} = \langle \mathbb{R}; \dot{<}^{\mathfrak{B}} \rangle$, where $\dot{<}^{\mathfrak{A}}$ and $\dot{<}^{\mathfrak{B}}$ are to be interpreted as the standard "less than" relation. Find a sentence $\varphi$ in the language $\mathcal{L}$ that is true in one of the structures but false in the other.
8. Let $\mathcal{L} = \{\dot{\times}, \doteq\}$ be the language having equality and $\dot{\times}$, a 2-place function symbol. Let $\mathfrak{A} = \langle \mathbb{R}; \dot{\times}^{\mathfrak{A}} \rangle$ and $\mathfrak{B} = \langle \mathbb{R}^*; \dot{\times}^{\mathfrak{B}} \rangle$ be structures where $\mathbb{R}^*$ is the set of nonzero real numbers and $\dot{\times}^{\mathfrak{A}}$ and $\dot{\times}^{\mathfrak{B}}$ are both the usual multiplication operation. Find a sentence $\varphi$ in the language $\mathcal{L}$ that is true in one of the structures but false in the other.
*9. Let $\alpha, \psi, \varphi$ be wffs and let $\Gamma$ be a set of wffs in a language $\mathcal{L}$. Show that
   (a) $\Gamma \cup \{\alpha\} \models \varphi$ if and only if $\Gamma \models (\alpha \to \varphi)$,
   (b) $\varphi \models\!\!\!\!\dashv \psi$ if and only if $\models (\varphi \leftrightarrow \psi)$.
10. Show that the set of any two of the following sentences does not logically imply the third sentence:
   (a) $\forall x \forall y \forall z (Pxy \to Pyz \to Pxz)$,
   (b) $\forall x \forall y (Pxy \to Pyx \to x \doteq y)$,
   (c) $\forall x \exists y Pxy \to \exists y \forall x Pxy$.
*11. Let $\mathfrak{A}$ be a structure and let $s$ be an assignment such that $s(x) = s(y)$, where $x$ and $y$ are variables.
   (a) Prove that for all terms $t$, if $t'$ is obtained from $t$ by replacing some, none, or all of the occurrences of $x$ in $t$ with $y$, then $\bar{s}(t) = \bar{s}(t')$.

(b) Let $Pt_1t_2\cdots t_n$ be an atomic formula and let $Pt_1't_2'\cdots t_n'$ be the result of replacing some, none, or all of the occurrences of $x$ in $Pt_1t_2\cdots t_n$ with $y$. Show that $\mathfrak{A} \models Pt_1t_2\cdots t_n[s]$ if and only if $\mathfrak{A} \models Pt_1't_2'\cdots t_n'[s]$.

*12. Show that $\{\forall x(\alpha \to \beta), \forall x\alpha\} \models \forall x\beta$.

*13. Show that if $x$ does not occur free in $\alpha$, then $\alpha \models \forall x\alpha$.

*14. Show that a wff $\theta$ is logically valid if and only if $\forall x\theta$ is logically valid.

15. Let $\mathcal{L} = \{\dot{+}, \dot{\times}, \dot{=}\}$ and let $\mathcal{N} = (\mathbb{N}; +, \times)$ be an $\mathcal{L}$-structure, where $\dot{+}^{\mathcal{N}} = +$ and $\dot{\times}^{\mathcal{N}} = \times$ are the usual operations of addition and multiplication, respectively. Show that each of the following sets are definable over $\mathcal{N}$:
   (a) $\{0\}$,
   (b) $\{1\}$,
   (c) $\{\langle m, n\rangle : n \text{ is the successor of } m\}$,
   (d) $\{\langle m, n\rangle : m < n\}$.

16. Let $\mathcal{L} = \{\dot{+}, \dot{\times}, \dot{=}\}$ be the language having equality and 2-place function symbols $\dot{+}$ and $\dot{\times}$. Consider the $\mathcal{L}$-structure $\mathfrak{R} = (\mathbb{R}; +, \times)$, where $+$ and $\times$ are the usual addition and multiplication operations.
   (a) Show that $[0, \infty)$ is definable over $\mathfrak{R}$.
   (b) Show that $\{1\}$ is definable over $\mathfrak{R}$.
   (c) Show that $\{2\}$ is definable over $\mathfrak{R}$.

17. Let $\Sigma$ and $\Gamma$ be sets of sentences in a language $\mathcal{L}$. Show that:
   (a) if $\Gamma \subseteq \Sigma$, then $\text{Mod}(\Sigma) \subseteq \text{Mod}(\Gamma)$,
   (b) $\text{Mod}(\Gamma) \cap \text{Mod}(\Sigma) = \text{Mod}(\Gamma \cup \Sigma)$,
   (c) $\text{Mod}(\Gamma) \cup \text{Mod}(\Sigma) \subseteq \text{Mod}(\Gamma \cap \Sigma)$.

18. Let $\Sigma = \{\varphi_1, \varphi_2, \ldots, \varphi_n\}$ be a finite set of $\mathcal{L}$-sentences and let $\psi = \varphi_1 \wedge \cdots \wedge \varphi_n$. Show that $\text{Mod}(\Sigma) = \text{Mod}(\psi)$.

*19. Let $\mathfrak{A}$ and $\mathfrak{B}$ be $\mathcal{L}$-structures.
   (a) Show that for every $\mathcal{L}$-sentence $\psi$, either $\psi \in \text{Th}(\mathfrak{B})$ or $\neg\psi \in \text{Th}(\mathfrak{B})$.
   (b) Show that if $\mathfrak{A} \models \text{Th}(\mathfrak{B})$, then $\mathfrak{A}$ and $\mathfrak{B}$ are elementarily equivalent.

20. Let $\mathcal{L} = \{\dot{<}, \dot{=}\}$ and let $\mathfrak{R} = (\mathbb{R}; <)$, where $\mathbb{R}$ is the set of real numbers and $<$ is the standard "less than" relation on $\mathbb{R}$.
   (a) What subsets of $\mathbb{R}$ are definable over $\mathfrak{R}$?
   (b) Let $R \subseteq \mathbb{R}^2$ be such that $\langle a, a\rangle \in R$ and $\langle b, b\rangle \notin R$. Is $R$ definable over $\mathfrak{R}$?
   (c) Let $R \subseteq \mathbb{R}^2$ be such that $\langle a, b\rangle \in R$ and $\langle c, d\rangle \notin R$. Show that if $a < b$ and $c < d$, then $R$ is not definable over $\mathfrak{R}$.

21. Let $\mathcal{L} = \{\dot{<}, \dot{=}\}$, where $\dot{<}$ is a 2-place predicate symbol. Let $\mathcal{Z} = (\mathbb{Z}; <)$ be the structure, where $\mathbb{Z}$ is the set of integers and $<$ is the standard less than relation.
   (a) Prove that for all $n \in \mathbb{Z}$, the set $\{n\}$ is not definable over $\mathcal{Z}$.
   (b) Let $\theta$ be a wff with one free variable $v_1$. Suppose that $\mathcal{Z} \models \theta[\![2]\!]$. Using Theorem 3.2.38, prove that $\mathcal{Z} \models \forall v_1\theta$.

22. Let $\mathcal{L} = \{\dot{<}, \dot{\times}, \dot{=}\}$ be a language with a 2-place predicate symbol $\dot{<}$ and 2-place function symbol $\dot{\times}$. Let $\mathcal{R} = (\mathbb{P}; <, \times)$ be the structure, where $\mathbb{P}$ is the set of positive real

numbers, $<$ is the standard less than relation, and $\times$ is the standard multiplication function on the positive real numbers.

(a) Define the function $h\colon \mathbb{P} \to \mathbb{P}$ by $h(x) = x^2$. Prove that $h$ is an automorphism of the structure $\mathcal{R}$.

(b) Let $A = \{\langle x, y, z\rangle \in \mathbb{P}^3 : x + y = z\}$. Prove that $A$ is not definable over $\mathcal{R}$.

**Exercise Notes:** For Exercise 1, use Definition 3.2.6 and abbreviation 4 on page 62. For Exercise 3 and Exercise 4, use Remark 3.2.8. For Exercise 11, use induction on terms. For Exercise 21(b), show that $\mathcal{Z} \models \theta[\![n]\!]$ for all $n \in \mathbb{Z}$.

## 3.3 Deductions

### What is proof?

In Section 3.1 we described a formal language $\mathcal{L}$ and also defined what it means for a formula of the language to be grammatically correct. In addition, in Section 3.2 we defined the notion of "truth" in a structure, that is, $\mathfrak{A} \models \varphi$, where $\varphi$ is a sentence. Recall that a sentence $\varphi$ is logically implied by a set of sentences $\Gamma$, denoted by $\Gamma \models \varphi$, if *every* model of $\Gamma$ is a model of $\varphi$. In this section we shall define the notion of "proof" from a set of axioms (or formulas), that is, we shall define when $\varphi$ is *deducible* from the axioms in $\Gamma$, denoted by $\Gamma \vdash \varphi$.

Surely the most important discovery for mathematics by the ancient Greeks was the notion of *proof*, turning mathematics into a deductive science. Each theorem $\varphi$ must have a proof from a set $\Gamma$ of more or less explicitly stated assumptions, or *axioms*. The proof must demonstrate that the conclusion $\varphi$ follows from the axioms in $\Gamma$ by the laws of logic alone. The natural question is:

> Can the notions of "laws of logic" and "proof" be made mathematically precise?

A proof is an argument that you give to someone else which completely convinces him or her of the correctness of your assertion. Thus, a proof should be finitely long, as you cannot give an infinite argument to another person. If the set of axioms $\Gamma$ is infinite, that is fine, but they cannot all be used in one proof (otherwise, we would have an infinitely long proof). Another essential feature of a proof is that it must be possible for another person to check the proof to ascertain that it contains no fallacies.

In Section 2.5, we presented a system of deduction for propositional logic, where our axioms were the tautologies of propositional logic and our one rule of inference was modus ponens. Our system of deduction for first-order logic will be similar. We will select a set $\Lambda$ of wffs to be called logical axioms and we will use modus ponens as our one rule of inference. This will enable us to deduce a new wff from other wffs. Then for a set $\Gamma$ of wffs, the theorems of $\Gamma$ will be the wffs which can be deduced from $\Gamma \cup \Lambda$. A wff $\varphi$ will be a theorem of $\Gamma$ (written $\Gamma \vdash \varphi$) if and only if there a finite sequence of wffs which identifies how $\varphi$ was derived from $\Gamma \cup \Lambda$. Such a derivation, which uses the logical axioms and the rule of inference, will be called a *deduction* of $\varphi$ from $\Gamma$. The term deduction will be used to avoid confusion with our own mathematical proofs.

There are other deduction systems for first-order logic that are equivalent to the one we shall present. Each such system of deduction may have a different version of $\Lambda$ and different rules of inference, but each system will produce the same theorems.

Before we identify the set of logical axioms $\Lambda$, we must first discuss tautologies, generalizations, and substitutions in first-order logic.

### 3.3.1 Tautologies in first-order logic

In Remark 2.2.13, we defined the concept of a tautology in propositional logic. The definition of a tautology can be extended to wffs in first-order logic, where the wff may contain quantifiers—an attribute absent from wffs in propositional logic.

**Definition 3.3.1** (First-order tautologies). Let $\mathcal{L}$ be a first-order language and let $\mathbf{A}_1, \mathbf{A}_2, \mathbf{A}_3, \ldots$ be the sentence symbols of propositional logic. Let $\varphi$ be a tautology in propositional logic containing only the connectives $\neg$ and $\to$. Suppose that the sentence symbols in $\varphi$ are in the list $\mathbf{A}_1, \mathbf{A}_2, \ldots, \mathbf{A}_n$. Let $\varphi^*$ be the result of replacing $\mathbf{A}_1, \mathbf{A}_2, \ldots, \mathbf{A}_n$ with $\alpha_1, \alpha_2, \ldots, \alpha_n$, where each $\alpha_i$ is a first-order wff in $\mathcal{L}$. We shall then say that $\varphi^*$ is a *first-order tautology*.

First-order tautologies are simple generalizations of the tautologies of propositional logic. For example, the propositional wff

$$(\mathbf{A} \to \mathbf{B}) \to (\neg\mathbf{B} \to \neg\mathbf{A}) \tag{3.12}$$

is a tautology where $\mathbf{A}$ and $\mathbf{B}$ are two sentence symbols. We can use this tautology to create first-order tautologies. Let $\varphi$ and $\psi$ be wffs of first-order logic. Then

$$(\varphi \to \psi) \to (\neg\psi \to \neg\varphi)$$

is a first-order tautology. All the first-order tautologies are obtained in this manner, that is, from the tautologies of propositional logic.

For another example, consider the propositional tautology (3.12). Let $\alpha$ and $Pt$ be any two formulas of first-order logic, where $Pt$ is an atomic formula. Then

$$(\forall v\alpha \to Pt) \to ((\neg Pt) \to (\neg\forall v\alpha))$$

is a first-order tautology. We now present four more examples of first-order tautologies. In these examples, we reverse the replacement procedure and then determine if we get a propositional tautology. If so, we have a first-order tautology.

1. $(\forall zPz \vee \neg\forall zPz)$
   Replace the wff $\forall zPz$ with the sentence symbol **A**. Thus, item 1 is a first-order tautology, because $(\mathbf{A} \vee \neg\mathbf{A})$ is a tautology.
2. $(\exists zPz \to \forall xQx) \to (\neg\forall xQx \to \neg\exists zPz)$
   Replace $\exists zPz$ with the sentence symbol **A** and replace $\forall xQx$ with **B**. Item 2 is a first-order tautology, because $(\mathbf{A} \to \mathbf{B}) \to (\neg\mathbf{B} \to \neg\mathbf{A})$ is a tautology.
3. $\neg(\forall zPz \to Qx) \to \forall zPz$
   Replace $\forall zPz$ with **A** and replace $Qx$ with **B**. Item 3 is a first-order tautology because $\neg(\mathbf{A} \to \mathbf{B}) \to \mathbf{A}$ is a tautology.
4. $\neg(\forall zPz \to Qx) \to \neg Qx$
   Replace $\forall zPz$ with **A** and replace $Qx$ with **B**. Item 4 is a first-order tautology because $\neg(\mathbf{A} \to \mathbf{B}) \to \neg\mathbf{B}$ is a tautology,

### Propositional logic revisited

We will now apply certain concepts from propositional logic, covered in Chapter 2, to first-order logic. We will also refer to a first-order wff as being an $\mathcal{L}$-formula. In propositional logic, the concept of a truth assignment is based on having sentence symbols. Can the concept of a sentence symbol be extended to first-order logic?

**Definition 3.3.2** (Prime formulas). We divide the $\mathcal{L}$-formulas into two groups:
1. A wff is called *prime* if it is an atomic formula or has the form $\forall v\alpha$ for a wff $\alpha$.
2. A wff is called *nonprime* if it has the form $(\neg\alpha)$ or $(\alpha \to \beta)$ for wffs $\alpha$ and $\beta$.

Thus, $\neg\forall z\neg Pz$ is a nonprime formula, whereas $\forall z\neg Pz$ is a prime formula. The prime formulas of first-order logic can be viewed as analogues of the sentence symbols in propositional logic.

**Definition 3.3.3.** Let $\mathcal{S}$ be a set of prime formulas. Then we shall let $\overline{\mathcal{S}}$ be the set of wffs that can be built from the prime formulas in $\mathcal{S}$ by using the two formula building functions $\mathcal{E}_\neg$ and $\mathcal{E}_\to$.

Every wff of first-order logic can be built up from the prime formulas by the operations $\mathcal{E}_\neg$ and $\mathcal{E}_\to$. Thus, if one considers the prime formulas as "sentence symbols," then the wffs of first-order logic can also be seen, from a global point of view, as "formulas of propositional logic."

**Definition 3.3.4.** Let $\mathcal{S}$ be a set of prime formulas in a first-order language $\mathcal{L}$. A function $u: \mathcal{S} \to \{F, T\}$ is called an *$\mathcal{L}$-truth assignment* for $\mathcal{S}$.

Let $\mathcal{S}$ be a set of prime formulas and let $u: \mathcal{S} \to \{F, T\}$ be an $\mathcal{L}$-truth assignment for $\mathcal{S}$. Let $\mathcal{F} = \{\mathcal{E}_\neg, \mathcal{E}_\to\}$. Theorem 2.2.1 implies that the set $\overline{\mathcal{S}}$ of all wffs generated by $\mathcal{S}$ from the functions in $\mathcal{F}$ is freely generated. Thus, there is a unique function $\overline{u}: \overline{\mathcal{S}} \to \{F, T\}$ satisfying the analogous conclusions of Theorem 2.2.4(1)(4).

Let $\psi$ be a wff of first-order logic. Recall Definition 3.3.1. One can now show that $\psi$ is a first-order tautology if and only if for every $\mathcal{L}$-truth assignment $u$, defined on the prime formulas in $\psi$, we have $\overline{u}(\psi) = T$.

**Remark 3.3.5.** Let $\mathfrak{A}$ be an $\mathcal{L}$-structure with domain $A$ and also let $s\colon \mathcal{T} \to A$ be an assignment, where $\mathcal{T}$ is the set of all the variables and constant symbols of $\mathcal{L}$. Now let $u$ be an $\mathcal{L}$-truth assignment. It is possible that $u(\forall v\alpha) = F$, while $\mathfrak{A} \models \forall v\alpha[s]$. Thus, an $\mathcal{L}$-truth assignment may disagree with the satisfaction relation of Definition 3.2.6.

**Definition 3.3.6.** Let $\varphi$ be an $\mathcal{L}$-formula and let $u$ be an $\mathcal{L}$-truth assignment that is defined on the prime formulas in $\varphi$. Then $u$ *satisfies* $\varphi$ if and only if $\overline{u}(\varphi) = T$.

Let $\varphi$ and $\psi$ be two $\mathcal{L}$-wffs. Then $\varphi$ *tautologically implies* $\psi$ if and only if for every $\mathcal{L}$-truth assignment $u$ defined on all of the prime formulas that appear in $\varphi$ and $\psi$, if $u$ satisfies $\varphi$, then $u$ satisfies $\psi$. Moreover, $\varphi$ and $\psi$ are *tautologically equivalent* if $\varphi$ tautologically implies $\psi$ and the converse holds as well.

**Definition 3.3.7.** Let $\Sigma$ be a set of $\mathcal{L}$-wffs. Then $\Sigma$ is *$\mathcal{L}$-satisfiable* if there is an $\mathcal{L}$-truth assignment $u$ that is defined on the prime formulas in every $\varphi \in \Sigma$ such that $\overline{u}(\varphi) = T$ for every $\varphi \in \Sigma$.

**Definition 3.3.8.** A set of $\mathcal{L}$-formulas $\Sigma$ is *finitely $\mathcal{L}$-satisfiable* if and only if every finite subset of $\Sigma$ is $\mathcal{L}$-satisfiable.

The notion of tautological implication can now be applied to first-order logic.

**Definition 3.3.9.** Let $\Sigma$ be a set of $\mathcal{L}$-formulas and let $\varphi$ be an $\mathcal{L}$-formula. Then $\Sigma$ *tautologically implies* $\varphi$ if and only if for every $\mathcal{L}$-truth assignment $u$, defined on the prime formulas occurring in formulas in $\Sigma$ and in $\varphi$, if $u$ satisfies $\Sigma$, then $\overline{u}(\varphi) = T$.

Theorem 2.4.2, the compactness theorem of propositional logic, and its corollaries now extend to first-order logic.

**Theorem 3.3.10.** *Let $\Sigma$ be a set of $\mathcal{L}$-formulas. If $\Sigma$ is finitely $\mathcal{L}$-satisfiable, then $\Sigma$ is $\mathcal{L}$-satisfiable.*

The above extension of the propositional compactness theorem holds even when the set of prime formulas is uncountable (see Remark 2.4.3). The proof of Corollary 2.4.5 establishes the following result.

**Corollary 3.3.11.** *If a set of $\mathcal{L}$-formulas $\Sigma$ tautologically implies an $\mathcal{L}$-formula $\varphi$, then $\Sigma_0$ tautologically implies $\varphi$ for some finite $\Sigma_0 \subseteq \Sigma$.*

### 3.3.2 Generalization and substitution

Given an $\mathcal{L}$-formula $\psi$, a generalization of $\psi$ is the result of putting universal quantifiers prior to $\psi$.

**Definition 3.3.12** (Generalization). Let $\mathcal{L}$ be a first-order language. Let $\psi$ be a wff. A wff $\varphi$ is a *generalization* of $\psi$ if and only if for some $n \geq 0$ and some variables $x_1, x_2, \ldots, x_n$,

$$\varphi = \forall x_1 \forall x_2 \cdots \forall x_n \psi.$$

In particular, when $n = 0$, any wff is a generalization of itself.

### Substitution

In mathematics, one often replaces a variable with some expression that represents a possible value of the variable. We will also need to do this in first-order logic. In particular, we will need to substitute a free variable appearing in a wff $\alpha$ with a term. We shall write $\alpha_t^x$ to denote the wff obtained by replacing the free occurrences of the variable $x$ in the wff $\alpha$ with the term $t$.

Before we give a mathematical definition of the operation $\alpha_t^x$, we give an example of what the operation $\alpha_t^x$ should *not do*. Let $\mathcal{L} = \{\dot{+}, \dot{0}\}$, where $\dot{+}$ is a 2-place function symbol and $\dot{0}$ is a constant symbol. Consider the wff $\forall x(x \dot{+} \dot{0} \doteq x)$. If we replace the variable $x$ with the term $\dot{0}$, then we obtain the expression $\forall \dot{0}(\dot{0} \dot{+} \dot{0} \doteq \dot{0})$, which is not a wff. One cannot "quantify over a constant." So, the definition of the operation $\alpha_t^x$ must avoid replacing a quantified variable with the term $t$ (see item 4 of the following definition).

**Definition 3.3.13** (The operation $\alpha_t^x$). Let $\mathcal{L}$ be a language where $x$ is a variable and $t$ is a term. Let $\alpha$ be a wff. The wff $\alpha_t^x$ is defined recursively as follows:
1. $\alpha_t^x$ is the result of replacing all occurrences of $x$ in $\alpha$ with $t$, when $\alpha$ is atomic;
2. $(\neg\alpha)_t^x = \neg\alpha_t^x$;
3. $(\alpha \to \beta)_t^x = \alpha_t^x \to \beta_t^x$;
4. $(\forall v\alpha)_t^x = \begin{cases} \forall v\alpha, & \text{if } x = v, \\ \forall v\alpha_t^x, & \text{if } x \neq v. \end{cases}$

In the right hand side of the equalities in items 2 and 3 of Definition 3.3.13, we have dropped the outermost parentheses (see 7 on page 63). Exercise 5 shows that Definition 3.3.13 is an application of Theorems 1.1.27 and 3.1.16.

We offer some observations and examples that illustrate Definition 3.3.13:
(a) $\varphi_x^x = \varphi$ for all wffs $\varphi$;
(b) $\varphi_t^x = \varphi$ when $x$ is not a free variable in $\varphi$;
(c) observe that

$$(Qx \to \forall yPy)_y^x = Qx_y^x \to (\forall yPy)_y^x = Qy \to \forall yPy$$

by items 3, 1, and 4 of Definition 3.3.13;
(d) note that

$$(\neg\forall yLxy)_z^x = \neg(\forall yLxy)_z^x = \neg\forall yLxy_z^x = \neg\forall yLzy$$

by items 2, 4, and 1 of Definition 3.3.13;

(e) observe that

$$\forall x \neg \forall y Lxy \to (\neg \forall y Lxy)^x_z = \forall x \neg \forall y Lxy \to \neg \forall y Lzy$$

by item (c). This example has the form $\forall x \alpha \to \alpha^x_t$, where $t$ is a term.

One might wonder why we have the two-case distinction in item 4 of Definition 3.3.13. Suppose that instead, item 4 was defined to be $(\forall v \alpha)^x_t = \forall v \alpha^x_t$ even if $v = x$. Then this version of Definition 3.3.13 would allow us to conclude that

$$(\exists x(x \not\doteq y))^x_y = \exists x(y \not\doteq y),$$

which is false in all structures and all assignments. This illustrates why item 4 is defined as it is in Definition 3.3.13.

The next two examples (i) and (ii) illustrate the concept of being and not being "substitutable," respectively. Let $\mathcal{L} = \{L, f, c\}$, where $L$ is a 2-place predicate symbol, $f$ is a 1-place function symbol, and $c$ is a constant symbol. Consider the wff $\alpha = \forall v Lvx$. Note that $\alpha$ contains the two distinct variables $v$ and $x$. Using the terms $fy$ and $fv$, let us evaluate the new formulas $\alpha^x_{fy}$ and $\alpha^x_{fv}$:

(i) $\alpha^x_{fy} = (\forall v Lvx)^x_{fy} = \forall v Lvfy$
So, when the term $fy$ is substituted for the variable $x$, the quantifier "$\forall v$" *does not* "capture" the variable $y$ in $fy$.

(ii) $\alpha^x_{fv} = (\forall v Lvx)^x_{fv} = \forall v Lvfv$
Thus, when the term $fv$ is substituted for the variable $x$, the quantifier "$\forall v$" *does* "capture" the variable $v$ in $fv$.

In (i) we shall say that $fy$ is "substitutable" for $x$, whereas in (ii) we shall say that $fv$ is *not* "substitutable" for $x$. We want to avoid substitutions, as in (ii), that result in a variable in a term being "captured" by a quantifier. Why? See Example 3.3.14, below.

**Example 3.3.14.** The wff $\forall x \alpha \to \alpha^x_t$ asserts that "if $\alpha$ is true of everything, then $\alpha$ is true of $t$." This appears to be obviously true. Let $\mathcal{L} = \{\doteq\}$ and let $\mathfrak{A}$ be an $\mathcal{L}$-structure whose domain contains at least two elements. Now let $\alpha$ be the formula $(\neg \forall y(x \doteq y))$. Observe that $\alpha^x_y = (\neg \forall y(x \doteq y))^x_y = \neg \forall y(y \doteq y)$. Thus, the conditional $\forall x \alpha \to \alpha^x_y$ is

$$\forall x \neg \forall y(x \doteq y) \to \neg \forall y(y \doteq y). \tag{3.13}$$

The hypothesis $\forall x \neg \forall y(x \doteq y)$ in (3.13) is equivalent to $\forall x \exists y(x \not\doteq y)$, which is true in $\mathfrak{A}$. On the other hand, the conclusion $\neg \forall y(y \doteq y)$ in (3.13) is equivalent to $\exists y(y \not\doteq y)$, so it is false in $\mathfrak{A}$. Thus, sentence (3.13) is also false in $\mathfrak{A}$. The problem here is that the substitution $(\neg \forall y(x \doteq y))^x_y$ resulted in a variable being "captured" by a quantifier.

We conclude from Example 3.3.14 that when performing a substitution, we must avoid capturing a variable by a quantifier. Thus, we must identify a specific restriction that will prevent a variable from being so captured. A term will be said to be *substitutable* for a variable in a formula if its substitution does not produce a captured variable. The formal definition follows.

**Definition 3.3.15** (Substitutable). Let $\mathcal{L}$ be a language with variable $x$ and term $t$. Then "$t$ is *substitutable* for $x$ in $\alpha$" is defined recursively as follows:

1. $t$ is always substitutable for $x$ in $\alpha$ when $\alpha$ is an atomic formula;[1]
2. $t$ is substitutable for $x$ in $(\neg\alpha)$ iff $t$ is substitutable for $x$ in $\alpha$;
3. $t$ is substitutable for $x$ in $(\alpha \to \beta)$ iff $t$ is substitutable for $x$ in both $\alpha$ and $\beta$;
4. $t$ is substitutable for $x$ in $(\forall v\alpha)$ iff either
   (a) $x$ does not occur free in $(\forall v\alpha)$[2] or
   (b) $v$ does not occur in $t$ and $t$ is substitutable for $x$ in $\alpha$.[3]

It follows from Definition 3.3.15 that a variable $x$ is substitutable for $x$ in every wff.

**Remark.** Given a wff $\alpha$ and a term $t$, the operation $\alpha_t^x$ is always defined, but it may be the case that $t$ is not substitutable for $x$ in $\alpha$. However, if a term $t$ is a constant symbol or contains no variables, then $t$ will always be substitutable for $x$ in $\alpha$.

**Problem 3.3.16.** Let $\mathcal{L} = \{P, Q, f, g, c\}$, where $P, Q$ are 2-place predicate symbols, $f, g$ are 1-place function symbols, and $c$ is a constant symbol. Let $\varphi$ be the wff $(Pxy \to \forall xQgxz)$ and consider the term $fx$. Evaluate the wff $\varphi_{fx}^z$ and decide if $fx$ is substitutable for $z$ in $\varphi$.

**Solution.** We evaluate $\varphi_{fx}^z$ as follows:

$$\begin{aligned}
\varphi_{fx}^z &= (Pxy \to \forall xQgxz)_{fx}^z && \text{by definition of } \varphi,\\
&= Pxy_{fx}^z \to (\forall xQgxz)_{fx}^z && \text{by Definition 3.3.13(3)},\\
&= Pxy \to (\forall xQgxz)_{fx}^z && \text{by Definition 3.3.13(1)},\\
&= Pxy \to \forall xQgxfx && \text{by Definition 3.3.13(4)}.
\end{aligned}$$

Thus, $\varphi_{fx}^z$ is the wff $Pxy \to \forall xQgxfx$. However, the term $fx$ is not substitutable for $z$ in $\varphi$ because $fx$ is not substitutable for $z$ in $\forall xQgxz$ by condition 4(b) of Definition 3.3.15 ($x$ in the term $fx$ is captured by $\forall x$).

---

1 There are no quantifiers in an atomic formula, so no variable in $t$ can be captured.

2 In this case, $(\forall v\alpha)_t^x = \forall v\alpha$.

3 This ensures that the quantifier "$\forall v$" does not capture any variable in $t$ and that no variable in $t$ will get captured in $\alpha_t^x$.

### 3.3.3 The logical axioms

There are some formulas in a first-order language that are logically valid, that is, these formulas are satisfied by every structure and every assignment. Usually one selects a minimal set of such formulas to be identified as the logical axioms. From a selected set of logical axioms, one must be able to deduce all the other logically valid formulas. Our discussions on first-order tautologies, generalizations, and substitutions will now allow us to present a set $\Lambda$ of the logical axioms, which are arranged in six groups.

**Logical Axioms 3.3.17.** Let $\mathcal{L}$ be a language of first-order logic. Let $\Lambda$ be the set of all *generalizations* of wffs having the following forms, where $x, y$ are any variables, $\alpha, \beta$ are any wffs, and $t$ is any term:
1. first-order tautologies;
2. $\forall x\alpha \to \alpha_t^x$, where $t$ is substitutable for $x$ in $\alpha$;
3. $\forall x(\alpha \to \beta) \to (\forall x\alpha \to \forall x\beta)$;
4. $\alpha \to \forall x\alpha$, where $x$ does not occur free in $\alpha$.

If the language $\mathcal{L}$ includes the equality symbol $\doteq$, then we add the following forms:
5. $x \doteq x$;
6. $x \doteq y \to (\alpha \to \alpha_y^x)$, where $\alpha$ is an atomic formula.

We shall call the resulting set $\Lambda$ the *set of logical axioms.*

We prove in Section 4.1 that every logical axiom in $\Lambda$ is logically valid (see Definition 3.2.17), that is, each logical axiom holds in all structures and for all assignments.

**Remark** (On the logical axioms). In Logical Axiom 1, the first-order tautologies will also be called tautologies. They are included to handle the propositional connective symbols. Logical Axiom 2 reflects the intended meaning of the quantifier symbol, that is, if a statement is true of all objects in the domain, then the statement is true of an individual denoted by $t$. Logical Axioms 3 and 4 are used to prove the generalization theorem (Theorem 3.3.29) later in this section. Logical Axioms 1 and 2 are used to deduce the important properties of equality (see Theorem 3.3.54). Logical Axiom 6 is an application of Leibniz's law: If two things are equal, then whatever is true of one is true of the other. Generalizations of logical axioms are also logical axioms.

In the following problem, one is asked to identify whether or not a given wff is a logical axiom.

**Problem 3.3.18.** Let $\mathcal{L} = \{P, Q, R, c\}$, where $P, Q$ are 1-place predicate symbols, $R$ is a 2-place predicate symbol, and $c$ is a constant symbol. To which logical axiom groups, if any, do each of the following wffs belong?
1. $(\exists zPz \to \forall xQx) \to (\neg\forall xQx \to \neg\exists zPz)$.
2. $\forall x(\forall zPz \vee \neg\forall zPz)$.

3. $\forall zQz \to Qc$.
4. $\forall x\forall zQz \to Qc$.
5. $\forall x\forall yRxy \to \forall yRcy$.
6. $\forall x\forall yRxy \to \forall yRyy$.
7. $\forall x(\exists zPz \to \forall zPz) \to (\forall x\exists zPz \to \forall x\forall zPz)$.
8. $\exists zPz \to \forall x\exists zPz$.

**Solution.**

1. $(\exists zPz \to \forall xQx) \to (\neg\forall xQx \to \neg\exists zPz)$
   The above wff belongs to axiom group 1. To verify this, let $\alpha$ be the wff $\exists zPz$ and let $\beta$ be the wff $\forall xQx$. The above can then be written as
   $$(\alpha \to \beta) \to (\neg\beta \to \neg\alpha),$$
   which is a tautology.
2. $\forall x(\forall zPz \vee \neg\forall zPz)$
   The above wff belongs to axiom group 1, since it is a generalization of a tautology. To see this, let $\alpha$ be the wff $\forall zPz$. The above can then be written as a generalization of $(\alpha \vee \neg\alpha)$, which is a tautology.
3. $\forall zQz \to Qc$
   This wff belongs to axiom group 2. To affirm this, let $\alpha$ be the wff $Qz$. The above can then be written as $\forall z\alpha \to \alpha^z_c$ and $c$ is substitutable for $z$ in $\alpha$.
4. $\forall x\forall zQz \to Qc$
   The above wff is not a logical axiom. Note that the above wff is not a generalization of an axiom in group 2.
5. $\forall x\forall yRxy \to \forall yRcy$
   The given wff belongs to axiom group 2. To see this, let $\alpha$ be the wff $\forall yRxy$. The above can then be written as $\forall x\alpha \to \alpha^x_c$ and $c$ is substitutable for $x$ in $\alpha$.
6. $\forall x\forall yRxy \to \forall yRyy$
   The above wff is not a logical axiom, for the following reason. Let $\alpha$ be the wff $\forall yRxy$. Then the above wff has the form $\forall x\alpha \to \alpha^x_y$. However, $y$ is not substitutable for $x$ in $\alpha$. So, the above wff is *not* a logical axiom.
7. $\forall x(\exists zPz \to \forall zPz) \to (\forall x\exists zPz \to \forall x\forall zPz)$
   The above wff belongs to axiom group 3. To confirm this, let $\alpha$ be the wff $\exists zPz$ and let $\beta$ be the wff $\forall zPz$. The above wff can then be written as
   $$\forall x(\alpha \to \beta) \to (\forall x\alpha \to \forall x\beta).$$
8. $\exists zPz \to \forall x\exists zPz$
   This wff belongs to axiom group 4. To see this, let $\alpha$ be the wff $\exists zPz$. The above can then be written as $\alpha \to \forall x\alpha$ and $x$ is not free in $\alpha$.

### 3.3.4 Formal deductions

A deduction in first-order logic, as in propositional logic, is going to be a finite list of formulas that satisfies certain conditions.

**Definition 3.3.19** (An inference rule). Let $\mathcal{L}$ be a language of first-order logic. Let $\alpha$ and $\beta$ be wffs in the language $\mathcal{L}$. *Modus ponens* is the rule of inference: *From the formulas $\alpha$ and $\alpha \to \beta$ we can infer $\beta$*, that is,

$$\begin{array}{l} \alpha \to \beta \\ \alpha \\ \hline \therefore \beta \end{array} \quad . \qquad \text{(modus ponens)}$$

**Definition 3.3.20.** Let $\Gamma$ be a set of formulas. A *deduction of $\varphi$ from $\Gamma$* is a sequence $\langle \alpha_1, \ldots, \alpha_n \rangle$ of formulas such that $\alpha_n = \varphi$ and for all $1 \leq k \leq n$, either
(a) $\alpha_k$ is in $\Gamma \cup \Lambda$, or
(b) $\alpha_k$ is obtained by modus ponens from two earlier wffs in the sequence, that is, for some $i$ and $j$ less than $k$, the wffs $\alpha_i$ and $\alpha_j = (\alpha_i \to \alpha_k)$ are in the sequence.

**Definition 3.3.21.** Let $\Gamma$ be a set of formulas. A formula $\varphi$ is a *theorem* of $\Gamma$, denoted by $\Gamma \vdash \varphi$, if there is a deduction of $\varphi$ from $\Gamma$. When $\Gamma = \{\psi\}$ we shall write $\psi \vdash \varphi$ to denote that there is a deduction of $\varphi$ from the wff $\psi$. When $\Gamma = \varnothing$ we shall write $\vdash \varphi$ to denote that there is a deduction of $\varphi$ from the logical axioms alone.

In Definition 3.3.21, $\Gamma$ can be viewed as a set of assumptions. The notation $\Gamma \vdash \varphi$ means that one can deduce $\varphi$ from the assumptions and the logical axioms.

**Example 3.3.22.** Let $\mathcal{L}$ contain two 1-place predicate symbols $P, Q$ and a constant symbol $c$. Let $\Gamma = \{\forall x(Px \to Qx), \forall z Pz\}$. Show that $\Gamma \vdash Qc$.

**Solution.** A deduction of $Qc$ from $\Gamma$ (with explanation) is given below:

1. $\forall x(Px \to Qx)$ — in $\Gamma$,
2. $\forall z Pz$ — in $\Gamma$,
3. $\forall x(Px \to Qx) \to (Pc \to Qc)$ — by Logical Axiom 2,
4. $\forall z Pz \to Pc$ — by Logical Axiom 2,
5. $Pc \to Qc$ — from 1 and 3, by modus ponens,
6. $Pc$ — from 2 and 4, by modus ponens,
7. $Qc$ — from 5 and 6, by modus ponens.

Therefore, $\Gamma \vdash Qc$.

**Example 3.3.23.** Let $P$ be a 1-place predicate symbol. Show that $\vdash Px \to \exists x Px$.

**Solution.** A deduction of $Px \to \neg\forall x \neg x Px$ (with explanation) is given below:

| | | |
|---|---|---|
| 1. | $\forall x \neg Px \to \neg Px$ | by Logical Axiom 2, |
| 2. | $(\forall x \neg Px \to \neg Px) \to (Px \to \neg\forall x \neg Px)$ | by Logical Axiom 1, |
| 3. | $Px \to \neg\forall x \neg Px$ | from 1 and 2, by modus ponens. |

Since $\neg\forall x \neg Px$ is abbreviated by $\exists x Px$, we conclude that $\vdash Px \to \exists x Px$.

In the above step 1 we used $\neg Px$ as $\alpha$ and $t = x$ in Logical Axiom 2, that is, $\forall x\alpha \to \alpha_x^x$ is the instance of Logical Axiom 2 that we used. Moreover, in step 2 we used the tautology $(\mathbf{A} \to \neg\mathbf{B}) \to (\mathbf{B} \to \neg\mathbf{A})$.

**Example 3.3.24.** Let $P$ be a 1-place predicate symbol. Show that $\vdash \forall x(Px \to \exists y Py)$.

**Solution.** A deduction of $\forall x(Px \to \neg\forall y \neg Py)$ (with explanation) is given below:

| | | |
|---|---|---|
| 1. | $\forall x(\forall y \neg Py \to \neg Px)$ | Logical Axiom 2 (gen), |
| 2. | $\forall x(\,(\forall y \neg Py \to \neg Px) \to (Px \to \neg\forall y \neg Py)\,)$ | Logical Axiom 1 (gen), |
| 3. | $\forall x(\,(\forall y \neg Py \to \neg Px) \to (Px \to \neg\forall y \neg Py)\,)$ | |
| | $\to (\,\forall x(\forall y \neg Py \to \neg Px) \to \forall x(Px \to \neg\forall y \neg Py)\,)$ | Logical Axiom 3, |
| 4. | $\forall x(\forall y \neg Py \to \neg Px) \to \forall x(Px \to \neg\forall y \neg Py)$ | modus ponens: 2 and 3, |
| 5. | $\forall x(Px \to \neg\forall y \neg Py)$ | modus ponens: 1 and 4. |

Thus, $\vdash \forall x(Px \to \exists y Py)$.

In the above solution, “gen” means “generalization.” In step 1 we used $\neg Py$ as $\alpha$ and $t = x$ in Logical Axiom 2, that is, $\forall y\alpha \to \alpha_x^y$ is the instance of Logical Axiom 2 we applied. In step 2 we used the tautology $(\mathbf{A} \to \neg\mathbf{B}) \to (\mathbf{B} \to \neg\mathbf{A})$. In step 3 we used $(\forall y \neg Py \to \neg Px)$ as $\alpha$ and $(Px \to \neg\forall y \neg Py)$ as $\beta$ in Logical Axiom 3.

**Example 3.3.25.** Let $\mathcal{L}$ have a 1-place predicate symbol $Q$. Show that $\forall y Qy \vdash \forall x Qx$.

**Solution.** A deduction of $\forall x Qx$ from $\forall y Qy$ (with explanation) is given below:

| | | |
|---|---|---|
| 1. | $\forall y Qy$ | given, |
| 2. | $\forall x(\forall y Qy \to Qx)$ | Logical Axiom 2 (gen), |
| 3. | $\forall y Qy \to \forall x\forall y Qy$ | Logical Axiom 4, |
| 4. | $\forall x(\forall y Qy \to Qx) \to (\forall x\forall y Qy \to \forall x Qx)$ | Logical Axiom 3, |
| 5. | $\forall x\forall y Qy \to \forall x Qx$ | modus ponens: 2 and 4, |
| 6. | $\forall x\forall y Qy$ | modus ponens: 1 and 3, |
| 7. | $\forall x Qx$ | modus ponens: 5 and 6. |

So $\forall y Qy \vdash \forall x Qx$.

All of the standard proof techniques that are used in mathematics hold for deductions as well; for example, “proof by contradiction” can be used to show that a deduction exists in first-order logic (see Corollary 3.3.37 below).

We will now present and prove lemmas and theorems about deductions. In our first such result we identify some useful observations about deductions. Each item in the following lemma follows directly from Definition 3.3.20.

**Lemma 3.3.26.** *Let $\Gamma$ be a set of formulas and let $\alpha$, $\beta$, $\delta$, and $\gamma$ be formulas.*
1. *If $\alpha \in \Gamma$ or $\alpha \in \Lambda$, then $\Gamma \vdash \alpha$.*
2. *If $\vdash \alpha$, then $\Gamma \vdash \alpha$.*
3. *If $\alpha \vdash \beta$ and $\beta \vdash \gamma$, then $\alpha \vdash \gamma$.*
4. *If $\Gamma \vdash \alpha$ and $\alpha \vdash \beta$, then $\Gamma \vdash \beta$.*
5. *If $\Gamma \vdash \delta$ and $\Gamma \vdash (\delta \to \beta)$, then $\Gamma \vdash \beta$.*

The proof of our next result can be seen as an application of the compactness theorem of propositional logic.

**Theorem 3.3.27.** *Let $\Gamma$ be a set of wffs of first-order logic and let $\varphi$ be a wff of first-order logic. Then $\Gamma \vdash \varphi$ if and only if $\Gamma \cup \Lambda$ tautologically implies $\varphi$.*

*Proof.* ($\to$): Assume that $\Gamma \vdash \varphi$. We shall show that $\Gamma \cup \Lambda$ tautologically implies $\varphi$. Let $\langle \alpha_1, \alpha_2, \ldots, \alpha_n \rangle$ be a deduction of $\varphi$ from the set $\Gamma$ of wffs. We shall prove the following statement: *For all $k \leq n$, $\Gamma \cup \Lambda$ tautologically implies $\alpha_k$.* We shall use strong induction on the natural number variable $k$.

*Base step:* Let $k = 1$. Since $\alpha_1$ is the first step in the deduction, we must have $\alpha_1 \in \Gamma \cup \Lambda$. Thus, $\Gamma \cup \Lambda$ tautologically implies $\alpha_1$.

*Inductive step:* Assume the strong induction hypothesis

$$\Gamma \cup \Lambda \text{ tautologically implies } \alpha_i \text{ for all } i < k. \tag{SIH}$$

We show that $\Gamma \cup \Lambda$ tautologically implies $\alpha_k$. As $\alpha_k$ is in the deduction, either (a) $\alpha_k \in \Gamma \cup \Lambda$ or (b) $\alpha_k$ is obtained by modus ponens from some $\alpha_i$ and $\alpha_j = (\alpha_i \to \alpha_k)$, where $i, j < k$. If (a) holds, then $\Gamma \cup \Lambda$ tautologically implies $\alpha_k$ just as in the above base step. If (b) holds, then by the strong induction hypothesis (SIH), we conclude that $\Gamma \cup \Lambda$ tautologically implies $\alpha_i$ and $\Gamma \cup \Lambda$ tautologically implies $(\alpha_i \to \alpha_k)$. Hence, $\Gamma \cup \Lambda$ tautologically implies $\alpha_k$ (see page 35).

($\Leftarrow$): Assume that $\Gamma \cup \Lambda$ tautologically implies $\varphi$. As $\Gamma \cup \Lambda$ tautologically implies $\varphi$, Corollary 3.3.11 implies that there exists a finite subset $\Gamma_0$ of $\Gamma \cup \Lambda$ such that $\Gamma_0$ tautologically implies $\varphi$. Since $\Gamma_0$ is finite, let $\Gamma_0 = \{\gamma_1, \gamma_2, \ldots, \gamma_n\}$. It follows (see Exercise 6 on page 51) that

$$\gamma_1 \to \gamma_2 \to \cdots \to \gamma_n \to \varphi$$

is a tautology. By repeated use of modus ponens, it easily follows that $\Gamma_0 \vdash \psi$. As $\Gamma_0 \subseteq \Gamma$, we conclude that $\Gamma \vdash \varphi$. □

Note that the above Theorem 3.3.27 was established by giving a mathematical proof in English. That is, we proved a theorem about deductions. Hence, one is now in a position to prove theorems about "mathematical proof." It is generally accepted that all proofs in classical mathematics can be expressed as a deduction in first-order logic. Thus, one can now ask and answer questions about the limits of mathematical proof.

### 3.3.5 Metatheorems about deductions

If $\Gamma \vdash \alpha$, then we say that $\alpha$ is a theorem of $\Gamma$. However, we have also been using the word "theorem" in a different way when proving theorems in English about deductions. These English theorems are sometimes referred to as being *metatheorems* to emphasize that they are statements in English about deductions and first-order logic.

We now pose a question about deductions.

> HOW CAN WE SHOW THAT A DEDUCTION EXISTS WITHOUT ACTUALLY GIVING ONE?
>
> **Answer:** *One must first prove metatheorems about deductions.*

#### Induction on deductions

Let $S$ be a set of wffs. Then $S$ is said to be *closed* under modus ponens if whenever $\alpha \in S$ and $(\alpha \to \beta) \in S$, we have $\beta \in S$. Since there is a procedure for building all the theorems of $\Gamma$ using the formulas in $\Gamma \cup \Lambda$ and applying modus ponens, we have the following *induction principle.*

**Theorem 3.3.28** (Induction principle). *Let $S$ be a set of wffs such that:*
(1) $\Gamma \cup \Lambda \subseteq S$,
(2) *$S$ is closed under modus ponens.*

*Then $\{\varphi : \Gamma \vdash \varphi\} \subseteq S$, that is, $S$ contains all the theorems of $\Gamma$.*

*Proof.* Let $S$ and $\Gamma$ be as stated. Assume that $\Gamma \vdash \varphi$. Now let $\langle \alpha_1, \alpha_2, \ldots, \alpha_n \rangle$ be a deduction of $\varphi$ from the set $\Gamma$ of wffs. Thus, $\alpha_n = \varphi$. We shall prove that for all $k \leq n$, $\alpha_k \in S$. We shall use strong induction on the natural number variable $k$.

*Base step:* Let $k = 1$. Since $\alpha_1$ is the first step in the deduction, we must have $\alpha_1 \in \Gamma \cup \Lambda$. Thus, $\alpha_1 \in S$ by (1).

*Inductive step:* Assume the strong induction hypothesis

$$\alpha_i \in S \text{ for all } i < k. \tag{SIH}$$

We show that $\alpha_k \in S$. As $\alpha_k$ is in the deduction, either (a) $\alpha_k \in \Gamma \cup \Lambda$ or (b) $\alpha_k$ is obtained by modus ponens from some $\alpha_i$ and $\alpha_j = (\alpha_i \to \alpha_k)$, where $i, j < k$. If (a) holds, then $\alpha_k \in S$.

If (b) holds, then by the induction hypothesis we see that $\alpha_i \in S$ and $(\alpha_i \to \alpha_k) \in S$. By (2), $S$ is closed under modus ponens. Hence, $\alpha_k \in S$. Therefore, for all $k \leq n$, $\alpha_k \in S$. In particular, $\varphi \in S$. □

Theorem 3.3.28 validates the following strategy, which will allow us to prove results about formulas $\varphi$ that are deducible from a set of wffs $\Gamma$.

**Proof Strategy.** Let $\mathbb{S}(\varphi)$ be a statement about a wff $\varphi$. In order to prove an assertion "for all wffs $\varphi$, if $\Gamma \vdash \varphi$, then $\mathbb{S}(\varphi)$" by induction, use the following diagram:

*Base step:* Prove $\mathbb{S}(\phi)$ for all formulas $\phi \in \Gamma \cup \Lambda$.
*Inductive step:* Let $\alpha$ and $\beta$ be wffs.
Assume $\mathbb{S}(\alpha)$ and $\mathbb{S}(\alpha \to \beta)$.
Prove $\mathbb{S}(\beta)$.

The above proof strategy is Theorem 3.3.28 applied to the set $S = \{\varphi \text{ is a wff} : \mathbb{S}(\varphi)\}$. We will apply this strategy in the proof of our next theorem, where $\mathbb{S}(\varphi)$ is "$\Gamma \vdash \forall x\varphi$." In this proof we will tacitly be using Lemma 3.3.26.

**Theorem 3.3.29** (Generalization theorem). *Suppose that $x$ does not occur free in any formula in $\Gamma$. For all wffs $\varphi$, if $\Gamma \vdash \varphi$, then $\Gamma \vdash \forall x\varphi$.*

*Proof.* Assume that the variable $x$ does not occur free in any formula in $\Gamma$. Using the induction principle, we shall prove the following: For all wffs $\varphi$, if $\Gamma \vdash \varphi$, then $\Gamma \vdash \forall x\varphi$.

*Base step:* Let $\phi$ be in $\Gamma \cup \Lambda$. If $\phi \in \Gamma$, then $x$ does not occur free in $\phi$. Thus, $(\phi \to \forall x\phi)$ is in axiom group 4 of the logical axioms (page 97). Since $\Gamma \vdash \phi$ and $\Gamma \vdash (\phi \to \forall x\phi)$, we conclude (by modus ponens) that $\Gamma \vdash \forall x\phi$. If $\phi \in \Lambda$, then $\phi$ is a logical axiom. Thus, the generalization $\forall x\phi$ is also a logical axiom. Therefore, $\Gamma \vdash \forall x\phi$.

*Inductive step:* Let $\alpha$ and $\beta$ be wffs. Assume the induction hypothesis

$$\Gamma \vdash \forall x\alpha \quad \text{and} \quad \Gamma \vdash \forall x(\alpha \to \beta). \tag{IH}$$

We need to prove that $\Gamma \vdash \forall x\beta$. Note that (▲) $\forall x(\alpha \to \beta) \to (\forall x\alpha \to \forall x\beta)$ is in axiom group 3 of the logical axioms. By (IH) we have $\Gamma \vdash \forall x(\alpha \to \beta)$ and $\Gamma \vdash \forall x\alpha$. Thus, by applying modus ponens twice to (▲), we have $\Gamma \vdash \forall x\beta$. □

A converse of the generalization theorem (Theorem 3.3.29) holds, whether or not $x$ occurs free in a formula in $\Gamma$.

**Theorem 3.3.30.** *For all wffs $\varphi$, if $\Gamma \vdash \forall x\varphi$, then $\Gamma \vdash \varphi$.*

*Proof.* Assume that $\Gamma \vdash \forall x\varphi$. By Logical Axiom 2, we have $\Gamma \vdash (\forall x\varphi \to \varphi)$. Thus, by modus ponens, we conclude that $\Gamma \vdash \varphi$. □

Given that certain deductions exist, the following "tautology rule" allows one to show that another deduction exists.

**Theorem 3.3.31** (Rule T). *If* $\Gamma \vdash \alpha_1, \ldots, \Gamma \vdash \alpha_n$ *and* $\{\alpha_1, \ldots, \alpha_n\}$ *tautologically implies* $\beta$, *then* $\Gamma \vdash \beta$.

*Proof.* Assume that $\Gamma \vdash \alpha_1, \ldots, \Gamma \vdash \alpha_n$ and that $\{\alpha_1, \ldots, \alpha_n\}$ tautologically implies $\beta$. Thus, $\alpha_1 \to \alpha_2 \to \cdots \to \alpha_n \to \beta$ is a tautology (see Exercise 6 on page 51), so it is in axiom group 1 of the logical axioms. Since $\Gamma \vdash \alpha_1, \ldots, \Gamma \vdash \alpha_n$, we conclude that $\Gamma \vdash \beta$ by applying modus ponens $n$ times. □

Note that $\neg(\psi \to \theta)$ tautologically implies $\psi$ and $\neg\theta$ and that $\{\psi, \neg\theta\}$ tautologically implies $\neg(\psi \to \theta)$. We therefore have the following corollary.

**Corollary 3.3.32.** *We have* $\Gamma \vdash \neg(\psi \to \theta)$ *if and only if* $\Gamma \vdash \psi$ *and* $\Gamma \vdash \neg\theta$.

In order to simplify our notation, we shall write $\Gamma; \psi$ as an abbreviation for $\Gamma \cup \{\psi\}$.

**Theorem 3.3.33** (Deduction theorem). *Let* $\Gamma$ *be a set of wffs and* $\gamma, \varphi$ *be wffs. Then* $\Gamma; \gamma \vdash \varphi$ *if and only if* $\Gamma \vdash (\gamma \to \varphi)$.

*Proof.* Let $\Gamma$ be a set of wffs and let $\gamma, \varphi$ be wffs. Then

| | | | |
|---|---|---|---|
| $\Gamma; \gamma \vdash \varphi$ | iff | $\Gamma; \gamma \cup \Lambda$ tautologically implies $\varphi$ | by Theorem 3.3.27, |
| | iff | $\Gamma \cup \Lambda$ tautologically implies $(\gamma \to \varphi)$ | by Exercise 9 on page 38, |
| | iff | $\Gamma \vdash (\gamma \to \varphi)$ | by Theorem 3.3.27. □ |

Thus, to deduce a conditional, one can assume the hypothesis and then deduce the conclusion, just like in mathematics.

**Corollary 3.3.34.** *Let* $\Gamma$ *be a set of wffs. If* $\alpha$ *and* $\beta$ *are tautologically equivalent, then:*
(1) $\Gamma \vdash \alpha$ *iff* $\Gamma \vdash \beta$,
(2) $\Gamma; \alpha \vdash \varphi$ *iff* $\Gamma; \beta \vdash \varphi$, *for any wff* $\varphi$.

*Proof.* Let $\Gamma$ be a set of wffs and let $\alpha$ and $\beta$ be wffs that are tautologically equivalent. Thus, $\alpha \to \beta$ and $\beta \to \alpha$ are tautologies. To prove (1), assume that $\Gamma \vdash \alpha$. Since $\alpha \to \beta$ is a tautology, we see that $\Gamma \vdash \beta$. The converse holds in a similar fashion.

To prove (2), let $\varphi$ be a wff. Since $\alpha$ and $\beta$ are tautologically equivalent, it follows that (▲) $\alpha \to \varphi$ and $\beta \to \varphi$ are tautologically equivalent. Thus,

| | | | |
|---|---|---|---|
| $\Gamma; \alpha \vdash \varphi$ | iff | $\Gamma \vdash (\alpha \to \varphi)$ | by Theorem 3.3.33, |
| | iff | $\Gamma \vdash (\beta \to \varphi)$ | by (1) and (▲), |
| | iff | $\Gamma; \beta \vdash \varphi$ | by Theorem 3.3.33. □ |

**Corollary 3.3.35** (Contraposition). *We have* $\Gamma; \varphi \vdash \neg\psi$ *iff* $\Gamma; \psi \vdash \neg\varphi$.

*Proof.* Let $\Gamma$ be a set of wffs and let $\varphi, \psi$ be wffs. Since

$$(\varphi \to \neg\psi) \text{ and } (\psi \to \neg\varphi) \text{ are tautologically equivalent,} \qquad (▲)$$

we have

$$\begin{aligned}
\Gamma;\varphi \vdash \neg\psi \quad &\text{iff} \quad \Gamma \vdash (\varphi \to \neg\psi) && \text{by Theorem 3.3.33,}\\
&\text{iff} \quad \Gamma \vdash (\psi \to \neg\varphi) && \text{by Corollary 3.3.34(1) and } (\blacktriangle),\\
&\text{iff} \quad \Gamma;\psi \vdash \neg\varphi && \text{by Theorem 3.3.33.}
\end{aligned}$$

□

**Definition 3.3.36.** A set $\Gamma$ of wffs is *consistent* if there is no formula $\beta$ such that $\Gamma \vdash \beta$ and $\Gamma \vdash \neg\beta$; $\Gamma$ is *inconsistent* if $\Gamma \vdash \beta$ and $\Gamma \vdash \neg\beta$ for some formula $\beta$.

If a set of wffs $\Gamma$ is inconsistent, then for any wff $\alpha$, we have $\Gamma \vdash \alpha$. The reason for this follows. Let $\beta$ such that $\Gamma \vdash \beta$ and $\Gamma \vdash \neg\beta$. Since $\beta \to (\neg\beta \to \alpha)$ is a tautology, it follows that $\Gamma \vdash \alpha$ via modus ponens.

Proof by contradiction is often used in mathematical proofs. It can also be used in first-order logic to show that a deduction exists.

**Corollary 3.3.37** (Reductio ad absurdum). *If $\Gamma;\varphi$ is inconsistent, then $\Gamma \vdash \neg\varphi$.*

*Proof.* Assume that $\Gamma;\varphi$ is inconsistent. So, there is a formula $\beta$ such that $\Gamma;\varphi \vdash \beta$ and $\Gamma;\varphi \vdash \neg\beta$. Thus, $\Gamma \vdash (\varphi \to \beta)$ and $\Gamma \vdash (\varphi \to \neg\beta)$ by the deduction theorem. Since $\{\varphi \to \beta, \varphi \to \neg\beta\}$ tautologically implies $\neg\varphi$, we conclude that $\Gamma \vdash \neg\varphi$ by Rule T. □

**Theorem 3.3.38.** *If a set of formulas $\Gamma$ is inconsistent, then a finite subset of $\Gamma$ is inconsistent.*

*Proof.* See Exercise 12. □

### Strategies to show that deductions exist

To show that a deduction exists without having to explicitly present a deduction, one may be able to use the following eight *deduction strategies*:

(S1) To show that $\Gamma \vdash (\psi \to \theta)$, it is sufficient to show that $\Gamma;\psi \vdash \theta$, by the deduction theorem.

(S2) To show that $\Gamma;\neg\psi \vdash \neg\theta$, it is sufficient to show that $\Gamma;\theta \vdash \psi$, by contraposition. Also, $\Gamma;\alpha \vdash \neg\forall x\psi$ iff $\Gamma;\forall x\psi \vdash \neg\alpha$.

(S3) To show that $\Gamma \vdash \forall x\psi$, it is sufficient to show that $\Gamma \vdash \psi$ when $x$ does not occur free in $\Gamma$, by the generalization theorem (Theorem 3.3.29).

(S4) To show that $\Gamma \vdash \neg(\psi \to \theta)$, it is sufficient to show that $\Gamma \vdash \psi$ and $\Gamma \vdash \neg\theta$, by Rule T and the fact that $\{\psi, \neg\theta\}$ tautologically implies $\neg(\psi \to \theta)$.

(S5) To show that $\Gamma \vdash \neg\neg\psi$, it is sufficient to show that $\Gamma \vdash \psi$, by Rule T and the fact that $\{\psi\}$ tautologically implies $\neg\neg\psi$.

(S6) To show that $\Gamma \vdash \neg\psi$, it is sufficient to show that $\Gamma;\psi$ is inconsistent, by reductio ad absurdum.

(S7) To show that $\Gamma \vdash \neg\forall x\psi$, it is sufficient to show that $\Gamma \vdash \neg\psi^x_t$ for some term $t$. Note that $\Gamma \vdash (\neg\psi^x_t \to \neg\forall x\psi)$ by Logical Axiom 2 and Corollary 3.3.34(1). Thus, $\Gamma \vdash \neg\forall x\psi$ would follow by modus ponens. (If this is not useful, try (S6).)

(S8) To show that $\Gamma; \forall y\alpha \vdash \neg\forall x\psi$, try to show that $\Gamma; \forall x\psi \vdash \neg\alpha$. Thus, $\Gamma; \alpha \vdash \neg\forall x\psi$, and since $\forall y\alpha \to \alpha$ is a logical axiom, we have $\Gamma; \forall y\alpha \vdash \neg\forall x\psi$. Now apply contraposition, Corollary 3.3.35, to conclude that $\Gamma; \forall y\alpha \vdash \neg\forall x\psi$.

**Problem 3.3.39.** Using the above strategies, show that:
(1) $\vdash \forall xPx \to \exists xPx$,
(2) $\forall x\forall yPxy \vdash \forall y\forall zPzy$,
(3) $\forall xPx \vdash \forall xQx \to \forall x\neg(Px \to \neg Qx)$.

**Solution.** For each of the above items, we will show that such a deduction exists.
(1) To show that $\vdash \forall xPx \to \exists xPx$, we will apply (S1). By the deduction theorem, we just need to show that $\forall xPx \vdash \exists xPx$. Theorem 3.3.30 implies that $\forall xPx \vdash Px$. By Example 3.3.23 on page 99 and the deduction theorem, we have $Px \vdash \exists xPx$. Since $\forall xPx \vdash Px$ and $Px \vdash \exists xPx$, Lemma 3.3.26(4) implies that $\forall xPx \vdash \exists xPx$. Therefore, $\vdash \forall xPx \to \exists xPx$.
(2) We shall show that $\forall x\forall yPxy \vdash \forall y\forall zPzy$. Two applications of Logical Axiom 2 and modus ponens shows that $\forall x\forall yPxy \vdash Pzy$. Since $z$ is not free in $\forall x\forall yPxy$, we see that $\forall x\forall yPxy \vdash \forall zPzy$, by applying (S3). Similarly, as $y$ is not free in $\forall x\forall yPxy$, we conclude that $\forall x\forall yPxy \vdash \forall y\forall zPzy$.
(3) We will show that $\forall xPx \vdash \forall xQx \to \forall x\neg(Px \to \neg Qx)$. By the deduction theorem, it is sufficient to show that $\{\forall xPx, \forall xQx\} \vdash \forall x\neg(Px \to \neg Qx)$. So by strategy (S3), we just need to show that $\{\forall xPx, \forall xQx\} \vdash \neg(Px \to \neg Qx)$. By Logical Axiom 2 and modus ponens, we see that

$$\{\forall xPx, \forall xQx\} \vdash Px \text{ and } \{\forall xPx, \forall xQx\} \vdash Qx.$$

Thus, $\{\forall xPx, \forall xQx\} \vdash Px$, and $\{\forall xPx, \forall xQx\} \vdash \neg\neg Qx$ by Corollary 3.3.34(1). Applying strategy (S4), we conclude that $\{\forall xPx, \forall xQx\} \vdash \neg(Px \to \neg Qx)$. Therefore, $\forall xPx \vdash \forall xQx \to \forall x\neg(Px \to \neg Qx)$.

**Problem 3.3.40.** Show that $\vdash \exists x\forall y\varphi \to \forall y\exists x\varphi$.

**Solution.** We will show that a deduction of $\exists x\forall y\varphi \to \forall y\exists x\varphi$ exists. By the deduction theorem it is sufficient to show that $\exists x\forall y\varphi \vdash \forall y\exists x\varphi$. Hence, by the generalization theorem (Theorem 3.3.29), it is sufficient to show that $\exists x\forall y\varphi \vdash \exists x\varphi$. Removing the abbreviations, we need to show that $\neg\forall x\neg\forall y\varphi \vdash \neg\forall x\neg\varphi$. By contraposition, this reduces to showing that $\forall x\neg\varphi \vdash \forall x\neg\forall y\varphi$ (see Corollary 3.3.35). So, by the generalization theorem (Theorem 3.3.29), we must show that $\forall x\neg\varphi \vdash \neg\forall y\varphi$ and thus, by reductio ad absurdum, it is now sufficient to show that $\Sigma = \{\forall x\neg\varphi, \forall y\varphi\}$ is inconsistent. Note that $\Sigma \vdash \neg\varphi$ and $\Sigma \vdash \varphi$ by Logical Axiom 2 and modus ponens. Hence, $\Sigma$ is inconsistent. Therefore, it follows that there is a deduction of $\exists x\forall y\varphi \to \forall y\exists x\varphi$.

**Proposition 3.3.41.** *Let $\alpha$ be a wff. Then* $\vdash \neg\forall x\alpha \leftrightarrow \exists x\neg\alpha$ *and* $\vdash \neg\exists x\alpha \leftrightarrow \forall x\neg\alpha$.

**Proposition 3.3.42.** *If $x$ does not occur free in $\alpha$, then*

$$\vdash (\alpha \to \forall x\beta) \leftrightarrow \forall x(\alpha \to \beta).$$

*Proof.* By Rule T, it is sufficient to show that

$$\vdash (\alpha \to \forall x\beta) \to \forall x(\alpha \to \beta), \qquad (\star)$$
$$\vdash \forall x(\alpha \to \beta) \to (\alpha \to \forall x\beta). \qquad (\star\star)$$

To prove $(\star)$, by the deduction theorem, we must show that

$$(\alpha \to \forall x\beta) \vdash \forall x(\alpha \to \beta).$$

By assumption, $x$ does not occur free in $\alpha$. Hence, $x$ does not occur free in $(\alpha \to \forall x\beta)$. Therefore, by Theorem 3.3.29, it is now enough to show that $(\alpha \to \forall x\beta) \vdash (\alpha \to \beta)$. Again, by the deduction theorem, it is enough to show that $\{(\alpha \to \forall x\beta), \alpha\} \vdash \beta$. Let $\Gamma = \{(\alpha \to \forall x\beta), \alpha\}$. We argue that $\Gamma \vdash \beta$ as follows:

(1) $\Gamma \vdash \alpha \to \forall x\beta$ in $\Gamma$,
(2) $\Gamma \vdash \alpha$ in $\Gamma$,
(3) $\Gamma \vdash \forall x\beta$ by (1), (2), and modus ponens,
(4) $\Gamma \vdash \forall x\beta \to \beta$ by Logical Axiom 2,
(5) $\Gamma \vdash \beta$ by (3), (4), and modus ponens.

This completes the proof of $(\star)$. Note that the above list (1)–(5) is not a deduction. It is just part of a proof showing that a deduction of $(\star)$ exists.

To establish $(\star\star)$, we need to show, by the deduction theorem, that

$$\forall x(\alpha \to \beta) \vdash (\alpha \to \forall x\beta).$$

So, again by the deduction theorem, we must show that $\{\forall x(\alpha \to \beta), \alpha\} \vdash \forall x\beta$. Since $x$ does not occur free in $\forall x(\alpha \to \beta)$ or in $\alpha$, the generalization theorem (Theorem 3.3.29) implies that we just need to show $\{\forall x(\alpha \to \beta), \alpha\} \vdash \beta$. Let $\Gamma = \{\forall x(\alpha \to \beta), \alpha\}$. We show that $\Gamma \vdash \beta$ as follows:

(1) $\Gamma \vdash \forall x(\alpha \to \beta)$ in $\Gamma$,
(2) $\Gamma \vdash \alpha$ in $\Gamma$,
(3) $\Gamma \vdash \forall x(\alpha \to \beta) \to (\alpha \to \beta)$ by Logical Axiom 2,
(4) $\Gamma \vdash \alpha \to \beta$ by (1), (3), and modus ponens,
(5) $\Gamma \vdash \beta$ by (2), (4), and modus ponens. □

**Proposition 3.3.43.** *If $x$ is not free in $\alpha$, then $\vdash (\alpha \to \exists x\beta) \leftrightarrow \exists x(\alpha \to \beta)$.*

**Proposition 3.3.44.** *If $x$ is not free in $\beta$, then* $\vdash (\forall x\alpha \to \beta) \leftrightarrow \exists x(\alpha \to \beta)$.

**Proposition 3.3.45.** *If $x$ is not free in $\beta$, then* $\vdash (\exists x\alpha \to \beta) \leftrightarrow \forall x(\alpha \to \beta)$.

*Proof.* By Rule T, it is sufficient to show that

$$\vdash (\exists x\alpha \to \beta) \to \forall x(\alpha \to \beta), \tag{$\star$}$$

$$\vdash \forall x(\alpha \to \beta) \to (\exists x\alpha \to \beta). \tag{$\star\star$}$$

We shall first prove ($\star$). It is sufficient to show that

$$(\exists x\alpha \to \beta) \vdash \forall x(\alpha \to \beta), \tag{3.14}$$

by the deduction theorem. By assumption, $x$ does not occur free in $\beta$. Thus, $x$ does not occur free in $(\exists x\alpha \to \beta)$. To establish (3.14), the generalization theorem (Theorem 3.3.29) implies that it is enough to verify that $(\exists x\alpha \to \beta) \vdash (\alpha \to \beta)$. Again, by the deduction theorem, we now just need to show that $\{(\exists x\alpha \to \beta), \alpha\} \vdash \beta$. Let $\Gamma = \{(\exists x\alpha \to \beta), \alpha\}$. Thus, as $\exists x\alpha$ is an abbreviation for $\neg\forall x\neg\alpha$, we have

| | | |
|---|---|---|
| (1) | $\Gamma \vdash \neg\forall x\neg\alpha \to \beta$ | in $\Gamma$, |
| (2) | $\Gamma \vdash \alpha$ | in $\Gamma$, |
| (3) | $\Gamma \vdash \forall x\neg\alpha \to \neg\alpha$ | by Logical Axiom 2, |
| (4) | $\Gamma \vdash \alpha \to \neg\forall x\neg\alpha$ | by (3) and Corollary 3.3.34(1), |
| (5) | $\Gamma \vdash \neg\forall x\neg\alpha$ | by (2), (4), and modus ponens, |
| (6) | $\Gamma \vdash \beta$ | by (1), (5), and modus ponens. |

In (4) we used the fact that $(\forall x\neg\alpha \to \neg\alpha)$ and $(\alpha \to \neg\forall x\neg\alpha)$ are tautologically equivalent.

We shall now prove ($\star\star$). By the deduction theorem we need to show that

$$\forall x(\alpha \to \beta) \vdash (\exists x\alpha \to \beta).$$

So, by the deduction theorem and contraposition, it is sufficient to show that

$$\{\forall x(\alpha \to \beta), \neg\beta\} \vdash \neg\exists x\alpha.$$

By Corollary 3.3.34(1), we only need to show that

$$\{\forall x(\alpha \to \beta), \neg\beta\} \vdash \forall x\neg\alpha.$$

Since $x$ does not occur free in $\beta$, we see that $x$ does not occur free in any formula in the set $\{\forall x(\alpha \to \beta), \neg\beta\}$. So, by the generalization theorem (Theorem 3.3.29), it is enough to show that

$$\{\forall x(\alpha \to \beta), \neg\beta\} \vdash \neg\alpha.$$

Furthermore, by contraposition, we just need to show that

$$\{\forall x(\alpha \to \beta), \alpha\} \vdash \beta.$$

Let $\Gamma = \{\forall x(\alpha \to \beta), \alpha\}$. We show that $\Gamma \vdash \beta$ as follows:

| | | |
|---|---|---|
| (1) | $\Gamma \vdash \forall x(\alpha \to \beta)$ | in $\Gamma$, |
| (2) | $\Gamma \vdash \alpha$ | in $\Gamma$, |
| (3) | $\Gamma \vdash \forall x(\alpha \to \beta) \to (\alpha \to \beta)$ | by Logical Axiom 2, |
| (4) | $\Gamma \vdash \alpha \to \beta$ | by (1), (3), and modus ponens, |
| (5) | $\Gamma \vdash \beta$ | by (2), (4), and modus ponens. |

□

The following theorem can be useful to show that certain deductions exist.

**Theorem 3.3.46.** *Let $\alpha$ and $\beta$ be wffs. If $\vdash \alpha \to \beta$, then $\vdash \forall x\alpha \to \forall x\beta$.*

### 3.3.6 Equality

As you may recall, the logical axioms have two groups of axioms that concern equality. However, as yet, we have not discussed deductions that involve the equality symbol $\doteq$. In this section, we will examine deductions in languages $\mathcal{L}$ that contain the equality symbol. Using the logical axioms, we will show that all of the common properties of equality can be deduced from the axioms. One should go to page 97 and review the logical axioms and the equality axioms before continuing.

**Proposition 3.3.47.** *We have $\vdash \forall x(x \doteq x)$.*

*Proof.* Since $x \doteq x$ is a logical axiom, $\forall x(x \doteq x)$ is also a logical axiom by generalization (a generalization of a logical axiom is a logical axiom). □

**Proposition 3.3.48.** *Let $\Gamma$ be any set of wffs and let $t$ be a term. Then $\Gamma \vdash t \doteq t$.*

*Proof.* The following holds for any set $\Gamma$, even the empty set:

| | | |
|---|---|---|
| (1) | $\Gamma \vdash \forall x(x \doteq x)$ | by Proposition 3.3.47, |
| (2) | $\Gamma \vdash \forall x(x \doteq x) \to t \doteq t$ | by Logical Axiom 2, |
| (3) | $\Gamma \vdash t \doteq t$ | by (1), (2), and modus ponens. |

In (2) we are using $x \doteq x$ as $\alpha$ in Logical Axiom 2 and $(x \doteq x)_t^x$ is $t \doteq t$. □

Before we prove our next result, we introduce some temporary notation. Let $\alpha$ be an atomic formula. The formula $\alpha$ may have multiple occurrences of a particular variable, say $x$. To identify all of the occurrences of $x$ that appear in $\alpha$, we shall use the notation $\alpha(x, \ldots, x)$. In our next proposition, we will discuss the result of replacing some of these occurrences of $x$ with another variable. To distinguish the occurrences of the variable $x$ that are to be kept from those that are to be replaced, we shall use the notation $\alpha(x,\dots,x|x,\dots,x)$, where the occurrences of $x$ on the left of | are to be kept and the occurrences on the right are to be replaced. Again, this somewhat ambiguous notation is only temporary. Recall that $\Gamma; (x \doteq y)$ is the set $\Gamma \cup \{x \doteq y\}$.

**Proposition 3.3.49.** *Suppose that $\Gamma; (x \doteq y) \vdash \alpha$, where $\alpha$ is an atomic formula and $x, y$ are variables. Let $\alpha'$ be the result of replacing some or all of the occurrences of $x$ in $\alpha$ with $y$. Then $\Gamma; (x \doteq y) \vdash \alpha'$.*

*Proof.* We are given that $\Gamma; (x \doteq y) \vdash \alpha$. As discussed above, we shall use the notation $\alpha(x,\dots,x|x,\dots,x)$ to indicate that the occurrences of $x$ on the right of | are to be replaced by $y$. Thus, $\alpha' = \alpha(x,\dots,x|y,\dots,y)$. Now let $z$ be a variable that is distinct from $x$, $y$, and all of the variables that appear in $\alpha$. Consider the new atomic formula $\alpha(z,\dots,z|x,\dots,x)$. We observe that the formula

$$x \doteq y \to \alpha(z,\dots,z|x,\dots,x) \to \alpha(z,\dots,z|x,\dots,x)^x_y \tag{3.15}$$

is in logical axiom group 6. Since $\alpha(z,\dots,z|x,\dots,x)^x_y = \alpha(z,\dots,z|y,\dots,y)$, we see that (3.15) is the same as

$$x \doteq y \to \alpha(z,\dots,z|x,\dots,x) \to \alpha(z,\dots,z|y,\dots,y). \tag{3.16}$$

Let $\beta$ denote the formula in (3.16). Therefore, by generalization, $\forall z\beta$ is in logical axiom group 6 and $\forall z\beta \to \beta^z_x$ is in logical axiom group 2. By modus ponens, we infer that

$$\Gamma; (x \doteq y) \vdash \beta^z_x.$$

Since $\alpha(z,\dots,z|x,\dots,x)^z_x = \alpha$ and $\alpha(z,\dots,z|y,\dots,y)^z_x = \alpha'$, we see that $\beta^z_x$ is the formula $x \doteq y \to \alpha \to \alpha'$. Hence,

$$\Gamma; (x \doteq y) \vdash x \doteq y \to \alpha \to \alpha'.$$

As $\Gamma; (x \doteq y) \vdash \alpha$, by applying modus ponens twice, we see that $\Gamma; (x \doteq y) \vdash \alpha'$. □

**Proposition 3.3.50.** *We have $\vdash \forall x\forall y(x \doteq y \to y \doteq x)$.*

*Proof.* By Theorem 3.3.29, it is sufficient to show that $\vdash (x \doteq y \to y \doteq x)$, and by the deduction theorem, we just need to show that $\{x \doteq y\} \vdash y \doteq x$. We do this as follows:

(1) $\{x \doteq y\} \vdash x \doteq x$ as $x \doteq x$ is a logical axiom,

(2) $\{x \doteq y\} \vdash y \doteq x$ by Proposition 3.3.49.

In (2) we are using $x \doteq x$ as $\alpha$ in Proposition 3.3.49. □

The proof of the following result is requested in Exercise 19.

**Proposition 3.3.51.** *We have* $\vdash \forall x \forall y \forall z(x \doteq y \to y \doteq z \to x \doteq z)$.

**Proposition 3.3.52.** *We have* $\vdash \forall x_1 \forall x_2 \forall y_1 \forall y_2(x_1 \doteq y_1 \to x_2 \doteq y_2 \to Px_1x_2 \to Py_1y_2)$.

*Proof.* By the generalization theorem (Theorem 3.3.29), it is sufficient to show that

$$\vdash x_1 \doteq y_1 \to x_2 \doteq y_2 \to Px_1x_2 \to Py_1y_2.$$

By the deduction theorem we need to prove that

$$\{x_1 \doteq y_1, x_2 \doteq y_2, Px_1x_2\} \vdash Py_1y_2.$$

Let $\Gamma = \{x_1 \doteq y_1, x_2 \doteq y_2, Px_1x_2\}$. We now show that $\Gamma \vdash Py_1y_2$ as follows:

(1) $\Gamma \vdash Px_1x_2$ in $\Gamma$,
(2) $\Gamma \vdash Py_1x_2$ by Proposition 3.3.49,
(3) $\Gamma \vdash Py_1y_2$ by Proposition 3.3.49. □

**Proposition 3.3.53.** *We have* $\vdash \forall x_1 \forall x_2 \forall y_1 \forall y_2(x_1 \doteq y_1 \to x_2 \doteq y_2 \to fx_1x_2 \doteq fy_1y_2)$.

*Proof.* By the generalization theorem (Theorem 3.3.29), it is sufficient to show that

$$\vdash x_1 \doteq y_1 \to x_2 \doteq y_2 \to fx_1x_2 \doteq fy_1y_2.$$

By the deduction theorem we need to prove that

$$\{x_1 \doteq y_1, x_2 \doteq y_2\} \vdash fx_1x_2 \doteq fy_1y_2.$$

Let $\Gamma = \{x_1 \doteq y_1, x_2 \doteq y_2\}$. We now prove that $\Gamma \vdash fx_1x_2 \doteq fy_1y_2$ as follows:

(1) $\Gamma \vdash fx_1x_2 \doteq fx_1x_2$ by Proposition 3.3.48,
(2) $\Gamma \vdash fx_1x_2 \doteq fy_1x_2$ by Proposition 3.3.49,
(3) $\Gamma \vdash fx_1x_2 \doteq fy_1y_2$ by Proposition 3.3.49. □

The following theorem follows directly from Propositions 3.3.47–3.3.53.

**Theorem 3.3.54.** *We have:*
1. $\vdash \forall x(x \doteq x)$,
2. $\vdash \forall x \forall y(x \doteq y \to y \doteq x)$,
3. $\vdash \forall x \forall y \forall z(x \doteq y \to y \doteq z \to x \doteq z)$,
4. $\vdash \forall x_1 \forall x_2 \forall y_1 \forall y_2(x_1 \doteq y_1 \to x_2 \doteq y_2 \to Px_1x_2 \to Py_1y_2)$, *where $P$ is a 2-place predicate symbol, and similarly for n-place predicate symbols,*
5. $\vdash \forall x_1 \forall x_2 \forall y_1 \forall y_2(x_1 \doteq y_1 \to x_2 \doteq y_2 \to fx_1x_2 \doteq fy_1y_2)$, *where $f$ is a 2-place function symbol, and similarly for n-place function symbols.*

Items 1–3 of Theorem 3.3.54 show that equality is reflexive, symmetric, and transitive, respectively. Items 4 and 5 show that the substitution property of equality holds, that is, if two quantities are equal, then one can replace one with the other. Thus, the important properties of equality can be deduced from the logical axioms alone.

### 3.3.7 More metatheorems about deductions

The next problem illustrates an ability to "generalize on a constant."

**Problem 3.3.55.** Let $\mathcal{L}$ contain a constant symbol $c$ and 1-place predicate symbols $P$ and $Q$. Let $\Gamma = \{\forall x(Px \to Qx), \forall zPz\}$. One can show that $\Gamma \vdash Qc$. Since $c$ does not appear in any formula in $\Gamma$, can one then show that $\Gamma \vdash \forall xQx$?

**Solution.** We show, in steps, that such a deduction exists.

(a) $\Gamma \vdash Qc$
A deduction confirming (a) is given in Example 3.3.22 on page 99.

(b) $\Gamma \vdash Qy$
To confirm (b), in the deduction given in Example 3.3.22, replace $c$ with $y$.

(c) $\Gamma \vdash \forall yQy$
This follows from (b) by Theorem 3.3.29, as $y$ does not occur in $\Gamma$.

(d) $\forall yQy \vdash \forall xQx$
This is proven in Example 3.3.25 on page 100.

(e) $\Gamma \vdash \forall xQx$
This follows from (c) and (d).

Our next theorem shows that the steps in the solution of the above problem can be generalized. Let $\alpha$ be a wff and let $y$ be a variable that does not occur in $\alpha$. Define $\alpha^c_y$ to be the wff obtained by replacing all occurrences of the constant symbol $c$ in $\alpha$ with $y$.

**Theorem 3.3.56** (Generalization on constants). *Suppose that $\Gamma \vdash \varphi$ and let $c$ be a constant symbol that occurs in no formula in $\Gamma$. Then there is a variable $y$ which does not appear in $\varphi$ such that $\Gamma \vdash \forall y\varphi^c_y$. Moreover, there is a deduction of $\forall y\varphi^c_y$ from $\Gamma$ in which $c$ does not occur.*

*Proof.* Assume $\Gamma \vdash \varphi$ and assume that $c$ is a constant symbol that occurs in *no* formula in $\Gamma$. Let $\langle \alpha_1, \ldots, \alpha_n \rangle$ be a deduction of $\varphi$ from $\Gamma$, where $\alpha_n = \varphi$. So for all $k \leq n$, either

(a) $\alpha_k$ is in $\Gamma \cup \Lambda$, or

(b) $\alpha_k$ is obtained by modus ponens from two earlier wffs in the sequence $\langle \alpha_1, \ldots, \alpha_k \rangle$.

Now, let $y$ be a variable that does not occur in $\alpha_k$ for all $1 \leq k \leq n$. We will show that $\langle \alpha^c_{1y}, \ldots, \alpha^c_{ny} \rangle$ is a deduction of $\varphi^c_y$ from $\Gamma$. Let $k$ be so that $1 \leq k \leq n$.

CASE (a): Suppose that $\alpha_k$ is in $\Gamma \cup \Lambda$. If $\alpha_k \in \Gamma$, then $c$ does not occur in $\alpha_k$. Thus, ${\alpha_k}^c_y = \alpha_k$, and therefore ${\alpha_k}^c_y \in \Gamma$. If $\alpha_k \in \Lambda$, then ${\alpha_k}^c_y$ is also a logical axiom (for each $\alpha$ that appears in a logical axiom, by replacing $\alpha$ with $\alpha^c_y$ we get another logical axiom). Therefore, ${\alpha_k}^c_y \in \Lambda$.

CASE (b): If $\alpha_k$ is obtained by modus ponens from two earlier wffs in the sequence $\langle \alpha_1, \dots, \alpha_k \rangle$, then for some $i$ and $j$ less than $k$, the wffs $\alpha_i$ and $\alpha_j = (\alpha_i \to \alpha_k)$ are in the sequence $\langle \alpha_1, \dots, \alpha_k \rangle$. Since

$$ {\alpha_j}^c_y = (\alpha_i \to \alpha_k)^c_y = ({\alpha_i}^c_y \to {\alpha_k}^c_y), $$

we see that ${\alpha_k}^c_y$ is obtained by modus ponens from two earlier wffs in the sequence $\langle {\alpha_1}^c_y, \dots, {\alpha_k}^c_y \rangle$.

So $\langle {\alpha_1}^c_y, \dots, {\alpha_n}^c_y \rangle$ is a deduction of $\varphi^c_y$ from $\Gamma$. Let

$$ \Phi = \{\alpha_i : \alpha_i \in \Gamma \text{ and } 0 \le i \le n\}. $$

Since ${\alpha_i}^c_y = \alpha_i$ for all $\alpha_i \in \Phi$, we see that $\Phi \vdash \varphi^c_y$. As $y$ does not occur in any formula in $\Phi$, the generalization theorem (Theorem 3.3.29) implies that $\Phi \vdash \forall y \varphi^c_y$. Since $\Phi \subseteq \Gamma$, we conclude that $\Gamma \vdash \forall y \varphi^c_y$.

Therefore, $\langle {\alpha_1}^c_y, \dots, {\alpha_n}^c_y \rangle$ is a deduction of $\varphi^c_y$ from $\Gamma$ that does not involve the constant $c$. As the proof of Theorem 3.3.29 adds no new symbols to a deduction, we conclude there is a deduction of $\forall y \varphi^c_y$ from $\Gamma$ in which $c$ does not occur. □

**Lemma 3.3.57** (Re-replacement). *Let $x$ be a variable and let $\varphi$ be a wff in which the variable $y$ does not appear. Then $x$ is substitutable for $y$ in the wff $\varphi^x_y$ and $\varphi^{xy}_{yx} = \varphi$.*

*Proof.* See Exercise 21. □

**Corollary 3.3.58.** *Assume $\Gamma \vdash \varphi^x_c$, where $c$ is a constant symbol that does not occur in $\varphi$ and $c$ does not occur in any formula in $\Gamma$. Then $\Gamma \vdash \forall x \varphi$. In addition, there is a deduction of $\forall x \varphi$ from $\Gamma$ in which $c$ does not occur.*

*Proof.* Assume $\Gamma \vdash \varphi^x_c$, where $c$ is a constant symbol that occurs neither in $\varphi$ nor in any formula in $\Gamma$. Theorem 3.3.56 implies that there is a variable $y$ which does not occur in $\varphi^x_c$ such that $\Gamma \vdash \forall y \varphi^{xc}_{cy}$. In addition, there is a deduction of $\forall y \varphi^{xc}_{cy}$ from $\Gamma$ in which $c$ does not occur. Since $c$ does not occur in $\varphi$, we see that $\varphi^{xc}_{cy} = \varphi^x_y$. Hence, (▲) $\Gamma \vdash \forall y \varphi^x_y$. Lemma 3.3.57 asserts that $x$ is substitutable for $y$ in $\varphi^x_y$. Thus,

$$ \forall y \varphi^x_y \to \varphi^{xy}_{yx} $$

is in axiom group 2. Since Lemma 3.3.57 also asserts that $\varphi^{xy}_{yx} = \varphi$, we conclude that

$$ \forall y \varphi^x_y \to \varphi $$

is a logical axiom. Thus, $\vdash \forall y\varphi^x_y \to \varphi$. So, by the deduction theorem, we have $\forall y\varphi^x_y \vdash \varphi$. Since $x$ does not occur free in $\forall y\varphi^x_y$, the generalization theorem (Theorem 3.3.29) implies that $\forall y\varphi^x_y \vdash \forall x\varphi$. Because $\Gamma \vdash \forall y\varphi^x_y$ by (▲), we conclude that $\Gamma \vdash \forall x\varphi$. □

In mathematics, suppose that one can prove a result by using an arbitrary constant. Afterwards, one can then ask: Is there a proof that does not use the constant? The next corollary positively addresses this question.

**Corollary 3.3.59.** *Let $c$ be a constant symbol that does not occur in $\varphi$, in $\psi$, or in any formula in $\Gamma$. If $\Gamma; \varphi^x_c \vdash \psi$, then $\Gamma; \exists x\varphi \vdash \psi$. Moreover, there is a deduction of $\psi$ from $\Gamma; \exists x\varphi$ in which $c$ does not occur.*

*Proof.* Let $c$ be a constant symbol that does not occur in $\varphi$, in $\psi$, or in any formula in $\Gamma$. Suppose that $\Gamma; \varphi^x_c \vdash \psi$. By contraposition (Corollary 3.3.35), we see that $\Gamma; \neg\psi \vdash \neg\varphi^x_c$. Corollary 3.3.58 implies that $\Gamma; \neg\psi \vdash \forall x\neg\varphi$ and this can be confirmed by a deduction in which $c$ does not appear. Thus, by contraposition, $\Gamma; \neg\forall x\neg\varphi \vdash \psi$. Hence, $\Gamma; \exists x\varphi \vdash \psi$ and this can be verified by a deduction in which $c$ does not occur. □

### Alphabetic variants

We begin with some terminology. Let $v$ be a variable and consider a wff of the form $\forall x\varphi$. If $v$ occurs in $\varphi$, then $v$ is said to be within the *scope* of the quantifier $\forall$. An *alphabetic variant* of wff $\alpha$ is a wff $\overline{\alpha}$ that is the result of a one-to-one replacement of some, none, or all of the quantified variables of $\alpha$ with variables that are not in the scope of a quantifier in $\alpha$. A precise definition of an alphabetic variant can be given by recursion (see the proof of Theorem 3.3.61 below). Here are some examples:

1. The wff $\forall w\forall xPwx$ is an alphabetic variant of $\forall y\forall zPyz$, because the change of quantified variables $y \mapsto w$, $z \mapsto x$ is one-to-one and $w, x$ do not occur in $\forall y\forall zPyz$.
2. The wff $\forall z\forall zPzz$ is not an alphabetic variant of $\forall y\forall zPyz$, because the change of the quantified variable $y \mapsto z$ is such that $z$ is in the scope of $\forall$ in $\forall y\forall zPyz$.

Here are two examples which indicate that one can deduce alphabetic variants:

1. $\forall yQy \vdash \forall xQx$ (see Example 3.3.25),
2. $\forall y\forall zPyz \vdash \forall x\forall wPxw$.

The above examples illustrate that one can deduce a "change of quantified variables." In Theorem 3.3.61 below, such a "change of variables" result is established in general. Consequently, when a term is not substitutable for a variable in a given wff, one can change the quantified variables so that the term will become substitutable.

The following is essentially a restatement of Theorem 3.3.46.

**Lemma 3.3.60.** *Let $\alpha$ and $\beta$ be wffs and let $x$ be any variable. If $\alpha \vdash \beta$, then $\forall x\alpha \vdash \forall x\beta$.*

Given a term $t$ and a wff $\varphi$, the term $t$ may not be substitutable for a specific variable in $\varphi$. The following theorem shows that one can change the quantified variables in $\varphi$ to get an equivalent formula $\overline{\varphi}$ in which $t$ is substitutable.

**Theorem 3.3.61** (Alphabetic variants). *Let $\varphi$ be a wff, $t$ a term, and $x$ a variable. There exists an alphabetic variant $\overline{\varphi}$ of $\varphi$ such that:*
(a) *$\varphi \vdash \overline{\varphi}$ and $\overline{\varphi} \vdash \varphi$;*
(b) *$t$ is substitutable for $x$ in $\overline{\varphi}$.*

*Proof.* Let the variable $x$ and let term $t$ be fixed. If $t$ is substitutable for $x$ in $\varphi$, then let $\overline{\varphi} = \varphi$. Otherwise, we will define $\overline{\varphi}$ and prove the following statement by induction on $\varphi$: $\varphi \vdash \overline{\varphi}$, $\overline{\varphi} \vdash \varphi$, and $t$ is substitutable for $x$ in $\overline{\phi}$.

*Base step:* Let $\phi = Pt_1t_2 \cdots t_n$ be an atomic sentence. Let $\overline{\phi} = \phi$. Clearly, $\phi \vdash \overline{\phi}$, $\overline{\phi} \vdash \phi$, and $t$ is substitutable for $x$ in $\overline{\phi}$.

*Inductive step:* Let $\alpha$ and $\beta$ be arbitrary wffs. Assume the induction hypothesis that there are wffs $\overline{\alpha}$ and $\overline{\beta}$ such that

$$\begin{aligned} &\alpha \vdash \overline{\alpha},\ \overline{\alpha} \vdash \alpha, \text{ and } t \text{ is substitutable for } x \text{ in } \overline{\alpha}, \\ &\beta \vdash \overline{\beta},\ \overline{\beta} \vdash \beta, \text{ and } t \text{ is substitutable for } x \text{ in } \overline{\beta}. \end{aligned} \tag{IH}$$

We must prove that the same holds for each of the following: $\neg\alpha, \alpha \to \beta, \forall y\alpha$.

CASE $\neg\alpha$: Let $\overline{\neg\alpha} = \neg\overline{\alpha}$. By (IH), contraposition, and Definition 3.3.15, we see that $\neg\alpha \vdash \neg\overline{\alpha}$, $\neg\overline{\alpha} \vdash \neg\alpha$, and $t$ is substitutable for $x$ in $\neg\overline{\alpha}$.

CASE $\alpha \to \beta$: Let $\overline{\alpha \to \beta} = \overline{\alpha} \to \overline{\beta}$. By (IH), Exercise 22, and Definition 3.3.15, we see that $(\alpha \to \beta) \vdash (\overline{\alpha} \to \overline{\beta})$, $(\overline{\alpha} \to \overline{\beta}) \vdash (\alpha \to \beta)$, and $t$ is substitutable for $x$ in $\overline{\alpha} \to \overline{\beta}$.

CASE $\forall y\alpha$ : Choose a variable $z \neq x$ that occurs neither in $\overline{\alpha}$ nor in $t$. Let $\overline{\forall y\alpha} = \forall z\overline{\alpha}^y_z$. We must prove that

$$\forall y\alpha \vdash \forall z\overline{\alpha}^y_z,\ \forall z\overline{\alpha}^y_z \vdash \forall y\alpha, \text{ and } t \text{ is substitutable for } x \text{ in } \forall z\overline{\alpha}^y_z.$$

We first show that $t$ is substitutable for $x$ in $\forall z\overline{\alpha}^y_z$. By the induction hypothesis (IH), the term $t$ is substitutable for $x$ in $\overline{\alpha}$. Since $x \neq z$ and $z$ does not occur in $t$, it follows that $t$ is substitutable for $x$ in $\overline{\alpha}^y_z$. Thus, by Definition 3.3.15(4)(b), we conclude that $t$ is substitutable for $x$ in $\forall z\overline{\alpha}^y_z$.

By the induction hypothesis we have $\alpha \vdash \overline{\alpha}$. Since $z$ does not occur in $\overline{\alpha}$, we see that $(\star)$ $z$ is substitutable for $y$ in $\overline{\alpha}$. Also, as $z$ does not occur in $\overline{\alpha}$, it follows that $z$ does not occur free in $\alpha$ (if it occurred free in $\alpha$, it would occur in $\overline{\alpha}$). Therefore, $(\star\star)$ $z$ does not occur free in $\forall y\alpha$. We now prove that $\forall y\alpha \vdash \forall z(\overline{\alpha})^y_z$ as follows:

| | |
|---|---|
| $\alpha \vdash \overline{\alpha}$ | by (IH), |
| $\forall y\alpha \vdash \forall y\overline{\alpha}$ | by Lemma 3.3.60, |

$\forall y\alpha \vdash \overline{\alpha}^y_z$ by modus ponens as $\forall y\overline{\alpha} \to \overline{\alpha}^y_z$ is a logical axiom (group 2) via ($\star$),

$\forall y\alpha \vdash \forall z\overline{\alpha}^y_z$ by ($\star\star$) and the generalization theorem (Theorem 3.3.29).

We now prove that $\forall z\overline{\alpha}^y_z \vdash \forall y\alpha$. Recall that $z$ does not appear in $\overline{\alpha}$. Lemma 3.3.57 thus implies that $y$ is substitutable for $z$ in $\overline{\alpha}^y_z$ and that $\overline{\alpha}^{yz}_{zy} = \overline{\alpha}$. Also, we note that (▲) $y$ does not occur free in $\forall z\overline{\alpha}^y_z$. Since $\forall z\overline{\alpha}^y_z \to \overline{\alpha}^{yz}_{zy}$ is a logical axiom, we see that (▼) $\forall z\overline{\alpha}^y_z \vdash \overline{\alpha}^{yz}_{zy}$. Finally, we prove that $\forall z\overline{\alpha}^y_z \vdash \forall y\alpha$ as follows:

$\forall z\overline{\alpha}^y_z \vdash \overline{\alpha}^{yz}_{zy}$ by (▼,)

$\forall z\overline{\alpha}^y_z \vdash \overline{\alpha}$ because $\overline{\alpha}^{yz}_{zy} = \overline{\alpha}$,

$\forall z\overline{\alpha}^y_z \vdash \alpha$ by (IH) $\overline{\alpha} \vdash \alpha$, and so Lemma 3.3.26(4) applies,

$\forall z\overline{\alpha}^y_z \vdash \forall y\alpha$ by (▲) and the generalization theorem (Theorem 3.3.29). □

**Exercises 3.3.**

*1. Let $\mathcal{S}$ be the set of all prime formulas in a language $\mathcal{L}$. Let $\mathcal{F} = \{\mathcal{E}_\neg, \mathcal{E}_\to\}$ and let $\overline{\mathcal{S}}$ be the set of all wffs generated by $\mathcal{S}$ from the functions in $\mathcal{F}$. Prove by induction on wffs that every wff is in $\overline{\mathcal{S}}$.

*2. Let $\mathfrak{A}$ be an $\mathcal{L}$-structure with domain $A$. Let $V$ be the set of all the variables of $\mathcal{L}$. Let $s\colon V \to A$ be a variable assignment. Define a truth assignment on the set of prime wffs $\alpha$ by

$$u(\alpha) = \begin{cases} T, & \text{if } \mathfrak{A} \models \alpha[s], \\ F, & \text{if } \mathfrak{A} \not\models \alpha[s]. \end{cases}$$

(a) Show that for all wffs $\theta$,

$$\overline{u}(\theta) = T \quad \text{iff} \quad \mathfrak{A} \models \theta[s].$$

(b) Let $\Gamma$ be a set of wffs and let $\varphi$ be a wff. Show that if $\Gamma$ tautologically implies $\varphi$, then $\Gamma$ logically implies $\varphi$.

3. Show that the wff $\forall v_1(v_1 \doteq v_1) \to v_1 \doteq v_1$ is logically valid and that it is not a first-order tautology.

4. Let $\mathcal{L} = \{c, f, \doteq\}$, where $c$ is a constant symbol and $f$ is a 1-place function symbol. For each of the following wffs $\varphi$ and terms $t$, evaluate $\varphi^x_t$ and decide if $t$ is substitutable for $x$ in $\varphi$:

(a) $\forall x(x \doteq y \to fx \doteq fy)$ and $t$ is $fc$,

(b) $\forall y(x \doteq y \to fx \doteq fy)$ and $t$ is $fy$,

(c) $(x \doteq y \to \forall x(fx \doteq fy))$ and $t$ is $fy$.

*5. Let $\mathcal{L}$ be a language, let $x$ be a variable in $\mathcal{L}$, and, in particular, let $x = v_{i_0}$. Now let $t$ be a term. Let $A$ be the set of all the atomic formulas and let $X$ be the set of wffs defined by

$$X = \{\forall v_{i_0}\alpha : \alpha \text{ is a wff}\}.$$

Thus, $X$ is the set of wffs of the form $\forall x\alpha$, where $\alpha$ is some wff. Let $\mathcal{S} = A \cup X$, a disjoint union. Let $Q = \{\mathcal{E}_{Q_i} : i \neq i_0\}$. Let $\mathcal{F} = \{\mathcal{E}_\neg, \mathcal{E}_\to\} \cup Q$ (see (3.4)) and let $\overline{\mathcal{S}}$ be the set generated from $\mathcal{S}$ by the functions in $\mathcal{F}$. Theorem 3.1.16 implies that $\overline{\mathcal{S}}$ is the set of all the wffs and is freely generated from the set $\mathcal{S}$ by the functions in $\mathcal{F}$. By Exercise 11 on page 65, for each term $\tau$, $\tau_t^x$ is the term obtained by replacing all occurrences of $x$ in $\tau$ with $t$. Define $h: \mathcal{S} \to \overline{\mathcal{S}}$ by

$$h(v) = \begin{cases} P\tau_{1t}^{\,x}\tau_{2t}^{\,x}\cdots\tau_{nt}^{\,x}, & \text{if } v = P\tau_1\tau_2\cdots\tau_n \text{ is in } A, \\ \forall x\alpha, & \text{if } v = \forall x\alpha \text{ is in } X. \end{cases} \tag{3.17}$$

Theorem 1.1.27 implies there is a unique function $\overline{h}: \overline{\mathcal{S}} \to \overline{\mathcal{S}}$ such that

(1) $\overline{h}(\alpha) = h(\alpha)$ if $\alpha \in S$,
(2) $\overline{h}((\neg\alpha)) = \neg\overline{h}(\alpha)$,
(3) $\overline{h}((\alpha \to \beta)) = \overline{h}(\alpha) \to \overline{h}(\beta)$,
(4) $\overline{h}(\forall v\alpha) = \forall v\overline{h}(\alpha)$ whenever $v \neq x$.

Prove, by induction on wffs, that for all wffs $\alpha$, $\overline{h}(\alpha) = \alpha_t^x$.

6. Let $\mathcal{L} = \{P, Q, c\}$, where $P$ is a 1-place predicate symbol, $Q$ is a 2-place predicate symbol, and $c$ is a constant symbol. To which logical axiom groups, if any, do each of the following wffs belong?
   (a) $((\forall xPx \to \forall yPy) \to Pz) \to (\forall xPx \to (\forall yPy \to Pz))$.
   (b) $\forall y(\forall x(Px \to Px) \to (Pc \to Px))$.
   (c) $\forall x\exists yQxy \to \exists yQyy$.
7. Using Definition 3.3.20, prove Lemma 3.3.26.
8. Suppose that $\Gamma$ is consistent and $\Gamma \vdash \varphi$. Show that $\Gamma \cup \{\varphi\}$ is consistent.
9. Let $\Gamma = \{\forall x\alpha, \forall x\neg\alpha\}$, where $\alpha$ is a wff. Show that $\Gamma$ is inconsistent.

*10. Suppose that the term $t$ is substitutable for $x$ in wff $\alpha$. Show that $\vdash \alpha_t^x \to \exists x\alpha$.

11. Show that $\vdash \forall x\varphi \to \exists x\varphi$ by presenting an explicit deduction.

*12. Prove Theorem 3.3.38.

13. Prove Proposition 3.3.41.
14. Prove Proposition 3.3.43.
15. Prove Proposition 3.3.44.
16. Prove Theorem 3.3.46.
17. Show the following:
    (a) $\vdash \exists x(Px \to \forall xPx)$,
    (b) $\{Qx, \forall y(Qy \to \forall zPz)\} \vdash \forall xPx$.
18. Show that $\forall x\forall yPxy \to \forall y\forall xPyx$.

*19. Prove Proposition 3.3.51.

*20. Let $f$ be a 1-place function symbol and let $t$ and $\tau$ be terms. Show that:
    (a) $\vdash \forall x\forall y(x \doteq y \to fx \doteq fy)$,
    (b) $\vdash (t \doteq \tau \to ft \doteq f\tau)$.

*21. Prove Lemma 3.3.57 by induction on $\varphi$.

*22. Suppose that:
   1. $\alpha \vdash \overline{\alpha}$ and $\overline{\alpha} \vdash \alpha$,
   2. $\beta \vdash \overline{\beta}$ and $\overline{\beta} \vdash \beta$.

   Prove that $(\alpha \to \beta) \vdash (\overline{\alpha} \to \overline{\beta})$ and $(\overline{\alpha} \to \overline{\beta}) \vdash (\alpha \to \beta)$.

23. Let $\Gamma$ and $\Delta$ be sets of wffs. Suppose that $\Gamma \vdash \varphi$ for all $\varphi \in \Delta$ and that $\Delta \vdash \sigma$. Prove that $\Gamma \vdash \sigma$.

*24. Prove that a set of wffs $\Gamma$ is consistent if and only if every finite subset of $\Gamma$ is consistent.

*25. Suppose that $\Gamma$ is a consistent set of wffs and let $\varphi$ be a wff. Show that if $\Gamma \nvdash \varphi$, then $\Gamma \cup \{\neg\varphi\}$ is consistent.

26. Suppose $\Sigma$ is a set of sentences, $\varphi$ is any wff, and $x$ is any variable. Prove that $\Sigma \vdash \varphi$ if and only if $\Sigma \vdash \forall x\varphi$.

*27. Show that $\vdash \neg\forall v\alpha \to \neg\forall v\neg\neg\alpha$.

28. Show that $\vdash \exists x\beta \to \forall x\beta$ if $x$ does not occur free in $\beta$.

29. Show that $\vdash \exists x(\exists xPx \to Px)$.

30. Let $\Sigma_0 \subseteq \Sigma_1 \subseteq \Sigma_2 \subseteq \Sigma_3 \subseteq \cdots$ be sets of wffs. Suppose that each $\Sigma_i$ is consistent. Show that $\bigcup_{i\in\mathbb{N}} \Sigma_i$ is consistent.

**Exercise Notes:** For Exercise 11, $\forall x\varphi \to \varphi$ and $\forall x\neg\varphi \to \neg\varphi$ are logical axioms. In addition,

$$(\forall x\neg\varphi \to \neg\varphi) \to (\varphi \to \neg\forall x\neg\varphi) \text{ and}$$
$$(\forall x\varphi \to \varphi) \to (\varphi \to \neg\forall x\neg\varphi) \to (\forall x\varphi \to \neg\forall x\neg\varphi)$$

are tautologies. For Exercise 17, use reductio ad absurdum by showing that

$$\forall x\neg(Px \to \forall xPx) \vdash \forall xPx$$

(via Logical Axiom 2, a tautology, and generalization) and showing, via Logical Axiom 2 and a tautology, that $\forall x\neg(Px \to \forall xPx) \vdash \neg\forall xPx$. For Exercise 18, use Problem 3.3.39(2) on page 106. For Exercise 20, by Theorem 3.3.61 one can assume that $x$ and $y$ do not appear in $t$ or $\tau$.

# 4 Soundness and completeness

Is there a relationship between the concept of *truth* and the concept of *proof*? In the previous chapter we investigated what it means for a structure to satisfy a first-order formula and then we discussed deductions of first-order formulas. In this chapter we will establish two consequential relationships between satisfaction and deduction. The first relationship is called the soundness theorem, which implies that every first-order formula that is deducible, from the logical axioms alone, is logically valid. The second relationship, the completeness theorem, shows that every first-order formula that is logically valid is deducible.

## 4.1 The soundness theorem

The soundness theorem asserts that "deductions preserve truth," that is, if $\Gamma$ is a set of well-formed formulas (wffs) and each wff in $\Gamma$ holds in a structure, then any wff that is deducible from $\Gamma$ will also hold in this structure. The goal of this section is to prove the soundness theorem.

**Theorem** (Soundness). *Let $\Gamma$ be a set of formulas. If $\Gamma \vdash \varphi$, then $\Gamma \vDash \varphi$.*

The soundness theorem shows that if a formula $\varphi$ is deducible from a set of wffs $\Gamma$, then the set $\Gamma$ logically implies $\varphi$ (see Definition 3.2.15). Therefore, deductions produce valid conclusions, that is, these deduced conclusions will hold in every structure in which the assumptions hold. This confirms why proof in mathematics is so important. Before we present the proof of this theorem, we need to state and prove some technical lemmas.

### 4.1.1 Technical lemmas

Our first two technical lemmas are used to prove that all the logical axioms are logically valid, that is, every logical axiom is true in every structure. Recall that for a wff $\alpha$, we defined the wff $\alpha_t^x$ in Definition 3.3.13 so that $\alpha_t^x$ is the result of "replacing the variable $x$ in $\alpha$ with the term $t$," with one exception (item 4 of Definition 3.3.13). Now, given a term $\tau$, let us define $\tau_t^x$ to be the term obtained by replacing the variable $x$ in $\tau$ with the term $t$ (see Exercise 11 on page 65). Also recall that, given an assignment $s$, $s_{x|d}$ is the function which is exactly like $s$ except for one thing: at the variable $x$ it assumes the value $d$, that is,

$$s_{x|d}(v) = \begin{cases} s(v), & \text{if } v \neq x, \\ d, & \text{if } v = x. \end{cases}$$

https://doi.org/10.1515/9783110782073-004

Given an assignment $s$, we will show in our next lemma that the value of $\overline{s}(\tau_t^x)$ can be obtained by using the assignment $s_{x|\overline{s}(t)}$.

**Lemma 4.1.1.** *Let $\mathfrak{A}$ be a structure and let $s$ be an assignment. For terms $\tau, t$ and variable $x$ we have $\overline{s}(\tau_t^x) = \overline{s_{x|\overline{s}(t)}}(\tau)$.*

*Proof.* Let $\mathfrak{A}$ be a structure, $s$ an assignment, $t$ a term, and $x$ a variable. We shall prove that $\overline{s}(\tau_t^x) = \overline{s_{x|\overline{s}(t)}}(\tau)$ by induction on terms $\tau$.

*Base step:* Let $c$ and $v$ be a constant and a variable, respectively, of the language $\mathcal{L}$. Clearly, $c_t^x = c$, so $\overline{s}(c_t^x) = c^{\mathfrak{A}} = \overline{s_{x|\overline{s}(t)}}(c)$ by Definition 3.2.4. If $v \neq x$, then $v_t^x = v$, so $\overline{s}(v_t^x) = s(v) = \overline{s_{x|\overline{s}(t)}}(v)$. If $v = x$, then $x_t^x = t$, so $\overline{s}(x_t^x) = \overline{s}(t) = \overline{s_{x|\overline{s}(t)}}(x)$.

*Inductive step:* Let $f$ be an $n$-place function symbol in $\mathcal{L}$ and let $\tau_1, \ldots, \tau_n$ be terms. Assume the induction hypothesis

$$\text{For each } i \leq n \text{ we have } \overline{s}(\tau_{i\,t}^{\,x}) = \overline{s_{x|\overline{s}(t)}}(\tau_i). \tag{IH}$$

We must prove that the same holds for the term $f\tau_1 \cdots \tau_n$. Thus,

$$\begin{aligned}
\overline{s}((f\tau_1 \cdots \tau_n)_t^x) &= \overline{s}(f\tau_{1t}^{\,x} \cdots \tau_{nt}^{\,x}) && \text{Exercise 11(2), page 65,}\\
&= f^{\mathfrak{A}}(\overline{s}(\tau_{1t}^{\,x}), \ldots, \overline{s}(\tau_{nt}^{\,x})) && \text{by Theorem 3.2.5(3),}\\
&= f^{\mathfrak{A}}(\overline{s_{x|\overline{s}(t)}}(\tau_1), \ldots, \overline{s_{x|\overline{s}(t)}}(\tau_n)) && \text{by (IH),}\\
&= \overline{s_{x|\overline{s}(t)}}(f\tau_1 \cdots \tau_n) && \text{by Theorem 3.2.5(3).}
\end{aligned}$$

□

**Lemma 4.1.2** (Substitution). *If the term $t$ is substitutable for the variable $x$ in the wff $\varphi$, then for every structure $\mathfrak{A}$ and every assignment $s$, we have*

$$\mathfrak{A} \models \varphi_t^x[s] \quad \textit{iff} \quad \mathfrak{A} \models \varphi[s_{x|\overline{s}(t)}].$$

*Proof.* Let $\mathfrak{A}$ be an arbitrary structure. We shall prove the following statement by induction on wffs: *If $t$ is substitutable for $x$ in $\varphi$, then*

$$\mathfrak{A} \models \varphi_t^x[s] \quad \text{iff} \quad \mathfrak{A} \models \varphi[s_{x|\overline{s}(t)}] \textit{ for all assignments } s.$$

*Base step:* Let $P\tau_1 \cdots \tau_n$ be an atomic formula. Let $s$ be an assignment. We note that $(\star)$ $(P\tau_1 \cdots \tau_n)_t^x = P\tau_{1t}^{\,x} \cdots \tau_{nt}^{\,x}$. Thus,

$$\begin{aligned}
\mathfrak{A} \models (P\tau_1 \cdots \tau_n)_t^x[s] \quad &\text{iff} \quad \mathfrak{A} \models P\tau_{1t}^{\,x} \cdots \tau_{nt}^{\,x}[s] && \text{by } (\star),\\
&\text{iff} \quad \langle \overline{s}(\tau_{1t}^{\,x}), \ldots, \overline{s}(\tau_{nt}^{\,x})\rangle \in P^{\mathfrak{A}} && \text{by Definition 3.2.6(2),}\\
&\text{iff} \quad \langle \overline{s_{x|\overline{s}(t)}}(\tau_1), \ldots, \overline{s_{x|\overline{s}(t)}}(\tau_n)\rangle \in P^{\mathfrak{A}} && \text{by Lemma 4.1.1,}\\
&\text{iff} \quad \mathfrak{A} \models P\tau_1 \cdots \tau_n[s_{x|\overline{s}(t)}] && \text{by Definition 3.2.6(2).}
\end{aligned}$$

*Inductive step:* Let $\alpha$ and $\beta$ be arbitrary wffs. Assume the induction hypothesis

$$\begin{aligned}&\text{if } t \text{ is substitutable for } x \text{ in } \alpha, \text{ then } \mathfrak{A} \models \alpha_t^x[s] \text{ iff } \mathfrak{A} \models \alpha[s_{x|\bar{s}(t)}],\\&\text{if } t \text{ is substitutable for } x \text{ in } \beta, \text{ then } \mathfrak{A} \models \beta_t^x[s] \text{ iff } \mathfrak{A} \models \beta[s_{x|\bar{s}(t)}]\end{aligned} \tag{IH}$$

for *all* assignments $s$. We must prove that the same holds for each of the following:

$$(\neg\alpha), (\alpha \to \beta), \forall v\alpha.$$

CASE $(\neg\alpha)$: Let $s$ be an assignment and let $t$ be substitutable for $x$ in $(\neg\alpha)$. It follows that $t$ is substitutable for $x$ in $\alpha$ (see Definition 3.3.15) and that $(\neg\alpha)_t^x = \neg\alpha_t^x$. Hence

$$\begin{aligned}\mathfrak{A} \models (\neg\alpha)_t^x[s] \quad &\text{iff} \quad \mathfrak{A} \not\models \alpha_t^x[s] && \text{by Definition 3.2.6(3),}\\&\text{iff} \quad \mathfrak{A} \not\models \alpha[s_{x|\bar{s}(t)}] && \text{by (IH),}\\&\text{iff} \quad \mathfrak{A} \models (\neg\alpha)[s_{x|\bar{s}(t)}] && \text{by Definition 3.2.6(3).}\end{aligned}$$

CASE $(\alpha \to \beta)$: Let $s$ be an assignment and let $t$ be substitutable for $x$ in $(\alpha \to \beta)$. Thus, $t$ is substitutable for $x$ in $\alpha$ and in $\beta$, and $(\alpha \to \beta)_t^x = (\alpha_t^x \to \beta_t^x)$. Hence

$$\begin{aligned}\mathfrak{A} \models (\alpha \to \beta)_t^x[s] \quad &\text{iff} \quad \mathfrak{A} \models \alpha_t^x[s] \text{ implies } \mathfrak{A} \models \beta_t^x[s] && \text{by Definition 3.2.6(4),}\\&\text{iff} \quad \mathfrak{A} \models \alpha[s_{x|\bar{s}(t)}] \text{ implies } \mathfrak{A} \models \beta[s_{x|\bar{s}(t)}] && \text{by (IH),}\\&\text{iff} \quad \mathfrak{A} \models (\alpha \to \beta)[s_{x|\bar{s}(t)}] && \text{by Definition 3.2.6(4).}\end{aligned}$$

CASE $\forall v\alpha$: Let $s$ be an assignment and $t$ be substitutable for $x$ in $\forall v\alpha$. By Definition 3.3.15, either (a) $x$ does not occur free in $(\forall v\alpha)$ or (b) $v$ does not occur in $t$ and $t$ is substitutable for $x$ in $\alpha$.

(a) Suppose $x$ does not occur free in $\forall v\alpha$. Thus, (▲) $(\forall v\alpha)_t^x = \forall v\alpha$. Also, it follows that $s$ and $s_{x|\bar{s}(t)}$ agree on the free variables in $\forall v\alpha$. Thus, by Theorem 3.2.11 we have $\mathfrak{A} \models \forall v\alpha[s]$ if and only if $\mathfrak{A} \models \forall v\alpha[s_{x|\bar{s}(t)}]$. So, by (▲), $\mathfrak{A} \models (\forall v\alpha)_t^x[s]$ if and only if $\mathfrak{A} \models \forall v\alpha[s_{x|\bar{s}(t)}]$.

(b) Assume $v$ does not occur in $t$ and $t$ is substitutable for $x$ in $\alpha$. By Definition 3.3.13, we have

$$(\forall v\alpha)_t^x = \begin{cases}\forall v\alpha, & \text{if } x = v,\\ \forall v\alpha_t^x, & \text{if } x \neq v.\end{cases}$$

If $x = v$, then $x$ does not occur free in $\forall v\alpha$, and this is just case (a) above. If $x \neq v$, then $(\forall v\alpha)_t^x = \forall v\alpha_t^x$. Therefore, letting $A$ be the domain of $\mathfrak{A}$, we have

$$\begin{aligned}\mathfrak{A} \models (\forall v\alpha)_t^x[s] \quad &\text{iff} \quad \text{for every } d \in A, \mathfrak{A} \models \alpha_t^x[s_{v|d}] && \text{by Definition 3.2.6(5),}\\&\text{iff} \quad \text{for every } d \in A, \mathfrak{A} \models \varphi[s_{v|d,x|\bar{s}(t)}] && \text{by (IH),}\\&\text{iff} \quad \mathfrak{A} \models \forall v\alpha[s_{x|\bar{s}(t)}] && \text{by Definition 3.2.6(5).}\end{aligned}$$

This completes the proof of the substitution lemma. □

Recall that an alphabetic variant of wff $\varphi$ is obtained by replacing some, none, or all of the quantified variables of $\varphi$ with different variables (see page 114). Lemma 4.1.2 and Theorem 3.3.61 imply the following observation.

**Corollary 4.1.3.** *Let $t$ be a term, $x$ a variable, and $\varphi$ a wff. There is an alphabetic variant $\overline{\varphi}$ of $\varphi$ such that*

$$\varphi \vdash \overline{\varphi}, \overline{\varphi} \vdash \varphi, \textit{ t is substitutable for x in } \overline{\varphi},$$

*and for every structure $\mathfrak{A}$ and every assignment $s$, we have*

$$\mathfrak{A} \models \overline{\varphi}^x_t[s] \quad \textit{iff} \quad \mathfrak{A} \models \overline{\varphi}[s_{x|\overline{s}(t)}].$$

**Lemma 4.1.4.** *Every logical axiom is logically valid.*

*Proof.* By Exercise 14 on page 89, we know that any generalization of a logically valid formula is also logically valid. Therefore, we only need to prove the logical validity of the axioms described in the six groups of Logical Axioms 3.3.17. Let $\mathfrak{A}$ be a structure and let $s$ be an assignment. We shall show that $\mathfrak{A} \models \varphi[s]$ for every $\varphi$ that belongs to one of the six categories in Logical Axioms 3.3.17.

1. Suppose $\varphi$ is a tautology. Exercise 2(b) on page 116 (where $\Gamma = \varnothing$) implies that $\models \varphi$, and therefore $\mathfrak{A} \models \varphi[s]$.
2. Assume that $\varphi$ has the form $\forall x\alpha \to \alpha^x_t$, where $t$ is substitutable for $x$ in $\alpha$. We must show that $\mathfrak{A} \models (\forall x\alpha \to \alpha^x_t)[s]$. So, assume that $\mathfrak{A} \models \forall x\alpha[s]$. We conclude that $\mathfrak{A} \models \alpha[s_{x|\overline{s}(t)}]$. The substitution lemma (Lemma 4.1.2) implies that $\mathfrak{A} \models \alpha^x_t[s]$. Therefore, $\mathfrak{A} \models (\forall x\alpha \to \alpha^x_t)[s]$.
3. Let $\varphi$ have the form $\forall x(\alpha \to \beta) \to (\forall x\alpha \to \forall x\beta)$. To show that

$$\mathfrak{A} \models (\forall x(\alpha \to \beta) \to (\forall x\alpha \to \forall x\beta))[s],$$

   assume that $\mathfrak{A} \models \forall x(\alpha \to \beta)[s]$ and $\mathfrak{A} \models \forall x\alpha[s]$. Exercise 12 on page 89 implies that $\mathfrak{A} \models \forall x\beta[s]$.
4. Let $\varphi$ be of the form $\alpha \to \forall x\alpha$, where $x$ does not occur free in $\alpha$. To show that $\mathfrak{A} \models (\alpha \to \forall x\alpha)[s]$, assume $\mathfrak{A} \models \alpha[s]$. Thus, $\mathfrak{A} \models \forall x\alpha[s]$ by Exercise 13 on page 89.
5. Suppose $\varphi$ has the form $x \doteq x$. We must show that $\mathfrak{A} \models (x \doteq x)[s]$. This follows immediately from Definition 3.2.6(1), the definition of satisfaction.
6. Assume that $\varphi$ has the form $x \doteq y \to (\alpha \to \alpha^x_y)$, where $\alpha$ is an atomic formula. We must show that

$$\mathfrak{A} \models (x \doteq y \to (\alpha \to \alpha^x_y))[s].$$

   To affirm this, assume that $\mathfrak{A} \models (x \doteq y)[s]$. Thus, $s(x) = s(y)$. By Exercise 11 on page 88, we have $\mathfrak{A} \models \alpha[s]$ iff $\mathfrak{A} \models \alpha^x_y[s]$. It thus follows that $\mathfrak{A} \models (x \doteq y \to (\alpha \to \alpha^x_y))[s]$. □

### 4.1.2 Proof of the soundness theorem

**Theorem 4.1.5** (Soundness). *Let $\Gamma$ be a set of formulas. If $\Gamma \vdash \varphi$, then $\Gamma \models \varphi$.*

*Proof.* Assume that $\Gamma \vdash \varphi$. Let $\langle \alpha_1, \alpha_2, \dots, \alpha_n \rangle$ be a deduction from the set $\Gamma$ of wffs, where $\alpha_n = \varphi$. We shall prove the following statement: *For all $k \leq n$, $\Gamma \models \alpha_k$.* We shall use strong induction on the natural number variable $k$.

*Base step:* Let $k = 1$. We must show that $\Gamma \models \alpha_1$. Since $\alpha_1$ is the first item in the deduction, either (a) $\alpha_1 \in \Gamma$ or (b) $\alpha_1$ is a logical axiom. If $\alpha_1 \in \Gamma$, then clearly we have $\Gamma \models \alpha_1$. If $\alpha_1$ is a logical axiom, then Lemma 4.1.4 implies that $\Gamma \models \alpha_1$.

*Inductive step:* Let $k \leq n$. Assume the strong induction hypothesis

$$\Gamma \models \alpha_i \quad \text{for all } i < k. \tag{SIH}$$

We must show that $\Gamma \models \alpha_k$. Since $\alpha_k$ is in the deduction, either (a) $\alpha_k \in \Gamma$, or (b) $\alpha_k$ is a logical axiom, or (c) $\alpha_k$ is obtained by modus ponens from an $\alpha_i$ and $\alpha_j = (\alpha_i \to \alpha_k)$, where $i, j < k$. If (a) or (b) hold, then we can conclude that $\Gamma \models \alpha_k$ just as in the base step. So, suppose that (c) holds. By the strong induction hypothesis (SIH), we conclude that $\Gamma \models \alpha_i$ and $\Gamma \models (\alpha_i \to \alpha_k)$. It now follows that $\Gamma \models \alpha_k$. □

**Corollary 4.1.6.** *Let $\Gamma$ be a set of formulas. If $\Gamma \not\models \varphi$, then $\Gamma \nvdash \varphi$.*

**Corollary 4.1.7.** *Let $\varphi$ be a wff. If $\vdash \varphi$, then $\models \varphi$.*

**Corollary 4.1.8.** *If $\vdash (\varphi \leftrightarrow \psi)$, then the wffs $\varphi$ and $\psi$ are logically equivalent.*

*Proof.* Assume $\vdash (\varphi \leftrightarrow \psi)$. By the soundness theorem, we conclude that $\models (\varphi \leftrightarrow \psi)$. Exercise 9(b) on page 88 implies that $\varphi$ and $\psi$ are logically equivalent. □

Recalling Theorem 3.3.61, a wff $\overline{\varphi}$ is an alphabetic variant of $\varphi$ when $\overline{\varphi}$ differs from $\varphi$ only in the choice of quantified variables. For example, $\forall x \forall w Pxw$ is an alphabetic variant of $\forall x \forall z Pxz$.

**Corollary 4.1.9.** *Suppose that $\overline{\varphi}$ is an alphabetic variant of $\varphi$. Then $\varphi$ and $\overline{\varphi}$ are logically equivalent.*

*Proof.* Let $\overline{\varphi}$ be an alphabetic variant of $\varphi$. It follows from Theorem 3.3.61 that $\overline{\varphi} \vdash \varphi$ and $\varphi \vdash \overline{\varphi}$. By the soundness theorem, we have $\overline{\varphi} \models \varphi$ and $\varphi \models \overline{\varphi}'$. Therefore, $\varphi$ and $\overline{\varphi}$ are logically equivalent. □

**Definition 4.1.10.** A set of wffs $\Gamma$ is *satisfiable* if for some structure $\mathfrak{A}$ and some assignment $s$ we have $\mathfrak{A} \models \varphi[s]$ for all $\varphi \in \Gamma$.

**Corollary 4.1.11.** *If $\Gamma$ is satisfiable, then $\Gamma$ is consistent.*

*Proof.* Assume that $\Gamma$ is satisfiable. We will show that $\Gamma$ is consistent. Suppose, for a contradiction, that there is a wff $\beta$ such that $\Gamma \vdash \beta$ and $\Gamma \vdash \neg\beta$. Since $\Gamma$ is satisfiable, let $\mathfrak{A}$

be a structure and let $s$ be an assignment that satisfies every wff in $\Gamma$. The soundness theorem now implies that $\mathfrak{A} \models \beta[s]$ and $\mathfrak{A} \not\models \beta[s]$, which is a contradiction. □

**Exercises 4.1.**

1. Let $c$ be a constant symbol and let $\alpha$ be a wff. So $c$ is substitutable for $x$ in $\alpha$. Let $\mathfrak{A}$ be a structure and let $s$ be an assignment. Let $s(c) = d$. Show that
$$\mathfrak{A} \models \alpha^x_c[s] \quad \text{iff} \quad \mathfrak{A} \models \alpha[s_{x|d}].$$
Conclude that if $\mathfrak{A} \models \alpha^x_c[s]$, then $\mathfrak{A} \models \exists x\alpha[s]$.
2. Let $c$ be a constant symbol, let $\Gamma$ be a set of wffs, and let $\psi$ be a wff. Suppose that $\Gamma \models \psi^x_c$. Using Exercise 1, show that $\Gamma \models \exists x\psi$.
*3. Let $t$ be a term that is substitutable for $x$ in the wff $\varphi$. Show that $\{\varphi^x_t, x \doteq t\} \models \varphi$.
4. Show that $Px \not\models \forall xPx$. Conclude that $\{Px, \neg\forall xPx\}$ is consistent.
5. Let $\Gamma = \{\neg\forall v_1 Pv_1, Pv_2, Pv_3, Pv_4, \dots\}$. Is $\Gamma$ is consistent?
6. A set of wffs $\Gamma$ is said to be *independent* if for every wff $\varphi \in \Gamma$, $\Gamma \setminus \{\varphi\} \nvdash \varphi$. Suppose that for every $\varphi \in \Gamma$, the set of wffs $(\Gamma \setminus \{\varphi\}) \cup \{\neg\varphi\}$ is satisfiable. Show that $\Gamma$ is independent.
7. Let $\mathcal{L} = \{\doteq\}$ and let $\alpha$ be the sentence $\exists x\forall y(y \doteq x)$. Show that $\nvdash \alpha$ and $\nvdash \neg\alpha$.

## 4.2 The completeness theorem

The completeness theorem is the converse of the soundness theorem, but it is a much deeper result. Given a set $\Gamma$ of wffs, the completeness theorem shows that if $\Gamma$ logically implies a formula, then that formula is deducible from $\Gamma$. The completeness theorem and its proof first appeared in the 1930 doctoral dissertation of Kurt Gödel. The proof of the completeness theorem presented below is due to Leon Henkin.

**Theorem** (Completeness). *Let $\Gamma$ be a set of first-order formulas and let $\varphi$ be a wff.*
(a) *If $\Gamma \models \varphi$, then $\Gamma \vdash \varphi$.*
(b) *If $\Gamma$ is consistent, then $\Gamma$ is satisfiable.*

The basic idea behind the proof of (b) is as follows: Given that $\Gamma$ is consistent, we first construct a set of wffs $\Delta$ such that:
(1) for each wff $\varphi$ and variable $x$, the wff $\exists x\varphi \to \varphi^x_c$ is in $\Delta$ for a constant symbol $c$;
(2) $\Gamma \subseteq \Delta$;
(3) $\Delta$ is consistent;
(4) for every wff $\alpha$, either $\alpha \in \Delta$ or $(\neg\alpha) \in \Delta$.

We shall then use $\Delta$ to build a structure that satisfies $\Gamma$. Before we present the proof of the completeness theorem, we need to state and prove some technical lemmas that will allow us to complete the above steps (1)–(4). In the proof of the completeness theorem, we will then focus on the construction of the structure that satisfies Γ.

### 4.2.1 Technical lemmas

On page 2 we discussed indexed sets, that is, sets of the form $\{x_i : i \in I\}$, where $I$ is an unspecified set. Such notation will be used in this section.

**Definition 4.2.1.** Let $\mathcal{L}$ be a first-order language. Let $\mathcal{L}' = \mathcal{L} \cup \{c_i : i \in I\}$, where each $c_i$ is a new constant symbol that does not appear in $\mathcal{L}$ and $c_i \neq c_j$ whenever $i, j \in I$ are distinct. Then $\mathcal{L}'$ is an *extension of $\mathcal{L}$ by new constants.*

**Lemma 4.2.2** (Adding constants). *Suppose $\Gamma$ is a consistent set of wffs in a language $\mathcal{L}$. Let $\mathcal{L}'$ be an extension of $\mathcal{L}$ by new constants. Then $\Gamma$ remains a consistent set of wffs in the extended language $\mathcal{L}'$.*

*Proof.* Assume that $\Gamma$ is a consistent set of wffs in the language $\mathcal{L}$. Suppose, for a contradiction, that $\Gamma$ is an inconsistent set of wffs in the language $\mathcal{L}'$. Therefore, there is an $\mathcal{L}'$-formula $\varphi$ such that $\Gamma \vdash \varphi$ and $\Gamma \vdash \neg\varphi$, where the deductions take place in the language $\mathcal{L}'$. Let $\langle \alpha_1, \alpha_2, \ldots, \varphi \rangle$ be an $\mathcal{L}'$-deduction of $\varphi$ and let $\langle \beta_1, \beta_2, \ldots, \neg\varphi \rangle$ be an $\mathcal{L}'$-deduction of $\neg\varphi$. Since these two deductions are finite, there is only a finite number of the new constant symbols that appear in these deductions. Suppose that the finite list $c_1, \ldots, c_n$ contains all of the new constant symbols that appear in these deductions. The proof of Theorem 3.3.56 implies that there are variables $y_1, \ldots, y_n$ that do not appear in the two deductions $\langle \alpha_1, \alpha_2, \ldots, \varphi \rangle$ and $\langle \beta_1, \beta_2, \ldots, \neg\varphi \rangle$ such that the replacement of each new constant symbol $c_i$ in the deductions with the variable $y_i$ results in $\mathcal{L}$-deductions $\langle \alpha_1', \alpha_2', \ldots, \varphi' \rangle$ and $\langle \beta_1', \beta_2', \ldots, \neg\varphi' \rangle$ from $\Gamma$. We conclude that $\Gamma$ is inconsistent in $\mathcal{L}$. This contradiction completes the proof. □

In our next lemma, we will be adding constants to $\mathcal{L}$ that act as "witnesses." They are called Henkin constants and will allow us to construct a structure in which each existential formula that holds in the structure can be verified by a witness among the new constants. Recall that $\exists v\alpha$ is an abbreviation of $\neg\forall v\neg\alpha$.

**Lemma 4.2.3** (Ensuring witnesses). *Let $\mathcal{L}$ be a first-order language and let $\Gamma$ be a consistent set of wffs in the language $\mathcal{L}$. Let $\mathcal{L}' = \mathcal{L} \cup \{c_i' : i \in I\}$ be an extension of $\mathcal{L}$ by new constants where $\mathcal{L}$ and $I$ have the same cardinality. Then there is a set $\Theta$ of wffs in the language $\mathcal{L}'$ such that:*

(1) $\Gamma \subseteq \Theta$;
(2) *for every $\mathcal{L}'$-wff $\varphi$ and every variable $x$, the wff $\exists x\varphi \to \varphi_c^x$ is in $\Theta$ for some new constant symbol $c$;*
(3) *$\Theta$ is consistent in $\mathcal{L}'$.*

Before we present the proof, we make a comment. Let $S = \{c_1', c_2', c_3', \dots\}$ be an infinite set of new constant symbols. Given a finite subset $A$ of $S$ we shall be referring to the "first new constant symbol $c$ not in $A$". This will mean that $c = c_k'$, where $k$ is the smallest $k$ such that $c_k' \notin A$. For example, if $A = \{c_1', c_2', c_3', c_6'\}$, then $c = c_4'$ is the first new constant symbol not in $A$.

*Proof.* Let $\mathcal{L}$ be a countable first-order language and let $\mathcal{L}' = \mathcal{L} \cup \{c_1', c_2', c_3', \dots\}$ be an extension of $\mathcal{L}$ by new constants. Consider the set

$$P = \{\langle \varphi, v\rangle : \varphi \text{ is an } \mathcal{L}'\text{-wff and } v \text{ is a variable of the language}\}.$$

Theorem 1.1.30 (on page 12) implies that the set of *all* wffs in the language $\mathcal{L}'$ is a countable set, because the set of all wffs is a subset of the set of all finite sequences of symbols in the countable language $\mathcal{L}'$. From this it follows that the set $P$ of pairs is also a countable set. Thus, let

$$\langle \varphi_1, x_1\rangle, \langle \varphi_2, x_2\rangle, \langle \varphi_3, x_3\rangle, \dots, \langle \varphi_n, x_n\rangle, \dots \tag{4.1}$$

be a fixed enumeration of all the pairs in $P$. Let $\theta_1$ be the wff

$$\exists x_1 \varphi_1 \to \varphi_1{}^{x_1}_{c_1},$$

where $c_1$ is the first new constant symbol not occurring in $\varphi_1$. In general, let us define (by recursion on $\mathbb{N}$) $\theta_n$ to be the wff

$$\exists x_n \varphi_n \to (\varphi_n)^{x_n}_{c_n},$$

where $c_n$ is the first new constant symbol not occurring in $\varphi_n$ or in $\theta_k$ for all $k < n$. Let

$$\Theta = \Gamma \cup \{\theta_1, \theta_2, \dots, \theta_n, \dots\}.$$

Clearly $\Gamma$ satisfies conditions (1) and (2) of the statement of this lemma. We thus need to show that $\Theta$ satisfies condition (3), that is, we must show that $\Theta$ is consistent.

Suppose, for a contradiction, that $\Theta$ is inconsistent. Theorem 3.3.38 implies that a finite subset of $\Theta$ is inconsistent. Thus, there is a natural number $m$ such that

$$\Gamma \cup \{\theta_1, \theta_2, \dots, \theta_{m+1}\}$$

is inconsistent. *Assume that $m$ is the least such $m$.* Since $\Gamma$ is consistent, we must have $m \geq 0$. By the reductio ad absurdum corollary (Corollary 3.3.37), it follows that

$$\Gamma \cup \{\theta_1, \theta_2, \dots, \theta_m\} \vdash \neg\theta_{m+1}.$$

That is,

$$\Gamma \cup \{\theta_1, \theta_2, \ldots, \theta_m\} \vdash \neg(\exists x\varphi \to \varphi^x_c)$$

for some $\varphi, x, c$. As $\neg(\exists x\varphi \to \varphi^x_c) \to \exists x\varphi$ and $\neg(\exists x\varphi \to \varphi^x_c) \to \neg\varphi^x_c$ are tautologies, rule T (Theorem 3.3.31) implies that

$$\Gamma \cup \{\theta_1, \theta_2, \ldots, \theta_m\} \vdash \exists x\varphi, \tag{$\star$}$$

$$\Gamma \cup \{\theta_1, \theta_2, \ldots, \theta_m\} \vdash \neg\varphi^x_c. \tag{$\star\star$}$$

Since the constant symbol $c$ occurs neither in $\varphi$ nor in any formula in $\Gamma \cup \{\theta_1, \theta_2, \ldots, \theta_m\}$, Corollary 3.3.58 with $(\star\star)$ implies that

$$\Gamma \cup \{\theta_1, \theta_2, \ldots, \theta_m\} \vdash \forall x\neg\varphi.$$

However, $(\star)$ implies (after removing the abbreviation) that

$$\Gamma \cup \{\theta_1, \theta_2, \ldots, \theta_m\} \vdash \neg\forall x\neg\varphi.$$

So, $\Gamma \cup \{\theta_1, \theta_2, \ldots, \theta_m\}$ is inconsistent. This contradicts the leastness of $m$ if $m > 0$, and contradicts the consistency of $\Gamma$ if $m = 0$. □

**Remark 4.2.4.** The proof of Lemma 4.2.3 assumes that the language $\mathcal{L}$ is countable. The lemma still holds for uncountable languages by a modification of this proof, which requires knowledge of cardinal numbers. Suppose that $\mathcal{L}$ and $I$ both have cardinality $\kappa$, where $\kappa$ is an uncountable cardinal. We can presume that $I = \kappa$. In the above proof, replace (4.1) with the sequence of pairs

$$\langle \varphi_\gamma, x_\gamma \rangle_{\gamma \in \kappa}$$

indexed by the ordinals $\gamma \in \kappa$. For each $\gamma \in \kappa$, define (by recursion) $\theta_\gamma$ to be the wff

$$\exists x_\gamma \varphi_\gamma \to (\varphi_\gamma)^{x_\gamma}_{c_\gamma},$$

where $c_\gamma$ is the first new constant symbol not occurring in $\varphi_\gamma$ or in $\theta_\eta$ for all $\eta \in \gamma$. It then follows that $\Theta = \Gamma \cup \{\theta_\eta : \eta \in \kappa\}$ is consistent.

Let $\Theta$ be as in Lemma 4.2.3. Our next result will allow us to extend $\Theta$ to a maximal consistent set of wffs.

**Lemma 4.2.5** (Going to the max). *Let $\mathcal{L}$ be a first-order language and suppose that $\Theta$ is a consistent set of wffs in the language $\mathcal{L}$. Then there is a set $\Delta$ of wffs in the language $\mathcal{L}$ such that:*

(1) $\Theta \subseteq \Delta$;
(2) *$\Delta$ is consistent;*
(3) *for every wff $\varphi$, either $\varphi \in \Delta$ or $(\neg\varphi) \in \Delta$ but not both;*
(4) *for every wff $\varphi$, if $\Delta \vdash \varphi$, then $\varphi \in \Delta$.*

*Proof.* Let $\mathcal{L}$ be a countable first-order language. Theorem 1.1.30 (on page 12) implies that the set of *all* wffs is a countable set, because the set of all wffs is a subset of the set of all finite sequences of symbols in the language $\mathcal{L}$. Thus, let

$$\delta_1, \delta_2, \delta_3, \ldots, \delta_n, \ldots \tag{4.2}$$

be a fixed enumeration of all the wffs in the language $\mathcal{L}$. Define by recursion on $\mathbb{N}$ the following sets:

(i) $\Delta_0 = \Theta$

(ii) $\Delta_{n+1} = \begin{cases} \Delta_n \cup \{\delta_{n+1}\}, & \text{if } \Delta_n \cup \{\delta_{n+1}\} \text{ is consistent,} \\ \Delta_n \cup \{\neg\delta_{n+1}\}, & \text{otherwise.} \end{cases}$

We shall now establish four claims.

**Claim 1.** *For all $n \in \mathbb{N}$, $\Delta_n$ is consistent.*

*Proof of Claim* 1. We shall use induction on $n$.

*Base step:* Let $n = 0$. Then $\Delta_0 = \Theta$, which is consistent by assumption.

*Inductive step:* Let $n \in \mathbb{N}$ be arbitrary. Assume the induction hypothesis

$$\Delta_n \text{ is consistent.} \tag{IH}$$

We will prove that $\Delta_{n+1}$ is consistent. Suppose, for a contradiction, that $\Delta_{n+1}$ is not consistent. Thus, it follows from the above definition (ii) of $\Delta_{n+1}$ that $\Delta_n \cup \{\delta_{n+1}\}$ is inconsistent and $\Delta_n \cup \{\neg\delta_{n+1}\}$ is inconsistent. By the reductio ad absurdum corollary (Corollary 3.3.37), it follows that

$$\Delta_n \vdash \neg\delta_{n+1},$$
$$\Delta_n \vdash \delta_{n+1}.$$

Thus, $\Delta_n$ is inconsistent, contradicting the induction hypothesis (IH). (Claim 1) □

Let $\Delta = \bigcup_{n\in\mathbb{N}} \Delta_n$. Clearly, $\Theta \subseteq \Delta$ (as $\Theta = \Delta_0$), and (▲) $\Delta_n \subseteq \Delta_{n+1}$ for all $n \in \mathbb{N}$.

**Claim 2.** *The set $\Delta$ is consistent.*

*Proof of Claim* 2. Suppose, to the contrary, that $\Delta$ is inconsistent. By Theorem 3.3.38 there is a finite subset $\Pi$ of $\Delta$ that is inconsistent. Since $\Pi \subseteq \Delta = \bigcup_{n\in\mathbb{N}} \Delta_n$ and $\Pi$ is finite, it follows, from (▲), that $\Pi \subseteq \Delta_n$ for some $n$. Hence, $\Delta_n$ must be inconsistent, contradicting Claim 1. (Claim 2) □

**Claim 3.** *For every wff $\alpha$, either $\alpha \in \Delta$ or $(\neg\alpha) \in \Delta$, but not both.*

*Proof of Claim* 3. Let $\alpha$ be any wff. Since (4.2) lists all of the wffs, there is an $n \in \mathbb{N}$ such that $\alpha = \alpha_n$. So either $\alpha_n \in \Delta_n$ or $(\neg\alpha_n) \in \Delta_n$ by (ii) above. Since $\Delta_n \subseteq \Delta$, it follows that

$\alpha \in \Delta$ or $(\neg\alpha) \in \Delta$. If $\alpha \in \Delta$ and $(\neg\alpha) \in \Delta$, then $\Delta$ would be inconsistent, contradicting Claim 2. Hence, we cannot have $\alpha \in \Delta$ and $(\neg\alpha) \in \Delta$. (Claim 3) □

**Claim 4.** *For every wff $\varphi$, if $\Delta \vdash \varphi$, then $\varphi \in \Delta$.*

*Proof of Claim* 4. Let $\alpha$ be any wff. Assume that $\Delta \vdash \varphi$. To prove that $\varphi \in \Delta$, suppose to the contrary that $\varphi \notin \Delta$. By Claim 3, we see that $\neg\varphi \in \Delta$. Thus, $\Delta \vdash \neg\varphi$. Hence, $\Delta$ is inconsistent, contradicting Claim 2. (Claim 4) □

Therefore, $\Delta$ satisfies conditions (1)–(4) stated in the lemma. (Lemma) □

**Remark 4.2.6.** The proof of Lemma 4.2.5 assumes that the language $\mathcal{L}$ is countable. However, Lemma 4.2.5 also holds when $\mathcal{L}$ is uncountable. In this case, one can obtain the set $\Delta$ by applying Zorn's lemma (Lemma 1.1.36).

### 4.2.2 Proof of the completeness theorem

Given a consistent set of $\mathcal{L}$-wffs $\Gamma$, the upcoming proof of the completeness theorem constructs a structure $\mathfrak{A}$ and an assignment $s$. It is then shown that $\mathfrak{A} \models \alpha[s]$ for all $\alpha \in \Gamma$. To do this, one first lets $\mathcal{L}'$ be an extension of $\mathcal{L}$ by adding infinitely many new constants. The structure $\mathfrak{A}$ is then constructed, surprisingly, by using the terms of $\mathcal{L}'$.

**Definition 4.2.7.** Let $\mathcal{L}$ be a language. For each wff $\alpha$, let $n(\alpha)$ be the number of places at which connectives and quantifier symbols appear in $\alpha$.

For example, let $\alpha$ be the wff $\forall v(Pv \to \neg Qt)$. Then $n(\alpha) = 3$ and $n(Qt) = 0$. Now let $\alpha$ be any wff. For any variable $x$ and term $t$, note that the resulting substitution $\alpha^x_t$ does not add any connectives and quantifier symbols to $\alpha$. In fact, $n(\alpha) = n(\alpha^x_t)$.

**Theorem 4.2.8** (Completeness). *Let $\Gamma$ be a set of $\mathcal{L}$-wffs and let $\varphi$ be an $\mathcal{L}$-wff.*
(a) *If $\Gamma \models \varphi$, then $\Gamma \vdash \varphi$.*
(b) *If $\Gamma$ is consistent, then $\Gamma$ is satisfiable.*

*Proof.* We only provide a proof of (b), as (b) implies (a) (see Exercise 2). So, let $\Gamma$ be a consistent set of formulas in a language $\mathcal{L}$. We must prove that $\Gamma$ is satisfiable, that is, we must identify a structure and an assignment that satisfies every wff $\varphi$ in $\Gamma$. Let $\mathcal{L}' = \mathcal{L} \cup \{c_i : i \in I\}$ be an extension of $\mathcal{L}$ by new constants where $\mathcal{L}$ and $I$ have the same cardinality. By the "adding constants" Lemma 4.2.2, we know that $\Gamma$ remains consistent in the language $\mathcal{L}'$. By the "ensuring witnesses" Lemma 4.2.3, we know that there is a set $\Theta$ of wffs in the language $\mathcal{L}'$ such that
(1) $\Gamma \subseteq \Theta$;
(2) for every $\mathcal{L}'$-wff $\varphi$ and every variable $x$, the wff $\exists x\varphi \to \varphi^x_c$ is in $\Theta$ for some new constant symbol $c$;
(3) $\Theta$ is consistent in $\mathcal{L}'$.

By the “going to the max” Lemma 4.2.5, there is a set $\Delta$ of $\mathcal{L}'$-wffs such that:
(4) $\Theta \subseteq \Delta$;
(5) $\Delta$ is consistent;
(6) for every wff $\varphi$, either $\varphi \in \Delta$ or $(\neg\varphi) \in \Delta$, but not both;
(7) for every wff $\varphi$, if $\Delta \vdash \varphi$, then $\varphi \in \Delta$.

We will now construct a structure. There are two cases to consider. The first case is when the language $\mathcal{L}$ does not contain the equality symbol $\doteq$. The second case is when the language does contain $\doteq$.

**Case 1: The language $\mathcal{L}$ does not contain the equality symbol**

Note that $\Gamma \subseteq \Theta \subseteq \Delta$. We shall now use the set $\Delta$ to construct a structure $\mathfrak{A}$ for the language $\mathcal{L}'$ as follows:
(a) Let $A$, the domain of $\mathfrak{A}$, be the set of all terms in the language $\mathcal{L}'$.
(b) For each $n$-place predicate symbol $P$ define $P^{\mathfrak{A}}$ by

$$\langle t_1, t_2, \ldots, t_n\rangle \in P^{\mathfrak{A}} \quad \text{iff} \quad \text{the atomic formula } Pt_1t_2\cdots t_n \text{ belongs to } \Delta.$$

(c) For each $n$-place function symbol $f$ define $f^{\mathfrak{A}}$ by

$$f^{\mathfrak{A}}(t_1, t_2, \ldots, t_n) = ft_1t_2\cdots t_n.$$

(d) For each constant symbol $c$ define $c^{\mathfrak{A}}$ by $c^{\mathfrak{A}} = c$.

Now define an assignment $s$ by

$$s(v) = \begin{cases} v, & \text{if } v \text{ is a variable,} \\ c, & \text{if } v \text{ is a constant symbol } c. \end{cases} \tag{$\blacktriangle$}$$

Note that $s\colon \mathcal{T} \to \overline{\mathcal{T}}$, where $\mathcal{T}$ consists of the variables and constant symbols of $\mathcal{L}'$ and $\overline{\mathcal{T}}$ is the set of all the terms of $\mathcal{L}'$. Also note that $A = \overline{\mathcal{T}}$.

**Claim 1.** *For all terms $t$, we have $\bar{s}(t) = t$.*

*Proof.* One can prove this by using induction on terms using ($\blacktriangle$). (Claim 1) □

**Claim 2.** *For every wff $\varphi$, we have $\mathfrak{A} \vDash \varphi[s]$ if and only if $\varphi \in \Delta$.*

*Proof.* Let $\mathfrak{A}$ and $s$ be defined as above. We shall prove the following statement by strong induction on the natural number $k$:

$$\text{For all } \mathcal{L}'\text{-wffs } \varphi, \text{ if } n(\varphi) = k, \text{ then } \mathfrak{A} \vDash \varphi[s] \text{ iff } \varphi \in \Delta.$$

*Base step:* Let $k = 0$. If $n(\varphi) = 0$, then $\varphi = Pt_1 \cdots t_n$ for some atomic formula. Thus,

$$\begin{aligned}
\mathfrak{A} \models Pt_1 \cdots t_n[s] \quad &\text{iff} \quad \langle \bar{s}(t_1), \ldots, \bar{s}(t_n) \rangle \in P^{\mathfrak{A}} && \text{by Definition 3.2.6(2),} \\
&\text{iff} \quad \langle t_1, \ldots, t_n \rangle \in P^{\mathfrak{A}} && \text{by Claim 1,} \\
&\text{iff} \quad Pt_1 t_2 \cdots t_n \in \Delta && \text{by (b), above.}
\end{aligned}$$

*Inductive step:* Let $k > 0$. Assume the strong induction hypothesis:

$$\begin{aligned}
&\text{For all } \mathcal{L}'\text{-wffs } \varphi, \text{ if } n(\varphi) < k, \text{ then} \\
&\qquad \mathfrak{A} \models \varphi[s] \quad \text{iff} \quad \varphi \in \Delta.
\end{aligned} \tag{IH}$$

Let $\theta$ be an $\mathcal{L}'$-wff such that $n(\theta) = k$. We must prove that $A \models \theta[s]$ iff $\theta \in \Delta$. Since $k > 0$, $\theta$ must have one of the following forms:

$$(\neg\alpha), (\alpha \to \beta), \forall v\alpha,$$

where $n(\alpha) < k$ and $n(\beta) < k$. Thus, the induction hypothesis applies to $\alpha$ and $\beta$.

Case $(\neg\alpha)$: We show that $\mathfrak{A} \models (\neg\alpha)[s]$ iff $(\neg\alpha) \in \Delta$ as follows:

$$\begin{aligned}
\mathfrak{A} \models (\neg\alpha)[s] \quad &\text{iff} \quad \mathfrak{A} \not\models \alpha[s] && \text{by Definition 3.2.6(3),} \\
&\text{iff} \quad \alpha \notin \Delta && \text{by (IH),} \\
&\text{iff} \quad \neg\alpha \in \Delta && \text{by (6), above.}
\end{aligned}$$

Case $(\alpha \to \beta)$: We show that $\mathfrak{A} \models (\alpha \to \beta)[s]$ iff $(\alpha \to \beta) \in \Delta$ as follows:

$$\begin{aligned}
\mathfrak{A} \models (\alpha \to \beta)[s] \quad &\text{iff} \quad \text{if } \mathfrak{A} \models \alpha[s], \text{ then } \mathfrak{A} \models \beta[s] && \text{by Definition 3.2.6(4),} \\
&\text{iff} \quad \text{if } \alpha \in \Delta, \text{ then } \beta \in \Delta && \text{by (IH),} \\
&\text{iff} \quad (\alpha \to \beta) \in \Delta && \text{by Exercise 1.}
\end{aligned}$$

Case $\forall v\alpha$: We must show that $\mathfrak{A} \models \forall v\alpha[s]$ iff $\forall v\alpha \in \Delta$.

($\Rightarrow$). Assume that $\mathfrak{A} \models \forall v\alpha[s]$, that is, assume that $\mathfrak{A} \models \alpha[s_{v|t}]$, for all $t \in A$ (by Definition 3.2.6(5) and as $A = \overline{\mathcal{T}}$). Thus, in particular, we have

$$\mathfrak{A} \models \alpha[s_{v|c}], \text{ for all constant symbols } c \in A. \tag{4.3}$$

We shall prove that $\forall v\alpha \in \Delta$. Suppose, for a contradiction, that $\forall v\alpha \notin \Delta$. By (6) we infer that $\neg\forall v\alpha \in \Delta$. By Exercise 27 on page 118, we have $\vdash \neg\forall v\alpha \to \neg\forall v\neg\neg\alpha$. Thus, $\Delta \vdash \exists v\neg\alpha$ (using the abbreviation). So, by (2), (4), and (7), we conclude that

$$\neg\alpha_c^v \in \Delta \quad \text{for some constant symbol } c. \tag{4.4}$$

As $n(\alpha_c^v) = n(\alpha) < k$, the induction hypothesis (IH) and (4.4) imply that $\mathfrak{A} \models \neg\alpha_c^v[s]$. Thus, $\mathfrak{A} \models \neg\alpha[s_{v|\bar{s}(c)}]$ by Lemma 4.1.2. As $\bar{s}(c) = c$ by Claim 1, we conclude that $\mathfrak{A} \models \neg\alpha[s_{v|c}]$ and this contradicts (4.3). Therefore, $\forall v\alpha \in \Delta$.

($\Leftarrow$). Assume that $\forall v\alpha \in \Delta$. We will show that $\mathfrak{A} \models \forall v\alpha[s]$, that is, we will show that $\mathfrak{A} \models \alpha[s_{v|t}]$, for all $t \in A$. To do this, let $t \in A$. Corollary 4.1.3 implies that there is a wff $\overline{\alpha}$ such that $n(\overline{\alpha}) = n(\alpha)$, $\overline{\alpha} \vdash \alpha$, $\alpha \vdash \overline{\alpha}$, $t$ is substitutable for $v$ in $\overline{\alpha}$, and

$$\mathfrak{A} \models \overline{\alpha}^v_t[s] \quad \text{iff} \quad \mathfrak{A} \models \overline{\alpha}[s_{v|t}], \tag{4.5}$$

as $\overline{s}(t) = t$ by Claim 1. We now show that $\mathfrak{A} \models \overline{\alpha}[s_{v|t}]$. Since $\alpha \vdash \overline{\alpha}$, Lemma 3.3.60 implies that $\forall v\alpha \vdash \forall v\overline{\alpha}$. As $\forall v\alpha \in \Delta$, we conclude that $\forall v\overline{\alpha} \in \Delta$ by (7). Because $t$ is substitutable for $v$ in $\overline{\alpha}$, we see that $\forall v\overline{\alpha} \to \overline{\alpha}^v_t$ is a logical axiom. Hence, $\Delta \vdash \overline{\alpha}^v_t$ so $\overline{\alpha}^v_t \in \Delta$. Since $n(\overline{\alpha}^v_t) = n(\alpha) < k$, the induction hypothesis (IH) implies that $\mathfrak{A} \models \overline{\alpha}^v_t[s]$. By (4.5), we conclude that $\mathfrak{A} \models \overline{\alpha}[s_{v|t}]$. Since $\overline{\alpha} \vdash \alpha$, the soundness theorem (Theorem 4.1.5) implies that $\mathfrak{A} \models \alpha[s_{v|t}]$. Therefore, $\mathfrak{A} \models \forall v\alpha[s]$. (Claim 2) □

Thus, the structure $\mathfrak{A}$ and the assignment $s$ satisfy every formula in $\Delta$ and, because $\Gamma \subseteq \Delta$, $\mathfrak{A}$ and $s$ satisfy every formula in $\Gamma$. Recall, however, that $\mathfrak{A}$ is a structure for the language $\mathcal{L}'$. By restricting the structure $\mathfrak{A}$ to the language $\mathcal{L}$ (that is, by removing the interpretations of new constant symbols from $\mathfrak{A}$), we shall then have our desired structure for the language $\mathcal{L}$.

### Case 2: The language $\mathcal{L}$ does contain the equality symbol

Recall that $\Gamma \subseteq \Theta \subseteq \Delta$. By ignoring the equality symbol, we can construct the structure $\mathfrak{A}$ just as was done in case 1. Thus, the domain of the structure $\mathfrak{A}$ is the set $A$ of all the terms of the language $\mathcal{L}'$. We can now use $\mathfrak{A}$ to build a new structure whose domain is a set of equivalence classes of $A$, also called a *quotient structure*.

Before we construct our new structure, we first define an equivalence relation on the set $A$ of terms. For any $t, \tau \in A$, define the relation $t \sim \tau$ as follows:

$$\tau \sim t \quad \text{iff} \quad (\tau \doteq t) \in \Delta. \tag{4.6}$$

One can now prove that $\sim$ is a symmetric relation on $A$ as follows. Assume that $t_1 \sim t_2$. Then $(t_1 \doteq t_2) \in \Delta$, and thus $\Delta \vdash t_1 \doteq t_2$. Using Logical Axiom 2, Theorem 3.3.54(2) implies that $\Delta \vdash t_2 \doteq t_1$, and thus $(t_2 \doteq t_1) \in \Delta$. So $t_2 \sim t_1$. Similar reasoning will confirm the following claim.

**Claim 3.** *Let $\sim$ be the relation on $A$ defined by (4.6). Then:*
(i) *$\sim$ is an equivalence relation on $A$;*
(ii) *for all terms $t_1, t_2$ and $\tau_1, \tau_2$, if $t_1 \sim \tau_1$ and $t_2 \sim \tau_2$, then*

$$Pt_1t_2 \in \Delta \quad \textit{iff} \quad P\tau_1\tau_2 \in \Delta,$$

*where $P$ is a 2-place predicate symbol, and similarly for n-place predicate symbols;*
(iii) *for all terms $t_1, t_2$ and $\tau_1, \tau_2$, if $t_1 \sim \tau_1$ and $t_2 \sim \tau_2$, then $ft_1t_2 \sim f\tau_1\tau_2$, where $f$ is a 2-place function symbol, and similarly for n-place function symbols.*

*Proof Sketch.* Note that for all $\tau, t \in A$, we have

$$\tau \sim t \quad \text{iff} \quad (\tau \doteq t) \in \Delta \quad \text{iff} \quad \Delta \vdash \tau \doteq t.$$

One proves items (i), (ii), and (iii) as follows:

(i) To show that $\sim$ is an equivalence relation on $A$, one uses items (1), (2), and (3) of Theorem 3.3.54 and Logical Axiom 2.

(ii) One applies (4) of Theorem 3.3.54 to show that for all terms $t_1, t_2$ and $\tau_1, \tau_2$, if $t_1 \sim \tau_1$ and $t_2 \sim \tau_2$, then

$$Pt_1t_2 \in \Delta \quad \text{iff} \quad P\tau_1\tau_2 \in \Delta,$$

where $P$ is a 2-place predicate symbol, and similarly, for $n$-place predicate symbols.

(iii) One employs (5) of Theorem 3.3.54 to show that for all terms $t_1, t_2$ and $\tau_1, \tau_2$, if $t_1 \sim \tau_1$ and $t_2 \sim \tau_2$, then $ft_1t_2 \sim f\tau_1\tau_2$, where $f$ is a 2-place function symbol, and similarly for $n$-place function symbols. (Claim 3) □

Since $\mathcal{L}$ contains equality, we use the set $\Delta$, together with the equivalence relation $\sim$, to construct a structure $\mathfrak{B}$ for the language $\mathcal{L}'$ (see Section 1.1.2) as follows:

(a) Let the domain of $\mathfrak{B}$ be $B = A/{\sim} = \{[t] : t \in A\}$, the set of all equivalence classes.

(b) For each $n$-place predicate symbol $P$, define $P^{\mathfrak{B}}$ by

$$\langle [t_1], [t_2], \ldots, [t_n] \rangle \in P^{\mathfrak{B}} \quad \text{iff} \quad \text{the atomic formula } Pt_1t_2\cdots t_n \text{ belongs to } \Delta.$$

(c) For each $n$-place function symbol $f$, define $f^{\mathfrak{B}}$ by

$$f^{\mathfrak{B}}([t_1], [t_2], \ldots, [t_n]) = [ft_1t_2\cdots t_n].$$

(d) For each constant symbol $c$, define $c^{\mathfrak{B}}$ by $c^{\mathfrak{B}} = [c]$.

Claim 3 implies that every $P^{\mathfrak{B}}$ and $f^{\mathfrak{B}}$ are "compatible" with $\sim$. Thus, each $P^{\mathfrak{B}}$ and $f^{\mathfrak{B}}$ is "well-defined." Now define an assignment $s$ by

$$s(v) = \begin{cases} [v], & \text{if } v \text{ is a variable,} \\ [c], & \text{if } v \text{ is a constant symbol } c. \end{cases} \tag{4.7}$$

Using Claim 3, the proofs of the following two claims are very similar to the proofs of the corresponding Claims 1 and 2 in case 1. For the proof of Claim 5 below, in the "base step" one must include the atomic formula $t_1 \doteq t_2$ for terms $t_1$ and $t_2$.

**Claim 4.** *For all terms $t$, we have $\bar{s}(t) = [t]$.*

**Claim 5.** *For every wff $\varphi$, we have $\mathfrak{B} \models \varphi[s]$ if and only if $\varphi \in \Delta$.*

As before, after restricting the structure $\mathfrak{B}$ to the language $\mathcal{L}$, we have our desired structure in the language $\mathcal{L}$ and an assignment that satisfies every wff $\varphi$ in $\Gamma$. (Theorem) □

So, given a consistent set of formulas $\Gamma$, it has a model. The early attempts to prove Euclid's parallel postulate by contradiction would have benefited from the knowledge of the completeness theorem. An historical remark: Girolamo Saccheri tried to prove the parallel postulate by reductio ad absurdum, that is, he denied the parallel postulate and assumed the other axioms of geometry, hoping to derive a contradiction. This failed attempt could make one wonder if the negation of the parallel postulate is consistent with the other axioms. In any case, Saccheri made no attempt to find a model of geometry in which the parallel postulate is false. He had no idea that the result of his investigations were theorems about non-Euclidean geometry. Saccheri gave up his study of the negation of the parallel postulate and, as a result, was not credited with the discovery of non-Euclidean geometry.

The soundness theorem (Theorem 4.1.5) and the completeness theorem (Theorem 4.2.8) clearly imply the following equivalence.

**Corollary 4.2.9.** *Let $\Gamma$ be a set of formulas. Then $\Gamma \vdash \varphi$ if and only if $\Gamma \vDash \varphi$.*

### 4.2.3 The compactness theorem

The next theorem is the key result that led the logician Abraham Robinson to discover nonstandard analysis. Nonstandard analysis revives the notion of an "infinitesimal"—a number that is infinitely small yet greater than zero (see Section 4.3.1). New theorems in analysis, functional analysis, and other areas of mathematics have been discovered as a result of Robinson's work. Apparently, mathematical logic is an applicable branch of mathematics.

**Theorem 4.2.10** (Compactness theorem). *Let $\Gamma$ be a set of wffs and let $\varphi$ be a wff.*
(a) *If $\Gamma \vDash \varphi$, then for some finite $\Gamma_0 \subseteq \Gamma$, we have $\Gamma_0 \vDash \varphi$.*
(b) *If every finite subset of $\Gamma$ is satisfiable, then $\Gamma$ is satisfiable.*

*Proof.* Let $\Gamma$ be a set of formulas.
(a) Assume that $\Gamma \vDash \varphi$. The completeness theorem thus implies that $\Gamma \vdash \varphi$. Since deductions are finite, there is a finite subset $\Gamma_0 \subseteq \Gamma$ such that $\Gamma_0 \vdash \varphi$. Thus, by the soundness theorem, we conclude that $\Gamma_0 \vDash \varphi$.
(b) Assume that every finite subset of $\Gamma$ is satisfiable. Thus, every finite subset of $\Gamma$ is consistent by Corollary 4.1.11. Hence, by Exercise 24 on page 118, $\Gamma$ is consistent. So $\Gamma$ is satisfiable by part (b) of the completeness theorem (Theorem 4.2.8). □

The following definition is presented here for some of the exercises.

**Definition 4.2.11.** A set of formulas $\Sigma$ is *maximally consistent* if $\Sigma$ is consistent but $\Sigma \cup \{\varphi\}$ is inconsistent for any $\varphi \notin \Sigma$.

**Exercises 4.2.**

*1. Let $\alpha$, $\beta$, and $\Delta$ be as in the proof of Theorem 4.2.8. Show that

$$\text{if } \alpha \in \Delta, \text{ then } \beta \in \Delta \quad \text{iff} \quad (\alpha \to \beta) \in \Delta.$$

*2. In Theorem 4.2.8, show that (b) implies (a).

3. Let $\Gamma$ be a set of $\mathcal{L}$-wffs and let $\varphi$ be a logically valid wff. Show that $\Gamma \vdash \varphi$.

4. Let $\Gamma$ be a set of $\mathcal{L}$-wffs, let $\mathfrak{A}$ be an $\mathcal{L}$-structure with domain $A$, and let $V$ be the set of variables in $\mathcal{L}$. We write $\mathfrak{A} \models^* \Gamma$ to mean that $\mathfrak{A} \models \varphi[s]$ for every assignment $s\colon V \to A$ and for every $\varphi \in \Gamma$. Now let $\Gamma$ and $\Sigma$ be sets of $\mathcal{L}$-wffs. Suppose that for every $\mathcal{L}$-structure $\mathfrak{A}$, we have

$$\mathfrak{A} \models^* \Gamma \quad \text{iff} \quad \mathfrak{A} \models^* \Sigma.$$

Show that

$$\text{for all wffs } \varphi, \Gamma \vdash \varphi \quad \text{iff} \quad \Sigma \vdash \varphi.$$

5. Let $\mathfrak{A}$ be a structure and let $s$ be an assignment. Prove that $\Sigma = \{\varphi : \mathfrak{A} \models \varphi[s]\}$ is maximally consistent.

6. Let $\Sigma$ be a maximally consistent set of formulas. Prove that if $\varphi$ is a formula and $\Sigma \vdash \varphi$, then $\varphi \in \Sigma$.

*7. Let $t$ be a term that is substitutable for $x$ in the wff $\varphi$. Show that $\{\varphi^x_t, x \doteq t\} \vdash \varphi$.

8. Suppose that $\Sigma$ is a maximally consistent set of formulas and $\varphi$ is a formula. Prove that $\neg\varphi \in \Sigma$ if and only if $\varphi \notin \Sigma$.

9. Let $\Gamma$ be a set of sentences in a language $\mathcal{L}$. Suppose that for every $\mathcal{L}$-structure $\mathfrak{A}$, there is a sentence $\sigma \in \Gamma$ such that $\mathfrak{A} \models \sigma$. Using the compactness theorem, show that there must be a finite number of sentences $\sigma_1, \sigma_2, \ldots, \sigma_n$ in $\Gamma$ such that the sentence $\sigma_1 \vee \sigma_2 \vee \cdots \vee \sigma_n$ is logically valid.

10. Let $\Gamma$ be a set of sentences in a language $\mathcal{L}$. Suppose that $\Gamma$ is finitely satisfiable. Let $\varphi$ be a wff. Show that either $\Gamma \cup \{\varphi\}$ or $\Gamma \cup \{\neg\varphi\}$ is finitely satisfiable.

11. Let $\Sigma_0 \subseteq \Sigma_1 \subseteq \Sigma_2 \subseteq \Sigma_3 \subseteq \cdots$ be sets of wffs where each $\Sigma_i$ is finitely satisfiable. Show that $\bigcup_{i\in\mathbb{N}} \Sigma_i$ is finitely satisfiable. Conclude that there is a structure $\mathfrak{A}$ and an assignment $s$ such that $\mathfrak{A} \models \varphi[s]$ for all $\varphi \in \bigcup_{i\in\mathbb{N}} \Sigma_i$.

12. Let $c$ be a constant symbol that does not appear in the wffs $\varphi$, $\psi$, or in any wff in $\Gamma$. Suppose that $\Gamma; \psi^x_c \models \varphi$. Show that $\Gamma; \exists x\psi \models \varphi$.

13. Let $\Gamma$ be a set of $\mathcal{L}$-wffs and let $\mathcal{L}'$ be an extension of $\mathcal{L}$ by new constants. Show that for every $\mathcal{L}$-wff $\varphi$, if there is a deduction of $\varphi$ from $\Gamma$ in the language $\mathcal{L}'$, then there is a deduction of $\varphi$ from $\Gamma$ in the language $\mathcal{L}$.

**Exercise Notes:** For Exercise 1, $\neg(\alpha \to \beta) \to \alpha$ and $\neg(\alpha \to \beta) \to \neg\beta$ are tautologies. For Exercise 2, see Exercise 25 on page 118. For Exercise 7, see Exercise 3 on page 124. For Exercise 12, see Theorem 3.3.56.

## 4.3 Applications

In this section, we will present some applications of the soundness, completeness, and compactness theorems. We first give an application of the compactness theorem. Recall that the compactness theorem implies that if every finite subset of a set of sentences $\Sigma$ has a model, then $\Sigma$ also has a model.

**Theorem 4.3.1.** *Let $\Sigma$ be a set of sentences in the language $\mathcal{L}$. If $\Sigma$ has arbitrarily large finite models, then $\Sigma$ has an infinite model.*

Before we prove this theorem, we discuss the idea behind the proof. One first identifies an infinite set $\Gamma$ of sentences which describes the desired properties. Then one shows that every finite subset of $\Gamma$ has a model. The compactness theorem then implies that $\Gamma$ has a model and this model will have all of the desired properties.

*Proof of Theorem* 4.3.1. Let $\Sigma$ be a set of sentences. Suppose that $\Sigma$ has arbitrarily large finite models. Let us assume that the language contains the equality symbol $\doteq$. For each $k \geq 2$, let $\lambda_k$ be an $\mathcal{L}$-sentence that asserts "there are at least $k$ things;" for example,

$$\lambda_3 = \exists v_1 \exists v_2 \exists v_3 (v_1 \not\doteq v_2 \land v_1 \not\doteq v_3 \land v_2 \not\doteq v_3)$$

asserts "there are at least three things." We show that $\Gamma = \Sigma \cup \{\lambda_2, \lambda_3, \dots\}$ has a model. Let $\Pi \subseteq \Gamma$ be finite. We can write $\Pi = \Pi_0 \cup \Pi_1$, where $\Pi_0 \subseteq \Sigma$ and $\Pi_1 \subseteq \{\lambda_2, \lambda_3, \dots\}$. Since $\Pi_1$ is finite, it has the form

$$\Pi_1 = \{\lambda_{i_1}, \dots, \lambda_{i_k}\}, \quad \text{where } k \in \mathbb{N}.$$

Let $n = \max\{i_1, \dots, i_k\}$. Thus, $n$ is the largest natural number so that $\lambda_n \in \Pi_1$. By our assumption, there is a model $\mathfrak{A}$ of $\Sigma$ with at least $n$ elements. Thus, $\mathfrak{A} \models \lambda_n$ and hence, $\mathfrak{A} \models \Pi_1$. Since $\Pi_0 \subseteq \Sigma$, $\mathfrak{A}$ satisfies all of the sentences in $\Pi_0$. It now follows that $\mathfrak{A}$ must also satisfy all of the sentences in $\Pi$. So every finite subset of $\Gamma$ has a model and thus, by the compactness theorem (Theorem 4.2.10), $\Gamma$ has a model $\mathfrak{B}$. Since $\Sigma \subseteq \Gamma$ and $\mathfrak{B} \models \lambda_k$ for all $k \geq 2$, it follows that $\mathfrak{B}$ is an infinite model of $\Sigma$.

If the language does not contain the equality symbol $\doteq$, then one can add it to the language $\mathcal{L}$ and then argue as above to get the model $\mathfrak{B}$ of $\Sigma$. Then we remove $\doteq^{\mathfrak{B}}$ from $\mathfrak{B}$ to get an infinite $\mathcal{L}$-model of $\Sigma$. □

### 4.3.1 Nonstandard models

In mathematical logic, a nonstandard model is a structure that is a proper elementary extension of a standard model.

**Definition 4.3.2.** Let $\mathfrak{A}$ and $\mathfrak{B}$ be $\mathcal{L}$-structures, where $\mathfrak{A}$ is a substructure of $\mathfrak{B}$. Then $\mathfrak{B}$ is said to be an *elementary extension* of $\mathfrak{A}$, denoted by $\mathfrak{A} \prec \mathfrak{B}$, if and only if for every assignment $s\colon \mathcal{T} \to A$ and every $\mathcal{L}$-formula $\varphi$, we have

$$\mathfrak{A} \models \varphi[s] \quad \text{iff} \quad \mathfrak{B} \models \varphi[s],$$

where $\mathcal{T}$ is the set of all the variables and constants in $\mathcal{L}$ and $A$ is the domain of $\mathfrak{A}$.

Consider the language $\mathcal{L} = \{\dot{<}, \dot{0}, \dot{S}, \dot{+}, \dot{\times}, \dot{E}, \dot{=}\}$ and the $\mathcal{L}$-structure

$$\mathcal{N} = \langle \mathbb{N}; 0, S, <, +, \times, E\rangle,$$

where $\dot{0}^{\mathcal{N}} = 0$, $\dot{S}^{\mathcal{N}} = S$ (successor function), $\dot{<}^{\mathcal{N}} = <$ (the usual "less than" relation), and $\dot{+}^{\mathcal{N}} = +$, $\dot{\times}^{\mathcal{N}} = \times$, $\dot{E}^{\mathcal{N}} = E$ are the usual operations of addition, multiplication, and exponentiation, respectively. The structure $\mathcal{N}$ is called the *standard model of arithmetic* because it is a number system that is commonly used in mathematics and its domain is the set of the standard natural numbers

$$\mathbb{N} = \{0, 1, 2, 3, 4, \dots\}.$$

A nonstandard model of arithmetic is one that contains nonstandard numbers. Using the compactness theorem, we can construct a nonstandard model of arithmetic.

Let $\Gamma = \mathrm{Th}(\mathcal{N})$ (see Definition 3.2.31) and let $\mathcal{L}' = \mathcal{L} \cup \{c\}$ be an extension of $\mathcal{L}$ by one new constant symbol. Let $\Sigma$ be the following infinite set of atomic $\mathcal{L}'$-formulas

$$\Sigma = \{\dot{0} \mathrel{\dot{<}} c,\ \dot{S}\dot{0} \mathrel{\dot{<}} c, \dot{S}\dot{S}\dot{0} \mathrel{\dot{<}} c, \dots, \dot{S}^n\dot{0} \mathrel{\dot{<}} c, \dots\},$$

where $\dot{S}^n\dot{0}$ denotes the term $\overbrace{\dot{S}\dot{S}\cdots\dot{S}}^{n\text{ times}}\,\dot{0}$. Note that $(\dot{S}^n\dot{0})^{\mathcal{N}} = n$ for each natural number $n \geq 1$. Let $\Pi \subseteq \Gamma \cup \Sigma$ be finite. Let $\Pi = \Pi_0 \cup \Pi_1$, where $\Pi_0 \subseteq \Gamma$ and $\Pi_1 \subseteq \Sigma$. Since $\mathcal{N} \models \Pi_0$, we just need to interpret $c$ so that we can use $\mathcal{N}$ to get a model of $\Pi_1$ as well. Since $\Pi_1$ is finite, it has the form

$$\Pi_1 = \{\dot{S}^{n_1}\dot{0} \mathrel{\dot{<}} c, \dots, \dot{S}^{n_i}\dot{0} \mathrel{\dot{<}} c\}, \quad \text{where } i \in \mathbb{N}.$$

Let $k = \max\{n_1, \dots, n_i\} + 1$. Thus, $(\dot{S}^{n_1}\dot{0})^{\mathcal{N}} < k$, ..., $(\dot{S}^{n_i}\dot{0})^{\mathcal{N}} < k$. Therefore, the structure

$$\mathcal{N}_k = \langle \mathbb{N}; 0, S, <, +, \times, E, k\rangle, \quad \text{where } c^{\mathcal{N}_k} = k,$$

is a model of $\Pi_1$ and thus it is a model of $\Pi$, because $\mathcal{N}_k \models \Pi_0$. By the compactness theorem (Theorem 4.2.10), $\Gamma \cup \Sigma$ has a model

$$\mathcal{M} = \langle M; \dot{0}^{\mathcal{M}}, \dot{S}^{\mathcal{M}}, \dot{<}^{\mathcal{M}}, \dot{+}^{\mathcal{M}}, \dot{\times}^{\mathcal{M}}, \dot{E}^{\mathcal{M}}, c^{\mathcal{M}} \rangle.$$

Now let $\overline{\mathcal{M}}$ be the following restriction of $\mathcal{M}$ to the original language $\mathcal{L}$, that is, let

$$\overline{\mathcal{M}} = \langle M; \dot{0}^{\mathcal{M}}, \dot{S}^{\mathcal{M}}, \dot{<}^{\mathcal{M}}, \dot{+}^{\mathcal{M}}, \dot{\times}^{\mathcal{M}}, \dot{E}^{\mathcal{M}} \rangle.$$

Thus, $\overline{\mathcal{M}}$ is a model of $\mathrm{Th}(\mathcal{N})$. As every $n \in \mathbb{N}$ can be identified with $(\dot{S}^n \dot{0})^{\overline{\mathcal{M}}}$, we can view $\mathbb{N}$ as being a subset of $M$. However, since $\overline{\mathcal{M}}$ contains the "infinite" number $c^{\overline{\mathcal{M}}}$, it follows that $\overline{\mathcal{M}}$ and $\mathcal{N}$ are not isomorphic. Nevertheless, Exercise 19(b) on page 89 implies that $\mathcal{N}$ and $\overline{\mathcal{M}}$ are elementarily equivalent. Moreover, one can show that $\overline{\mathcal{M}}$ is a proper elementary extension of the standard model $\mathcal{N}$ (see Definition 4.3.2).

The standard ordered field of real numbers is the structure $\langle \mathbb{R}; +, \times, <, 0, 1 \rangle$ for the language $\mathcal{L} = \{\dot{+}, \dot{\times}, \dot{<}, \dot{0}, \dot{1}, \doteq\}$, where $\dot{+}, \dot{\times}$ are two 2-place function symbols, $\dot{<}$ is a 2-place relation symbol, $\dot{0}, \dot{1}$ are constant symbols, and $\doteq$ is the equality symbol. We will now discuss how to construct a nonstandard "real" field. Let

$$\mathcal{L}' = \{\dot{+}, \dot{\times}, \dot{<}, \dot{0}, \dot{1}, \doteq\} \cup \{c_r : r \in \mathbb{R}\},$$

where $c_r$ is a constant symbol for each real number $r$. Consider the $\mathcal{L}$-structure

$$\mathcal{R} = \langle \mathbb{R}; +, \times, <, 0, 1, \{c_r^{\mathcal{R}} : r \in \mathbb{R}\} \rangle,$$

where $c_r^{\mathcal{R}} = r$ for each $r \in \mathbb{R}$. For each $n \geq 2$, let $\overline{n}$ denote the $\mathcal{L}'$-term $\overbrace{\dot{1} \dot{+} \dot{1} \dot{+} \cdots \dot{+} \dot{1}}^{n \text{ times}}$. Thus, $\overline{n}^{\mathcal{R}} = n$ for all $n \geq 2$. Now let $\Gamma = \mathrm{Th}(\mathcal{R})$ and let $\mathcal{L}^* = \mathcal{L}' \cup \{c\}$ be an extension of $\mathcal{L}'$ by the one new constant symbol $c$. Let $\Sigma$ be the following infinite set of atomic $\mathcal{L}^*$-formulas:

$$\Sigma = \{\overline{2} \dot{<} c, \overline{3} \dot{<} c, \overline{4} \dot{<} c, \ldots, \overline{n} \dot{<} c, \ldots\}.$$

Let $\Pi \subseteq \Gamma \cup \Sigma$ be finite. Let $\Pi = \Pi_0 \cup \Pi_1$, where $\Pi_0 \subseteq \Gamma$ and $\Pi_1 \subseteq \Sigma$. Since $\mathcal{R} \models \Pi_0$, we just need to interpret $c$ so that we can use $\mathcal{R}$ to get a model of $\Pi_1$ as well. Since $\Pi_1$ is finite, it has the form

$$\Pi_1 = \{\overline{n_1} \dot{<} c, \ldots, \overline{n_i} \dot{<} c\}, \quad \text{where } i \in \mathbb{N}.$$

Let $n = \max\{n_1, \ldots, n_i\}$. Thus, $n$ is the largest natural number so that $(\overline{n} \dot{<} c) \in \Pi_1$. Let $k = n + 1$. Thus, the structure

$$\mathcal{R}_k = \langle \mathbb{R}; +, \times, <, 0, 1, \{c_r^{\mathcal{R}} : r \in \mathbb{R}\}, k \rangle, \quad \text{where } c^{\mathcal{R}_k} = k,$$

is a model of $\Pi_1$ and therefore it is a model of $\Pi$, because $\mathcal{R}_k \vDash \Pi_0$. By the compactness theorem (Theorem 4.2.10), $\Gamma \cup \Sigma$ has a model

$$\mathcal{M} = \langle M; \dot{+}^{\mathcal{M}}, \dot{\times}^{\mathcal{M}}, \dot{<}^{\mathcal{M}}, \dot{0}^{\mathcal{M}}, \dot{1}^{\mathcal{M}}, \{c_r^{\mathcal{M}} : r \in \mathbb{R}\}, c^{\mathcal{M}} \rangle.$$

Since $\mathcal{M}$ is a model of Th($\mathcal{R}$), it follows that $\mathcal{M}$ is an ordered field that contains the "infinite" number $c^{\mathcal{M}}$. Moreover, $\mathcal{M}$ contains *infinitesimals*. Since $\mathcal{M} \vDash (\dot{0} \dot{<} \overline{n} \dot{<} c)$ for all $n \geq 2$, it follows that $\mathcal{M} \vDash (\dot{0} \dot{<} \frac{1}{c} \dot{<} \frac{1}{\overline{n}})$ for all $n \geq 2$, where $\frac{1}{c}$ and $\frac{1}{\overline{n}}$ denote the inverses, respectively, of $c$ and $\overline{n}$ in $\mathcal{M}$. Hence, the inverse of $c^{\mathcal{M}}$ in $\mathcal{M}$ is an infinitesimal. Now let $\overline{\mathcal{M}}$ be the restriction of $\mathcal{M}$ to the original language $\mathcal{L}$, namely,

$$\overline{\mathcal{M}} = \langle M; \dot{+}^{\mathcal{M}}, \dot{\times}^{\mathcal{M}}, \dot{<}^{\mathcal{M}}, \dot{0}^{\mathcal{M}}, \dot{1}^{\mathcal{M}} \rangle.$$

Clearly, $\overline{\mathcal{M}}$ is also an ordered field and $c^{\mathcal{M}} \in M$. As every $r \in \mathbb{R}$ can be identified with $c_r^{\mathcal{M}}$, we can view $\mathbb{R}$ as being a subset of $M$. Moreover, $\overline{\mathcal{M}}$ contains infinite numbers and infinitesimals. It thus follows that $\mathcal{R}$ and $\overline{\mathcal{M}}$ and are not isomorphic; but $\mathcal{R}$ and $\overline{\mathcal{M}}$ are elementarily equivalent. Moreover, one can show that $\overline{\mathcal{M}}$ is a proper elementary extension of the standard model $\langle \mathbb{R}; +, \times, <, 0, 1 \rangle$.

In this section we have shown that there is a structure that contains all the natural numbers and satisfies all of the first-order sentences that hold in the standard model $\mathcal{N}$, but this new structure also includes infinite numbers. Thus, we have constructed a new number system.

We have also shown that there exists a structure that contains all the real numbers and satisfies all of the first-order sentences that hold in the ordered field of real numbers. However, this structure contains infinitely large and infinitely small numbers. Again, we have constructed a new system of numbers.

### 4.3.2 Löwenheim–Skolem theorems

Suppose that a set of wffs $\Gamma$ in a countable language has an uncountable model. Thus, $\Gamma$ is satisfiable. Does $\Gamma$ have a countable model? This question will be addressed by our next theorem. Leopold Löwenheim and Thoralf Skolem were two mathematicians who asked such questions and, as a result, established theorems on the existence and cardinality of structures.

**Theorem 4.3.3** (Löwenheim–Skolem). *Let $\Gamma$ be a satisfiable set of wffs in a countable language $\mathcal{L}$. Then there is a countable model $\mathfrak{A}$ that satisfies $\Gamma$.*

*Proof.* Let $\Gamma$ be a satisfiable set of formulas in a countable language $\mathcal{L}$. First, we shall assume that $\mathcal{L}$ does not contain equality. Recall the proof of the completeness theorem (Theorem 4.2.8) in this case. In the proof, as $\mathcal{L}$ is countable, we considered the extension $\mathcal{L}' = \mathcal{L} \cup \{c_1, c_2, c_3, \dots\}$, where the $c_i$'s are new constant symbols. It thus follows that $\mathcal{L}'$

is countable. Consequently, the set of terms of the language $\mathcal{L}'$ is countable. Now, in the proof of case 1 of Theorem 4.2.8 we constructed a model $\mathfrak{A}$ of $\Gamma$ consisting of the terms in the language $\mathcal{L}'$. Therefore, $\mathfrak{A}$ is a countable model satisfying $\Gamma$.

Suppose that the language $\mathcal{L}$ does contain equality. Under case 2 in the proof of Theorem 4.2.8, we constructed a model $\mathfrak{B}$ of $\Gamma$ consisting of equivalence classes of the terms in the language $\mathcal{L}'$. Since the set of these equivalence classes cannot have more elements than the set of terms in the language $\mathcal{L}'$, it follows that the model $\mathfrak{B}$ is a countable model of $\Gamma$. □

**Corollary 4.3.4.** *Let $\mathfrak{A}$ be an uncountable structure for a countable language. Then there is a countable structure $\mathfrak{B}$ such that $\mathfrak{A} \equiv \mathfrak{B}$.*

*Proof.* Let $\Gamma = \mathrm{Th}(\mathfrak{A})$. Theorem 4.3.3 implies that there exists a countable structure $\mathfrak{B}$ that satisfies $\Gamma$. Exercise 19 on page 89 implies that $\mathfrak{A} \equiv \mathfrak{B}$. □

The proof of the above theorem adapts to establish our next result, which is referred to as the *downward Löwenheim–Skolem theorem.* The remainder of this section assumes familiarity with cardinal numbers.

**Theorem 4.3.5** (Löwenheim–Skolem). *Let $\Gamma$ be a satisfiable set of wffs, where $\mathcal{L}$ has cardinality $\kappa$. Then there is a model $\mathfrak{A}$ of cardinality $\leq \kappa$ that satisfies $\Gamma$.*

The following theorem is called the *upward Löwenheim–Skolem theorem.*

**Theorem 4.3.6** (Tarski). *Let $\mathcal{L}$ be of cardinality $\lambda$ and let $\Gamma$ be a set of $\mathcal{L}$-wffs. If an infinite structure satisfies $\Gamma$, then for every cardinal $\kappa \geq \lambda$, there is a structure $\mathfrak{B}$ of cardinality $\kappa$ that satisfies $\Gamma$.*

*Proof.* Let $\mathcal{L}$ be of cardinality $\lambda$ and let $\Gamma$ be a set of $\mathcal{L}$-wffs. Assume that $\Gamma$ is satisfiable in an infinite structure $\mathfrak{A}$ with domain $A$. Let $\kappa \geq \lambda$ and let $\mathcal{L}' = \mathcal{L} \cup \{c_i : i \in \kappa\}$ be an extension of $\mathcal{L}$ by new constant symbols. It follows that $\mathcal{L}'$ has cardinality $\kappa$. Let

$$\Sigma = \{c_i \not\approx c_j : i, j \in \kappa \text{ are distinct}\}.$$

We now show that $\Gamma \cup \Sigma$ is finitely satisfiable. Let $\Pi \subseteq \Gamma \cup \Sigma$ be finite. Let $\Pi = \Pi_0 \cup \Pi_1$, where $\Pi_0 \subseteq \Gamma$ and $\Pi_1 \subseteq \Sigma$. Clearly $\mathfrak{A}$ satisfies $\Pi_0$, and $\Pi_1$ is finite. Let $c_{i_1}, \ldots c_{i_n}$ be a finite list of all the new constant symbols that appear in $\Pi_1$. Since $\mathfrak{A}$ is infinite, we can assign the distinct constant symbols in $c_{i_1}, \ldots c_{i_n}$ to distinct elements $a_{i_1}, \ldots a_{i_n}$ in $A$. Now extend $\mathfrak{A}$ by adding $c_{i_1}^{\mathfrak{A}} = a_{i_1}, \ldots c_{i_n}^{\mathfrak{A}} = a_{i_n}$ to $\mathfrak{A}$ and get a structure that satisfies $\Pi$. Therefore, $\Gamma \cup \Sigma$ is finitely satisfiable. By Theorem 4.3.5, there is a structure $\mathfrak{B}$ of cardinality $\leq \kappa$ that satisfies $\Gamma \cup \Sigma$. Since any model of $\Sigma$ must have cardinality $\geq \kappa$, we conclude that $\mathfrak{B}$ has cardinality $\kappa$. □

**Corollary 4.3.7.** *Let $\mathfrak{A}$ be an infinite structure for a countable language $\mathcal{L}$. Then for every infinite cardinal $\kappa$, there is a structure $\mathfrak{B}$ of cardinality $\kappa$ such that $\mathfrak{A} \equiv \mathfrak{B}$.*

*Proof.* Let $\Gamma = \mathrm{Th}(\mathfrak{A})$ and let $\kappa$ be an infinite cardinal. As $\mathcal{L}$ is countable, it has cardinality $\leq \kappa$. Thus, by Theorem 4.3.6, there is a structure $\mathfrak{B}$ of cardinality $\kappa$ that satisfies $\Gamma$. By Exercise 19 on page 89, $\mathfrak{A} \equiv \mathfrak{B}$. □

### 4.3.3 Theories

A theory is a set of sentences in a language which contains all the sentences that are logically implied by the set.

**Definition 4.3.8.** Let $T$ be a set of sentences of a language. Then $T$ is said to be a *theory* if and only if $T$ is closed under logical implication, that is, for any sentence $\sigma$ of the language,

$$\text{if } T \models \sigma, \quad \text{then } \sigma \in T.$$

A theory $T$ is said to be *complete* if for every sentence $\varphi$, either $\varphi \in T$ or $\neg\varphi \in T$.

Let $\mathfrak{A}$ be a structure. Recalling Definition 3.2.31, $\mathrm{Th}(\mathfrak{A})$ is the set of all sentences $\varphi$ such that $\mathfrak{A} \models \varphi$. It thus follows that $\mathrm{Th}(\mathfrak{A})$ is a complete theory. The proof of the next result is established in Exercise 1.

**Theorem 4.3.9.** *Let $T$ be a theory in a language $\mathcal{L}$. Then $T$ is complete if and only if for all $\mathcal{L}$-structures $\mathfrak{A}$ and $\mathfrak{B}$, if $\mathfrak{A} \models T$ and $\mathfrak{B} \models T$, then $\mathfrak{A} \equiv \mathfrak{B}$.*

Theorem 4.3.9 provides a method for determining whether or not a theory is complete. We will present another such method that depends on the next definition. As you may recall, $\aleph_0$ is the cardinality of any countable infinite set.

**Definition 4.3.10.** Let $T$ be a theory in a language $\mathcal{L}$ and let $\kappa \geq \aleph_0$ be a cardinal. Then $T$ is *$\kappa$-categorical* if all models of $T$ having cardinality $\kappa$ are isomorphic.

In 1954, Jerzy Łoś and Robert L. Vaught independently proved our next theorem.

**Theorem 4.3.11** (Łoś–Vaught test). *Let $T$ be a theory in a countable language. If $T$ has no finite models and is $\kappa$-categorical for some cardinal $\kappa \geq \aleph_0$, then $T$ is complete.*

*Proof.* Let $\kappa \geq \aleph_0$ be a cardinal and let $T$ be a theory in a countable language. Assume that $T$ is $\kappa$-categorical and has no finite models. We shall apply Theorem 4.3.9. Let $\mathfrak{A}$ and $\mathfrak{B}$ be models of $T$. Because $T$ has no finite models, $\mathfrak{A}$ and $\mathfrak{B}$ must be infinite. Corollary 4.3.7 implies that there are structures $\mathfrak{A}'$ and $\mathfrak{B}'$ of cardinality $\kappa$ such that $\mathfrak{A}' \equiv \mathfrak{A}$ and $\mathfrak{B}' \equiv \mathfrak{B}$. Since $T$ is $\kappa$-categorical, we conclude that $\mathfrak{A} \equiv \mathfrak{A}' \cong \mathfrak{B}' \equiv \mathfrak{B}$. Theorem 3.2.39 now implies that $\mathfrak{A} \equiv \mathfrak{B}$. Thus, by Theorem 4.3.9, $T$ is complete. □

**Definition 4.3.12.** Let $\mathcal{K}$ be a class of structures of a given language. We define the *theory of $\mathcal{K}$*, denoted as $\mathrm{Th}(\mathcal{K})$, to be the set of sentences defined by

$$\mathrm{Th}(\mathcal{K}) = \{\varphi : \varphi \text{ is a sentence such that } \mathfrak{A} \models \varphi \text{ for all } \mathfrak{A} \in \mathcal{K}\}. \tag{4.8}$$

**Theorem 4.3.13.** *Let $\mathcal{K}$ be a class of structures of a given language. Then $Th(\mathcal{K})$ is a theory. Moreover, $Th(\mathcal{K})$ is a complete theory if and only if for all structures $\mathfrak{A}, \mathfrak{B} \in \mathcal{K}$, we have $\mathfrak{A} \equiv \mathfrak{B}$.*

*Proof.* Let $\sigma$ be a sentence of the language. Assume that $\mathrm{Th}(\mathcal{K}) \models \sigma$. We must prove that $\sigma \in \mathrm{Th}(\mathcal{K})$. In order to prove that $\sigma \in \mathrm{Th}(\mathcal{K})$, we must show that $\mathfrak{A} \models \sigma$ for all $\mathfrak{A} \in \mathcal{K}$. To do this, let $\mathfrak{A} \in \mathcal{K}$. Since $\mathfrak{A} \in \mathcal{K}$, it follows from (4.8) that $\mathfrak{A} \models \varphi$ for all $\varphi \in \mathrm{Th}(\mathcal{K})$. Because $\mathrm{Th}(\mathcal{K}) \models \sigma$, it now follows that $\mathfrak{A} \models \sigma$. Thus, $\mathrm{Th}(\mathcal{K})$ is a theory. For the remainder of the proof, see Exercise 2. □

For example, let $\mathcal{L} = \{e, *, \doteq\}$ be the language of groups (Example 3.1.2). So, if $\mathcal{G}$ is the class of all groups, then the theory $\mathrm{Th}(\mathcal{G})$ is the set of all $\mathcal{L}$-sentences that are true in all groups. Since some groups are Abelian and others are not, it follows that $\mathrm{Th}(\mathcal{G})$ is not complete.

Let $\Sigma$ be a set of sentences in a language $\mathcal{L}$. Recalling Definition 3.2.25, $\mathrm{Mod}(\Sigma)$ is the class of all $\mathcal{L}$-structures $\mathfrak{A}$ such that $\mathfrak{A} \models \varphi$ for all $\varphi \in \Sigma$. Thus, by Definition 4.3.12, $\mathrm{Th}(\mathrm{Mod}(\Sigma))$ is the set of all $\mathcal{L}$-sentences that are true in all models of $\Sigma$, that is, $\varphi \in \mathrm{Th}(\mathrm{Mod}(\Sigma))$ if and only if $\Sigma \models \varphi$.

**Definition 4.3.14.** Let $\Sigma$ be a set of sentences in a language $\mathcal{L}$. The *consequences* of $\Sigma$, denoted by $\mathrm{Cn}(\Sigma)$, are the set

$$\mathrm{Cn}(\Sigma) = \{\varphi : \varphi \text{ is a sentence and } \Sigma \models \varphi\}.$$

Of course, $\mathrm{Cn}(\Sigma) = \mathrm{Th}(\mathrm{Mod}(\Sigma))$ and $\mathrm{Cn}(\Sigma)$ is a theory. One can confirm that $\Sigma$ is a *theory* if and only if $\mathrm{Cn}(\Sigma) = \Sigma$ (see Exercise 3). For a single sentence $\varphi$, we write $\mathrm{Cn}(\varphi)$ for $\mathrm{Cn}(\{\varphi\})$.

**Definition 4.3.15.** A theory $T$ is *finitely axiomatizable* if $T = \mathrm{Cn}(\Sigma)$ for some finite set $\Sigma$ of sentences.

We will apply the compactness theorem in the proof of the following result.

**Theorem 4.3.16.** *Let $\Sigma$ be a set of sentences. If $Cn(\Sigma)$ is finitely axiomatizable, then there is a finite set $\Sigma_0 \subseteq \Sigma$ such that $Cn(\Sigma_0) = Cn(\Sigma)$.*

*Proof.* Let $\Sigma$ be a set of sentences such that $\mathrm{Cn}(\Sigma)$ is finitely axiomatizable. So there is a finite set $\Pi = \{\alpha_1, \alpha_2, \ldots, \alpha_n\}$ of sentences such that (▲) $\mathrm{Cn}(\Sigma) = \mathrm{Cn}(\Pi)$. Let

$$\varphi = \alpha_1 \wedge \alpha_2 \wedge \cdots \wedge \alpha_n.$$

It follows that $\mathrm{Cn}(\Pi) = \mathrm{Cn}(\varphi)$. Thus, by (▲), $\mathrm{Cn}(\varphi) = \mathrm{Cn}(\Sigma)$, and therefore $\Sigma \models \varphi$. The compactness theorem (Theorem 4.2.10(a)) implies that there is a finite $\Sigma_0 \subseteq \Sigma$ such that $\Sigma_0 \models \varphi$. Since $\Sigma_0 \models \varphi$ and $\Sigma_0 \subseteq \Sigma$, Exercise 7 implies that

$$\mathrm{Cn}(\varphi) \subseteq \mathrm{Cn}(\Sigma_0) \subseteq \mathrm{Cn}(\Sigma).$$

Since $\mathrm{Cn}(\varphi) = \mathrm{Cn}(\Sigma)$, we conclude that $\mathrm{Cn}(\Sigma_0) = \mathrm{Cn}(\Sigma)$. □

We now give an example of a finitely axiomatizable theory. Let $\mathcal{L} = \{\dot{<}, \dot{=}\}$ and let $\Psi$ be the set consisting of the following *axioms for a dense linear order without endpoints*:

1. Asymmetry: $\forall x \forall y (x \dot{<} y \to y \dot{\nless} x)$
2. Trichotomy: $\forall x \forall y (x \dot{<} y \vee x \dot{=} y \vee y \dot{<} x)$
3. Transitivity: $\forall x \forall y \forall z (x \dot{<} y \to y \dot{<} z \to x \dot{<} z)$
4. Density: $\forall x \forall y (x \dot{<} y \to \exists z (x \dot{<} z \dot{<} y))$
5. No endpoints: $\forall x \exists y \exists z (y \dot{<} x \dot{<} z)$.

The theory $\mathrm{Cn}(\Psi)$ is called the *theory of dense linear orders without endpoints*. Let $\mathcal{Q} = \langle \mathbb{Q}; < \rangle$ and $\mathcal{R} = \langle \mathbb{R}; < \rangle$, where $\mathbb{Q}$ is the set of rational numbers, $\mathbb{R}$ is the set of real numbers, and $<$ is the usual "less than" relation on $\mathbb{Q}$ and $\mathbb{R}$. Clearly, $\mathcal{Q} \vDash \Psi$ and $\mathcal{R} \vDash \Psi$. These two structures are not isomorphic because $\mathbb{Q}$ is countable and $\mathbb{R}$ is uncountable (see [3]); nevertheless, are the structures $\mathcal{Q}$ and $\mathcal{R}$ elementarily equivalent? The following theorem is due to Georg Cantor, and its proof appears below.

**Theorem 4.3.17.** *Let $\mathcal{L} = \{\dot{<}, \dot{=}\}$ and let $\Psi$ be the set of axioms for a dense linear order without endpoints. Any two countable models of $\Psi$ are isomorphic.*

Before proving Theorem 4.3.17, we need to make several preparatory remarks. First of all, Exercise 10 shows that any model of $\Psi$ must be infinite. Let $\mathfrak{A}$ and $\mathfrak{B}$ be countable models of $\Psi$ with the respective domains $A = \{a_1, a_2, \ldots\}$ and $B = \{b_1, b_2, \ldots\}$. Let $<$ denote $\dot{<}^{\mathfrak{A}}$ and let $\prec$ denote $\dot{<}^{\mathfrak{B}}$. Let $n \geq 1$ and

$$A' = \{a_1', a_2', \ldots, a_n'\} \subseteq A \quad \text{and} \quad B' = \{b_1', b_2', \ldots, b_n'\} \subseteq B$$

be such that

$$a_1' < a_2' < \cdots < a_i' < a_{i+1}' < \cdots < a_n',$$
$$b_1' \prec b_2' \prec \cdots \prec b_i' \prec b_{i+1}' \prec \cdots \prec b_n'.$$

Let $a \in A \setminus A'$ and $b \in B \setminus B'$. We will say that $b$ has the *same relationship* to $B'$ as $a$ does to $A'$ when the following items hold:

(1) if $a_i' < a < a_{i+1}'$, then $b_i' \prec b \prec b_{i+1}'$, whenever $1 \leq i < n$;
(2) if $a < a_1'$, then $b \prec b_1'$, and if $a_n' < a$, then $b_n' \prec b$.

Since $\mathfrak{A}$ and $\mathfrak{B}$ are models of $\Psi$, it follows that for all $a \in A \setminus A'$, there is a $b \in B \setminus B'$ such that $a$ and $b$ satisfy (1) and (2). Inversely, we shall say that $a$ has the *same relationship* to $A'$ as $b$ does to $B'$, when the following items hold:

(3) if $b_i' \prec b \prec b_{i+1}'$, then $a_i' < a < a_{i+1}'$, whenever $1 \leq i < n$;
(4) if $b \prec b_1'$, then $a < a_1'$, and if $b_n' \prec b$, then $a_n' < a$.

It also follows that for all $b \in B \setminus B'$, there exists an $a \in A \setminus A'$ such that $a$ and $b$ satisfy (3) and (4). We will say that $h'\colon A' \to B'$ is a *partial isomorphism* if $h'$ is a bijection such that

$$x < y \quad \text{if and only if} \quad h'(x) \prec h'(y), \text{ for all } x, y \in A'.$$

Finally, let $X = \{x_1, x_2, x_3, \dots\}$. For any nonempty $X' \subseteq X$, we will say that $x$ is the "least" element in $X'$ if $x = x_n$ and $n$ is the least natural number such that $x_n \in X'$.

*Proof of Theorem* 4.3.17. Let $\mathfrak{A}$ and $\mathfrak{B}$ be two countable models of $\Psi$ with respective domains $A = \{a_1, a_2, \dots\}$ and $B = \{b_1, b_2, \dots\}$. Let $<$ denote $\dot{<}^{\mathfrak{A}}$ and let $\prec$ denote $\dot{<}^{\mathfrak{B}}$. We will construct an isomorphism $h\colon A \to B$ by recursion on the natural numbers.

*Base step:* Let $A_1 = \{a_1\}$ and $B_1 = \{b_1\}$. Define $h_1\colon A_1 \to B_1$ by $h_1(a_1) = b_1$. The function $h_1$ is vacuously a partial isomorphism.

*Inductive step:* Let $n \geq 1$. Assume that $A_n$, $B_n$, and $h_n\colon A_n \to B_n$ have been defined such that $A_n \subseteq A$ and $B_n \subseteq B$, where $A_n$ and $B_n$ have $n$ elements, and $h_n\colon A_n \to B_n$ is a partial isomorphism. If $n$ is odd, then let $a$ be the "least" element in $a \in A \setminus A_n$. Now let $b \in B \setminus B_n$ be so that $b$ has the same relationship to $B_n$ as $a$ does to $A_n$. If $n$ is even, then let $b$ be the "least" element in $b \in B \setminus B_n$. Now let $a \in A \setminus A_n$ be so that $a$ has the same relationship to $A_n$ as $b$ does to $B_n$. Let $A_{n+1} = A_n \cup \{a\}$, $B_{n+1} = B_n \cup \{b\}$, and define $h_{n+1}\colon A_{n+1} \to B_{n+1}$ by

$$h_{n+1}(v) = \begin{cases} h_n(v), & \text{if } v \in A_n, \\ b, & \text{if } v = a. \end{cases} \tag{4.9}$$

Clearly, $h_{n+1}$ is a partial isomorphism.

One can now show that $A = \bigcup_{n\geq1} A_n$, $B = \bigcup_{n\geq1} B_n$, and that $h = \bigcup_{n\geq1} h_n$ is an isomorphism between $\mathfrak{A}$ and $\mathfrak{B}$. □

The method used in the above proof is referred to as Cantor's *back-and-forth method.*

**Corollary 4.3.18.** *The theory* $Cn(\Psi)$ *is complete and the structures* $\mathcal{Q} = \langle \mathbb{Q}; < \rangle$ *and* $\mathcal{R} = \langle \mathbb{R}; < \rangle$ *are elementarily equivalent.*

*Proof.* Theorem 4.3.17 implies $\mathrm{Cn}(\Psi)$ is $\aleph_0$-categorical. Theorems 4.3.17 and 4.3.11 imply that $\mathrm{Cn}(\Psi)$ is a complete theory. So by Theorem 4.3.9, $\mathcal{Q}$ and $\mathcal{R}$ are elementarily equivalent. □

Since $\mathcal{Q} \vDash \Psi$, Corollary 4.3.18 and Exercise 5 imply that $\mathrm{Th}(\mathcal{Q}) = \mathrm{Cn}(\Psi)$. Therefore, $\mathrm{Th}(\mathcal{Q})$ is finitely axiomatizable. The same holds for $\mathrm{Th}(\mathcal{R})$.

**Corollary 4.3.19.** *Let* $\mathcal{L} = \{\dot{<}, \dot{=}\}$ *and* $\mathcal{Q} = \langle \mathbb{Q}; < \rangle$. *Then for every* $\mathcal{L}$*-sentence* $\varphi$, *we have* $\mathcal{Q} \vDash \varphi$ *if and only if* $\Psi \vdash \varphi$.

*Proof.* Let $\varphi$ be an $\mathcal{L}$-sentence. Since $\mathrm{Th}(\mathcal{Q}) = \mathrm{Cn}(\Psi)$, we conclude that $\mathcal{Q} \vDash \varphi$ if and only if $\Psi \vDash \varphi$. Corollary 4.2.9 implies that $\mathcal{Q} \vDash \varphi$ if and only if $\Psi \vdash \varphi$. □

### 4.3.4 Prenex normal form

We establish a result that follows (indirectly) from the soundness theorem (Theorem 4.1.5). Given a wff $\alpha$ with quantifiers, is it the case that $\alpha$ is logically equivalent to a wff $\varphi$ where all of the quantifiers in $\varphi$ appear at the beginning of $\varphi$? In this section, we prove that this is the case. First we present a formal definition.

**Definition 4.3.20.** A *prenex formula* is a wff that has the form

$$Q_1x_1Q_2x_2\cdots Q_nx_n\beta,$$

where each $Q_i$ is $\forall$ or $\exists$ and $\beta$ is quantifier-free.

For example, the formula

$$\forall x\exists y\forall z(Pxy \to Qyz)$$

is a prenex formula.

Corollary 4.1.8 and Propositions 3.3.41–3.3.45 imply the following theorem, which concerns quantifier manipulation and logical equivalence. This theorem presents "rules" that can be applied to transform any wff into an equivalent formula that is in prenex form. In fact, we will use these "rules" to prove that every formula is logically equivalent to a prenex formula. However, in order to apply these "rules," one may have to first apply Theorem 3.3.61 (on alphabetic variants).

**Theorem 4.3.21.** *The following six logical equivalences identify valid operations that involve quantifier manipulation:*

1. $\neg\exists x\alpha \vDash\!\!\!\dashv \forall x\neg\alpha$,
2. $\neg\forall x\alpha \vDash\!\!\!\dashv \exists x\neg\alpha$,
3. $(\alpha \to \forall x\beta) \vDash\!\!\!\dashv \forall x(\alpha \to \beta)$ *if $x$ is not free in $\alpha$,*
4. $(\alpha \to \exists x\beta) \vDash\!\!\!\dashv \exists x(\alpha \to \beta)$ *if $x$ is not free in $\alpha$,*
5. $(\forall x\alpha \to \beta) \vDash\!\!\!\dashv \exists x(\alpha \to \beta)$ *if $x$ is not free in $\beta$,*
6. $(\exists x\alpha \to \beta) \vDash\!\!\!\dashv \forall x(\alpha \to \beta)$ *if $x$ is not free in $\beta$.*

**Theorem 4.3.22.** *Every wff is logically equivalent to a prenex formula.*

*Proof.* We prove the following statement by induction on wffs: *For all wffs $\varphi$, there exists a prenex formula $\psi$ such that $\varphi \vDash\!\!\!\dashv \psi$.*

*Base step:* Let $\phi = Pt_1t_2\cdots t_n$ be an atomic formula. Since $\phi$ has no quantifiers, $\phi$ is already a prenex formula and it is logically equivalent to itself.

*Inductive step:* Let $\alpha$ and $\beta$ be arbitrary wffs. Assume the induction hypothesis

$$\begin{aligned} \alpha &\vDash\!\!\!\dashv Q_1x_1Q_2x_2\cdots Q_nx_n\gamma, \\ \beta &\vDash\!\!\!\dashv Q_1'y_1Q_2'y_2\cdots Q_k'y_k\theta, \end{aligned} \tag{IH}$$

where $\gamma$ and $\theta$ are quantifier-free and $Q_i$ and $Q'_i$ are quantifiers. We must prove that the same holds for each of the following:

$$(\neg\alpha), (\alpha \to \beta), \forall v\alpha.$$

CASE $(\neg\alpha)$: From (IH), by repeatedly applying Theorem 4.3.21(1)(2), we obtain

$$\neg\alpha \vDash\!\dashv \neg Q_1x_1Q_2x_2\cdots Q_nx_n\gamma \vDash\!\dashv \widehat{Q}_1x_1\widehat{Q}_2x_2\cdots\widehat{Q}_nx_n\neg\gamma,$$

where $\widehat{\forall} = \exists$ and $\widehat{\exists} = \forall$. Since $\neg\gamma$ is quantifier-free, we see that $\neg\alpha$ is logically equivalent to a prenex formula.

CASE $(\alpha \to \beta)$: After applying Theorem 3.3.61 (if necessary), we can be assured that the variables $x_1, \ldots, x_n, y_1, \ldots y_k$ in (IH) are all distinct, that none of the variables $x_1, \ldots, x_n$ occur free in $\theta$, and that none of the variables $y_1, \ldots, y_k$ occur free in $\gamma$. From (IH), by repeatedly applying Theorem 4.3.21(3)–(6), we obtain

$$(\alpha \to \beta) \vDash\!\dashv \widehat{Q}_1x_1\widehat{Q}_2x_2\cdots\widehat{Q}_nx_nQ'_1y_1Q'_2y_2\cdots Q'_ky_k(\gamma \to \theta).$$

As $(\gamma \to \theta)$ is quantifier-free, $(\alpha \to \beta)$ is logically equivalent to a prenex formula.

CASE $\forall v\alpha$: From (IH), we conclude that

$$\forall v\alpha \vDash\!\dashv \forall vQ_1x_1Q_2x_2\cdots Q_nx_n\gamma,$$

and this completes the proof. □

**Exercises 4.3.**

*1. Prove Theorem 4.3.9.

*2. Let $\mathcal{K}$ be a class of structures of a given language. Prove that $\mathrm{Th}(\mathcal{K})$ is a complete theory if and only if for all structures $\mathfrak{A}, \mathfrak{B} \in \mathcal{K}$, we have $\mathfrak{A} \equiv \mathfrak{B}$.

*3. Let $\Sigma$ be a set of sentences in a language $\mathcal{L}$. Prove that $\Sigma$ is a *theory* if and only if $\mathrm{Cn}(\Sigma) = \Sigma$.

4. Let $\Sigma$ be a set of sentences in a language $\mathcal{L}$ and let $\mathfrak{A}$ be an $\mathcal{L}$-structure. Show that if $\mathfrak{A} \vDash \Sigma$, then $\mathfrak{A} \vDash \mathrm{Cn}(\Sigma)$.

*5. Let $\Sigma$ be a set of sentences in a language $\mathcal{L}$ and let $\mathfrak{A}$ be an $\mathcal{L}$-structure. Suppose that $\mathrm{Cn}(\Sigma)$ is a complete theory and $\mathfrak{A} \vDash \Sigma$. Show that $\mathrm{Th}(\mathfrak{A}) = \mathrm{Cn}(\Sigma)$.

6. Let $\Sigma$ be a set of sentences in a language $\mathcal{L}$ and let $\mathfrak{A}$ be an $\mathcal{L}$-structure such that $\mathrm{Th}(\mathfrak{A}) = \mathrm{Cn}(\Sigma)$. Show that $\mathfrak{A} \vDash \varphi$ if and only if $\Sigma \vdash \varphi$, for every $\mathcal{L}$-sentence $\varphi$. Now show that for any $\mathcal{L}$-sentence $\varphi$, either $\Sigma \vdash \varphi$ or $\Sigma \vdash \neg\varphi$, but not both.

*7. Let $\Sigma_1$ and $\Sigma_2$ be sets of sentences in a language $\mathcal{L}$.
   (a) Suppose that $\Sigma_1 \subseteq \Sigma_2$. Show that $\mathrm{Cn}(\Sigma_1) \subseteq \mathrm{Cn}(\Sigma_2)$.
   (b) Suppose that $\Sigma_1 \vDash \alpha$ for all $\alpha \in \Sigma_2$. Show that $\mathrm{Cn}(\Sigma_2) \subseteq \mathrm{Cn}(\Sigma_1)$.

8. Let $\Sigma_1, \Sigma_2, \Sigma$ be sets of $\mathcal{L}$-sentences and let $\mathcal{K}_1, \mathcal{K}_2, \mathcal{K}$ be classes of $\mathcal{L}$-structures.
   (a) Show that if $\Sigma_1 \subseteq \Sigma_2$, then $\mathrm{Mod}(\Sigma_2) \subseteq \mathrm{Mod}(\Sigma_1)$.

(b) Show that $\Sigma \subseteq \text{Th}(\text{Mod}(\Sigma))$.
(c) Show that if $\mathcal{K}_1 \subseteq \mathcal{K}_2$, then $\text{Th}(\mathcal{K}_2) \subseteq \text{Th}(\mathcal{K}_1)$.
(d) Show that $\mathcal{K} \subseteq \text{Mod}(\text{Th}(\mathcal{K}))$.
(e) Using (a), (b), and (d), show that $\text{Mod}(\Sigma) = \text{Mod}(\text{Th}(\text{Mod}(\Sigma)))$.
(f) Using (b), (c), and (d), show that $\text{Th}(\mathcal{K}) = \text{Th}(\text{Mod}(\text{Th}(\mathcal{K})))$.

9. Find prenex formulas that are logically equivalent to the following:
   (a) $(\exists x Px \to \forall x Qx)$,
   (b) $\neg(\exists x Px \to \exists x Qx)$,
   (c) $\forall v(\forall x Pxv \to \forall x Qvx)$.

*10. Let $\mathcal{L} = \{\dot{<}, \dot{=}\}$ and let $\Psi$ be the set of $\mathcal{L}$-sentences given on page 143. Show that $\Psi$ has no finite models.

11. Let $A = \{a_1, a_2, \ldots\}$ and let $\bigcup_{n\geq 1} A_n$ be as in the proof of Theorem 4.3.17. From the proof, we have $A_1 \subseteq A_2 \subseteq \cdots$ and $\bigcup_{n\geq 1} A_n \subseteq A$. Prove that $A \subseteq \bigcup_{n\geq 1} A_n$.

12. Let $\mathfrak{A}$ and $\mathfrak{B}$ be as in the proof of Theorem 4.3.17. Complete this proof by showing that $A = \bigcup_{n\geq 1} A_n$, $B = \bigcup_{n\geq 1} B_n$, and that $h = \bigcup_{n\geq 1} h_n$ is an isomorphism between $\mathfrak{A}$ and $\mathfrak{B}$.

13. Let $\mathcal{L} = \{\dot{<}\}$, where $\dot{<}$ is a 2-place relation symbol. Consider the $\mathcal{L}$-structure $\mathcal{N} = \langle \mathbb{N}; < \rangle$, where $<$ is the usual ordering. Show that there is a model $\mathcal{M}$ with domain $M$ that is elementarily equivalent to $\mathcal{N}$ such that there is a sequence $\langle a_i : i \in \mathbb{N} \rangle$ of elements in $M$ such that $\mathcal{M} \models (a_{i+1} \dot{<} a_i)$, for all $i \in \mathbb{N}$.

14. Suppose that the sentence $\varphi$ is true in all infinite models of a theory $T$. Show that there is a natural number $k \geq 1$ such that $\varphi$ is true in all models of $T$ whose domain has at least $k$ many elements.

**Exercise Notes:** For Exercise 11, use proof by contradiction. Let $k > 1$ be the least such that $a_k \in A$ and $a_k$ is not in $\bigcup_{n\geq 1} A_n$. Thus, $\{a_1, a_2, \ldots, a_{k-1}\} \subseteq \bigcup_{n\geq 1} A_n$. It follows that $\{a_1, a_2, \ldots, a_{k-1}\} \subseteq A_n$ for some odd $n$ (why?). Derive a contradiction. For Exercise 13, let $\mathcal{L}' = \mathcal{L} \cup \{c_1, c_2, c_3, \ldots\}$, where the $c_i$'s are new constant symbols. For each $k \geq 1$, let $\lambda_k$ be the sentence

$$c_k < c_{k-1} \wedge c_{k-1} < c_{k-2} \wedge \cdots \wedge c_1 < c_0.$$

Apply the compactness theorem. For Exercise 14, assume, for a contradiction, that for all $k \geq 2$ there is a model $\mathfrak{A}$ of $T$ such that $\mathfrak{A} \models \neg\sigma$ and $\mathfrak{A}$ has at least $k$ elements. Review the proof (and its notation) of Theorem 4.3.1. Show that from this assumption there is a model of $T \cup \{\neg\sigma\} \cup \{\lambda_2, \lambda_3, \ldots\}$.

# 5 Computability

**What is computability theory?**

Computability theory arose from the concept of an algorithm. Computability theory, also called recursion theory, is now a branch of mathematical logic that originated in the 1930s, before there were computers, with the study of computable functions and Turing degrees. The field has grown to include the study of generalized computability and definability. For the computer scientist, computability theory shows that there is a theoretical limit to what computer programs can actually do.

Let $\mathbb{N} = \{0, 1, 2, 3, 4, \dots\}$ be the set of natural numbers and let $f$ be a function of the form $f: \mathbb{N} \to \mathbb{N}$. What does it mean to say that $f$ is computable? One could say that $f$ is computable if there is an algorithm such that for each $n \in \mathbb{N}$, the algorithm with input $n$ will produce the output $f(n)$. This of course would require one to define the meaning of an algorithm.

**Questions**

1. Consider the function $f: \mathbb{N} \to \mathbb{N}$ defined by $f(n) = n + 123{,}572$. Is $f$ computable?
2. Consider the function $f: \mathbb{N} \to \mathbb{N}$ defined by $f(n) = 7n$. Is $f$ computable?
3. Consider the function $f: \mathbb{N} \to \mathbb{N}$ defined by $f(n) = r$, where $r$ is the remainder obtained after dividing $n$ by 5. Is $f$ computable?
4. Suppose that the functions $f: \mathbb{N} \to \mathbb{N}$ and $g: \mathbb{N} \to \mathbb{N}$ are computable. Define the function $h: \mathbb{N} \to \mathbb{N}$ by $h(n) = f(n) + g(n)$. Is $h$ computable?
5. Are all functions of the form $f: \mathbb{N} \to \mathbb{N}$ computable?
6. How many computable functions are there?
7. Suppose that the function $f: \mathbb{N} \to \mathbb{N}$ is computable and suppose that a function $g: \mathbb{N} \to \mathbb{N}$ satisfies $g(n) < f(n)$ for all $n \in \mathbb{N}$. Is $g$ computable?

Is it possible to give a mathematically precise definition of a computable function? Yes! Alan Turing was one of the first mathematicians to give such a definition.

## 5.1 The informal concept

Computability theory is the branch of mathematical logic that studies and identifies problems that are computationally solvable using different models of computation. A central question of computer science is to address the limits of computing devices by understanding the problems that computers can solve. In this section we shall discuss the basic concepts that appear in the theory of computation.

https://doi.org/10.1515/9783110782073-005

### 5.1.1 Decidable sets

A function $f : \mathbb{N}^k \to \mathbb{N}$ is computable if there is an effective (algorithmic) procedure that can evaluate $f$, that is, given any input $n_1, \ldots, n_k$ of $k$ natural numbers, the algorithm will evaluate $f(n_1, \ldots, n_k)$.

In this section, we will use pseudocode to describe our algorithms and procedures. As you may know, *pseudocode* is a compact and informal high-level description of an algorithm. Programmers use pseudocode to develop their algorithms.

The first computability concept that we discuss is the concept of a decidable set. A set is *decidable* if there is an *effective procedure* which will determine whether or not any legitimate candidate is a member of the set. An effective procedure is an algorithm that can be carried out by following the specific steps of an algorithm.

**Definition 5.1.1.** Given any set $S$ of natural numbers, we will say that $S$ is *decidable* if there is an effective procedure such that whenever one applies the procedure to any natural number $n$, the procedure will eventually end and respond "yes" if $n \in S$ and "no" if $n \notin S$.

Definition 5.1.1 applies only to subsets of $\mathbb{N}$, the set of natural numbers. This definition can be generalized to other sets as well.

**Example 5.1.2.** Let $S = \{n \in \mathbb{N} : n \text{ is a prime number}\}$. Show that $S$ is decidable.

**Solution.** Let $n$ be a natural number. The procedure can be described as follows:

```
Begin Procedure
If n = 0 or n = 1, then let A=No.
If n > 1, then perform the following:
  Let A=Yes.
  For j = 2 to n − 1;
    If j | n, then let A=No.
  End of For Loop
Return A
End of Procedure
```

The above algorithm is summarized as follows: Given a natural number $n$, if $n \leq 1$, then return No and halt. If $n > 1$, then search the numbers $2, 3, 4, \ldots, n-1$ for a number $j$ such that $j \mid n$ ($j$ evenly divides $n$). If you find one, then return No and halt. If you do not find such a number, then return Yes and halt. We note that $j \mid n$ if and only if $n = jk$ for some $k \in \mathbb{N}$.

In every step of an effective procedure, one must be able to use an algorithm to decide if a condition is true. Can the truth of the condition "$j \mid n$" be checked by an algorithm? One can effectively decide whether or not $j \mid n$ holds by carrying out the algorithm called long division. If the remainder is 0, then $j \mid n$. Otherwise, $j \nmid n$.

**Example 5.1.3.** Let $S = \{n \in \mathbb{N} : n = 7q + 3 \text{ for some natural number } q\}$. Show that $S$ is decidable.

**Solution.** Let $n$ be a natural number. The procedure can be described as follows:

---

Begin Procedure
Perform long division by dividing $n$ by 7.
If the remainder is 3, then return Yes.
If the remainder is not 3, then return No.
End of Procedure

---

In other words, given a number $n$, divide $n$ by 7. If the remainder is 3, then return Yes and halt. If the remainder is not 3, then return No and halt.

**Example 5.1.4.** Suppose that $S \subseteq \mathbb{N}$ is decidable. Show that $\mathbb{N} \setminus S$ is decidable.

**Solution.** Let $n$ be a natural number. The procedure can be described as follows:

---

Begin Procedure
Perform the procedure for deciding whether or not $n \in S$.
If $n \in S$, then return No.
If $n \notin S$, then return Yes.
End of Procedure

---

That is, given $n$, using the decision procedure for $S$, check to see if $n$ belongs to $S$ or not. If $n \in S$, then return No and halt. If $n \notin S$, then return Yes and halt.

**Example 5.1.5.** Let $A \subseteq \mathbb{N}$ and $B \subseteq \mathbb{N}$ be decidable. Show that $A \setminus B$ is decidable.

**Solution.** Let $n$ be a natural number. The procedure can be described as follows:

---

Begin Procedure
Perform the procedure for deciding whether or not $n \in A$.
Perform the procedure for deciding whether or not $n \in B$.
If $n \in A$ and $n \notin B$, then return Yes.
If $n \notin A$ or $n \in B$, then return No.
End of Procedure

---

Given $n$, using the decision procedure for $A$, check to see if $n$ is a member of $A$ or not. Using the decision procedure for $B$, check to see if $n$ belongs to $B$ or not. If $n \in A$ and $n \notin B$, then return Yes and halt. If $n \notin A$ or $n \in B$, then return No and halt.

**Example 5.1.6.** Let $A \subseteq \mathbb{N}$ be a finite set. Show that $A$ is decidable.

**Solution.** Let $A = \{k_1, k_2, \dots, k_\ell\}$ and $n \in \mathbb{N}$. The procedure is described as follows:

```
Begin Procedure
Let R=No.
If n = k_1, then let R=Yes.
If n = k_2, then let R=Yes.
⋮
If n = k_ℓ, then let R=Yes.
Return R.
End of Procedure
```

That is, given $n$, check to see if $n$ is in the finite list $k_1, k_2, \dots, k_\ell$ or not. If $n$ is not in the list, then return No and halt. If $n$ is in the list, then return Yes and halt.

The notion of an effective procedure is somewhat vague and we have not yet given a precise mathematical definition. We will shortly be giving just such a definition. There are at least two different but equivalent ways to define an effective procedure: Turing machines and register machines. There is also a more mathematical way of defining computable functions which involves the definition of partial recursive functions.

Are all subsets of $\mathbb{N}$ decidable? One can show that there are only countably many effective procedures. Since $\mathcal{P}(\mathbb{N})$ (power set of $\mathbb{N}$) is uncountable, it follows that the majority of sets of natural numbers are not decidable.

### 5.1.2 Computable functions

**Definition 5.1.7.** Let $k \geq 1$ be a natural number. A $k$-place *total function* has the form $f\colon \mathbb{N}^k \to \mathbb{N}$ and its domain is the entire set $\mathbb{N}^k$. A $k$-place *partial function* is a function whose domain is a subset of $\mathbb{N}^k$ and whose values are in $\mathbb{N}$. A total function will also be viewed as a partial function. The *empty function* is the partial function that is undefined for every input. When we say that $f\colon \mathbb{N}^k \to \mathbb{N}$ is a partial function, we mean that the domain of $f$ is a subset of $\mathbb{N}^k$.

**Example 5.1.8.** Define the 2-place total function $f\colon \mathbb{N}^2 \to \mathbb{N}$ and the 2-place partial function $g\colon \mathbb{N}^2 \to \mathbb{N}$ by

$$f(m,n) = \begin{cases} m - n, & \text{if } m \geq n, \\ 0, & \text{if } m < n, \end{cases} \qquad g(m,n) = \begin{cases} m - n, & \text{if } m \geq n, \\ \uparrow, & \text{if } m < n, \end{cases}$$

where $\uparrow$ means that $g(m,n)$ is undefined. So $g(5,2)$ is defined and $(5,2)$ is in the domain of $g$; but $g(2,5)$ is undefined and thus, $(2,5)$ is not in the domain of $g$.

We write $\vec{x} \in \mathbb{N}^k$ to indicate that $\vec{x}$ is a $k$-tuple of the form $x_1, \ldots, x_k$ or $\langle x_1, \ldots, x_k \rangle$, depending on the context.

**Definition 5.1.9.** Let $f$ be a $k$-place partial function. The function $f$ is *effectively computable* if there exists an effective procedure that satisfies the following:
1. Given a $k$-tuple $\vec{x}$ in the domain of $f$, the procedure eventually halts and returns the correct value for $f(\vec{x})$.
2. Given a $k$-tuple $\vec{x}$ *not* in the domain of $f$, the procedure will not halt and thereby will not return a value.

For example, the partial function for subtraction

$$g(m,n) = \begin{cases} m - n, & \text{if } m \geq n, \\ \uparrow, & \text{if } m < n \end{cases}$$

is effectively computable because there is a procedure for subtracting natural numbers, which we learned in elementary school. The procedure can be described as follows:

---

Begin Procedure
If $m \geq n$
  compute $r = m - n$, return $r$ and halt.
Else
  Infinite Loop.
End of Procedure

---

An infinite loop is a procedure that does not terminate, that is, it runs forever. The empty function is effectively computable by the following procedure:

---

Begin Procedure
Infinite Loop.
End of Procedure

---

The concept of a decidable set can now be described in terms of functions. Let $S \subseteq \mathbb{N}^k$. We say that $S$ is *decidable* if its *characteristic function* (which is total)

$$C_S(\vec{x}) = \begin{cases} 1, & \text{if } \vec{x} \in S, \\ 0, & \text{if } \vec{x} \notin S \end{cases}$$

is effectively computable. Thus, $S$ is decidable if its characteristic function $C_S$ is a computable total function.

**Remark 5.1.10.** If $S \subseteq \mathbb{N}^k$ is decidable, we can use the computable total function $C_S$ to construct another effectively computable partial function, namely, we can use the fact that $C_S(\vec{x}) = 1$ if $\vec{x}$ is in $S$ and $C_S(\vec{x}) = 0$ if $\vec{x}$ is not in $S$, for any $\vec{x} \in \mathbb{N}^k$.

**Definition 5.1.11.** A set $S \subseteq \mathbb{N}^k$ is *semi-decidable* if its *semi-characteristic* function

$$c_S(\vec{x}) = \begin{cases} 1, & \text{if } \vec{x} \in S, \\ \uparrow, & \text{if } \vec{x} \notin S \end{cases}$$

is an effectively computable partial function.

**Remark 5.1.12.** Thus, a set $S$ is semi-decidable if there is an effective procedure so that for any input $\vec{x} \in \mathbb{N}^k$ the procedure will halt and return "yes" if and only if $\vec{x} \in S$. Thus, if $\vec{x} \notin S$, then the procedure will not halt (that is, $c_S(\vec{x})$ is undefined).

Any decidable set is also semi-decidable; for example, suppose that $S$ is a decidable subset of $\mathbb{N}^k$. Then the following procedure shows that $S$ is semi-decidable.

```
Begin Procedure
Run C_S(x⃗)
If output = 1
   return 1 and halt.
Else
   Infinite Loop.
End of Procedure
```

**A universal 2-place partial function**

Now, let us just consider those effective procedures that have only one natural number as input and return at most one natural number as output. Suppose that we have adopted a fixed method of encoding each of these procedures $P$ by a single natural number $w$. Suppose, further, that each such code $w$ can be effectively decoded in such a way to retrieve the program $P$ (this can be done). Then the following 2-place "universal function"

$$\Phi(w, x) = \text{the result of applying the procedure coded by } w \text{ to the input } x$$

is an effectively computable partial function. It is to be understood that $\Phi(w, x)$ is undefined (that is, does not halt) when the procedure coded by $w$ fails to halt and return an output.

**Remark 5.1.13.** We present an effective procedure that can be used to show that $\Phi$ is effectively computable. The procedure has inputs $w$ and $x$.

```
Begin Procedure
Decode w into a procedure P.
If P is not a valid procedure, then infinite loop.
Execute procedure P with input x; set r = return value.
If r = empty
   Infinite Loop.
Else
   Return r and halt.
End of Procedure
```

The function $\Phi$ is not total for at least two reasons: Some natural numbers $w$ will not decode and produce a procedure. If $w$ does decode into a valid procedure, then this procedure may not yield an output. We shall write $\Phi(w, x)\downarrow$ to mean that the procedure in Remark 5.1.13 halts with inputs $w$ and $x$ and returns an output value. Thus, $\Phi(w, x)\downarrow$ means that the procedure coded by $w$ with input $x$ halts and returns an output value. The notation $\Phi(w, x)\uparrow$ means that the procedure fails to halt.

Using the universal function $\Phi$ we can enumerate all of the 1-place effectively computable partial functions. First, note that for any 1-place effectively computable partial function $f$ there is a natural number $e$ that encodes the procedure for $f$. Thus,

$$f(x) = \Phi(e, x) \quad \text{for all } x \in \mathbb{N}.$$

For this reason, we shall use the notation $[\![e]\!]$ to denote the corresponding function $f$. Hence, $[\![e]\!](x) = f(x) = \Phi(e, x)$ for all $x \in \mathbb{N}$.

Thus, for any natural number $e$, we will now let $[\![e]\!]$ be the partial computable function defined by $[\![e]\!](x) = \Phi(e, x)$. Of course, some natural numbers $e$ may not code a procedure and then $[\![e]\!]$ may be the empty function. In any case, we can conclude that every 1-place effectively computable partial function appears in the list

$$[\![0]\!], [\![1]\!], [\![2]\!], [\![3]\!], [\![4]\!], \dots. \tag{5.1}$$

Since an effectively computable function can be computed by more than one procedure, a computable function may have more than one representation in the list (5.1).

In computability theory, the *halting problem* can be stated as follows: Given a computer program, decide whether the program will eventually halt when it is executed with a given input or the program will run forever.

**Definition 5.1.14** (Turing). Define the *halting relation* $H$ on $\mathbb{N}$ by

$$\langle w, x\rangle \in H \quad \text{iff} \quad [\![w]\!](x)\downarrow,$$

where $[\![w]\!](x)\downarrow$ means that the procedure coded by $w$ with input $x$ halts and has an output value.

Of course, $[\![w]\!](x)\downarrow$ is equivalent to $\Phi(w,x)\downarrow$. Is the relation $H$ decidable? Is there a procedure that will determine whether or not a program coded by $w$ with input $x$ halts? Alan Turing proved in 1936 that no such procedure exists.

**Theorem 5.1.15** (Turing). *The halting relation H is not decidable.*

*Proof.* Suppose, to the contrary, that $H$ is decidable. Thus, the characteristic function $C_H$ defined by

$$C_H(w,x) = \begin{cases} 1, & \text{if } \langle w,x\rangle \in H, \\ 0, & \text{if } \langle w,x\rangle \notin H \end{cases} \tag{5.2}$$

is effectively computable. Define the partial function $f\colon \mathbb{N} \to \mathbb{N}$ by

$$f(n) = \begin{cases} 1, & \text{if } C_H(n,n) = 0, \\ \uparrow, & \text{if } C_H(n,n) = 1. \end{cases} \tag{5.3}$$

It follows that $f$ is an effectively computable partial function (see Remark 5.1.10). Thus, there is a procedure $P$ that evaluates $f$. So, if $C_H(n,n) = 1$, then the procedure $P$ with input $n$, which attempts to evaluate $f(n)$, will not halt. Since $f$ is an effectively computable partial function, there is a natural number $e$ such that $[\![e]\!] = f$. There are two cases to consider: Either $C_H(e,e) = 0$ or $C_H(e,e) = 1$.

Case 1. Suppose $C_H(e,e) = 0$. Thus, $f(e) = 1$ and so $[\![e]\!](e) = 1$. Hence, $\langle e,e\rangle \in H$ because the procedure coded by $e$ with input $e$ halts. Therefore, (5.2) implies that $C_H(e,e) = 1$, which is a contradiction.

Case 2. Suppose $C_H(e,e) = 1$. Thus, by (5.3), $f(e)$ is undefined because the procedure that attempts to evaluate $f(e)$ does not halt. Since $[\![e]\!] = f$, we conclude that $[\![e]\!](e)$ does not halt. Hence, $C_H(e,e) = 0$, which is a contradiction. □

Turing's proof presents an example of a *diagonal argument*, which was introduced by Georg Cantor in 1873 to prove that the set of real numbers is uncountable.

Even though the halting relation is not decidable, it is semi-decidable as $\langle w,x\rangle \in H$ if and only if $\Phi(w,x)\downarrow$. Thus, the semi-characteristic function $c_H$ satisfies

$$c_H(w,x) = \begin{cases} 1, & \text{if } \langle w,x\rangle \in H, \\ \uparrow, & \text{if } \langle w,x\rangle \notin H \end{cases}$$

and is an effectively computable partial function. The following psuedocode offers an effective procedure for evaluating $c_H(w,x)$ (see Remark 5.1.12 and Definition 5.1.9):

```
Begin Procedure
Execute procedure Φ(w, x).
Return 1 and halt.
End of Procedure
```

**Definition 5.1.16.** Define the subset $K$ on $\mathbb{N}$ by

$$K = \{x \in \mathbb{N} : \Phi(x,x)\downarrow\} = \{x \in \mathbb{N} : [\![x]\!](x)\downarrow\}.$$

The set $K$ is semi-decidable as the following procedure indicates:

```
Begin Procedure
Execute procedure Φ(x, x).
Return 1 and halt.
End of Procedure
```

**Theorem 5.1.17** (Kleene). *Let $S$ be a subset of $\mathbb{N}^k$. Then $S$ is decidable if and only if both $S$ and its complement $\overline{S} = \mathbb{N}^k \setminus S$ are semi-decidable.*

*Proof.* Let $S \subseteq \mathbb{N}^k$.

($\Rightarrow$). Assume that $S$ is decidable. Then, as we have seen before, $S$ is semi-decidable because the following effective procedure applies:

```
Begin Procedure
Run C_S(x⃗)
If output = 1
   return 1 and halt.
Else
   Infinite Loop.
End of Procedure
```

By replacing "output = 1" in the above procedure with "output = 0," we obtain an effective procedure that shows that $\overline{S}$ is semi-decidable.

($\Leftarrow$). Assume that $S$ and $\overline{S}$ are semi-decidable. Thus, the partial functions $c_S$ and $c_{\overline{S}}$ are effectively computable. We shall prove that $S$ is decidable, using a "system timer" as follows: Run the $c_S(\vec{x})$-procedure for one minute and then freeze it. Now, in another part of executable memory, run the $c_{\overline{S}}(\vec{x})$-procedure for one minute and then freeze it. Now continue the $c_S(\vec{x})$-procedure for one minute and then freeze it, and then continue the $c_{\overline{S}}(\vec{x})$-procedure for one minute and then freeze it. Keep alternating this "sharing of system time" until one of the procedures halts and yields an output of 1. If $c_S(\vec{x}) = 1$, then return 1. If $c_{\overline{S}}(\vec{x}) = 1$, then return 0. □

**Theorem 5.1.18.** *Let k be a natural number.*
1. *A k-place relation is semi-decidable if and only if the relation is the domain of some effectively computable partial function.*
2. *A partial function f is effectively computable if and only if* $\{\langle\vec{x},y\rangle : f(\vec{x}) = y\}$ *is a semi-decidable relation.*

*Proof.* Let $S \subseteq \mathbb{N}^k$.
1. We prove that a $k$-place relation is semi-decidable if and only if the relation is the domain of some effectively computable partial function.

   (⇒). Suppose that $S \subseteq \mathbb{N}^k$ is semi-decidable. Thus, by definition, the function $c_S$ is an effectively computable function. Since

   $$c_S(\vec{x}) = \begin{cases} 1, & \text{if } \vec{x} \in S, \\ \uparrow, & \text{if } \vec{x} \notin S, \end{cases}$$

   we see that $S$ is the domain of some effectively computable partial function.

   (⇐). Suppose that $S \subseteq \mathbb{N}^k$ is the domain of some effectively computable partial function, say $f$. We note that the following procedure evaluates $c_S$:

   ```
   Begin Procedure
   Run f(x⃗).
   Return 1 and halt.
   End of Procedure
   ```

   Since the procedure $f(\vec{x})$ halts if and only if $\vec{x} \in S$, we conclude that $c_S$ is an effectively computable function.
2. We shall prove that a partial function $f$ is effectively computable if and only if the set $G = \{\langle\vec{x},y\rangle : f(\vec{x}) = y\}$ is a semi-decidable relation.

   (⇒). Let $f$ be effectively computable. Consider the following procedure with inputs $\vec{x}$ and $y$:

   ```
   Begin Procedure
   Run f(x⃗).
   If output = y
      Return 1 and halt.
   Else
      Infinite Loop
   End of Procedure
   ```

Thus, the above effective procedure evaluates $c_G$. Therefore, $G$ is semi-decidable.

($\Leftarrow$). Let $G = \{\langle \vec{x}, y\rangle : f(\vec{x}) = y\}$ be a semi-decidable relation and let $c_G$ be the semi-characteristic function of $G$, which is a computable partial function. Given $\vec{x}$, execute the following procedure:

```
      Begin Procedure
      Let n = 0.
(L)   For j = 0 to n;
        Run c_G(x⃗, j) for n + 1 minutes.
        If c_G(x⃗, j) has halted and c_G(x⃗, j) = 1, then return j and go to (E).
      End for Loop
      Let n = n + 1 and go to (L).
(E)   Halt (End of Procedure)
```

If $f(\vec{x})\downarrow$, then this procedure will eventually halt and return the value $f(\vec{x})$. If $f(\vec{x})\uparrow$, then this procedure will not halt. Therefore, $f$ is effectively computable.

Let us explain what the above procedure does. We know that $G$ is semi-decidable. Thus, there is a procedure $P$ such that when given an input $\langle \vec{x}, y\rangle$, the procedure will halt and return "yes" if and only if $\langle \vec{x}, y\rangle \in G$. We now try to compute $f(\vec{x})$, if it is defined. The plan is to use the procedure $P$ to check $\langle \vec{x}, 0\rangle, \langle \vec{x}, 1\rangle, \langle \vec{x}, 2\rangle, \ldots$ for membership in $G$. To do this, we must budget our time by using a method called "dovetailing." We execute the following steps:

(1) Spend 1 minute testing whether $\langle \vec{x}, 0\rangle$ is in $G$.

(2) Spend 2 minutes each on $\langle \vec{x}, 0\rangle$ and $\langle \vec{x}, 1\rangle$ testing for membership in $G$.

(3) Spend 3 minutes each on $\langle \vec{x}, 0\rangle, \langle \vec{x}, 1\rangle, \langle \vec{x}, 2\rangle$ testing for membership in $G$.

$\vdots \quad \vdots$

If at a step we find an $\langle \vec{x}, k\rangle \in G$, then we return the value $k$ and halt. If $f(\vec{x})\downarrow$, then this process will eventually halt and return the correct value for $f(\vec{x})$. On the other hand, if $f(\vec{x})\uparrow$, then this process will not halt, that is, the process will run forever. Therefore, the partial function $f$ is effectively computable. □

**Exercises 5.1.**

1. Let $A \subseteq \mathbb{N}$ and $B \subseteq \mathbb{N}$ be decidable. Show that $A \cap B$ is decidable, that is, let $n \in \mathbb{N}$ and identify an effective procedure that decides whether or not $n \in A \cap B$.
2. Let $A$ and $B$ be decidable subsets of $\mathbb{N}$. Show that $A \cup B$ is decidable.
3. Let $A$ and $B$ be semi-decidable subsets of $\mathbb{N}$. Show that $A \cap B$ is semi-decidable.
4. Let $A$ and $B$ be semi-decidable subsets of $\mathbb{N}$. Show that $A \cup B$ is semi-decidable.
5. Let $R \subseteq \mathbb{N}^2$ be a decidable relation. Show that the set $\{x : \langle x, i\rangle \in \mathbb{R}\}$, domain of $R$, is semi-decidable. (Review the proof of Theorem 5.1.18(2).)
6. Suppose that $f: \mathbb{N} \to \mathbb{N}$ is a computable partial function. Show that the domain of $f$, $\{x \in \mathbb{N} : f(x)\downarrow\}$, is semi-decidable.

7. Let $f: \mathbb{N} \to \mathbb{N}$ be a total computable function. Show that the range $\{f(x) : x \in \mathbb{N}\}$ is semi-decidable.

## 5.2 Formalizations—an overview

In the preceding section, we used the term "effectively computable" to refer to an intuitive notion of an effective (algorithmic) procedure. In this section we present an overview of three different (but equivalent) methods that allow one to replace the notion of effectively computable with one that is mathematically precise. One of these methods will be developed in more detail in Sections 5.3 and 5.4. It is important to emphasize that these three approaches are equivalent (see Theorems 5.2.20 and 5.2.11).

### 5.2.1 Turing machines

A Turing machine is a device that manipulates symbols on a tape according to a set of instructions. Despite its simplicity, a Turing machine can be adapted to simulate the logic of any computer algorithm. The Turing machine is not intended as a practical method for implementing algorithms, but rather as a hypothetical device representing a computing machine. Turing machines help computer scientists understand the limits of mechanical computation.

A Turing machine has a potentially infinite tape, marked into squares and a tape head which has an arrow pointing at the square on the tape that is currently being addressed. Each square can hold a symbol or a blank space. The symbols must be taken from a given alphabet Σ. Initially, the tape contains only the input string (word) and is blank everywhere else (see Figure 5.1).

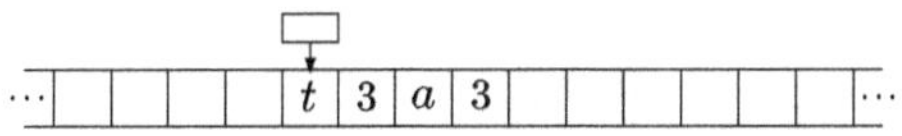


**Figure 5.1:** A tape with an input string from the alphabet $\Sigma = \{a, t, 3\}$.

A Turing machine instruction commands the machine to perform the simple steps indicated below:

(a) read the tape square under the tape head,
(b) write a symbol on the tape in that square,
(c) move the tape head to the left or right, and
(d) proceed to a new instruction.

Steps (b) through (d) all depend upon what symbol appears on the tape square being scanned before the instruction is executed. The machine can be in any one of finitely many "states" $q_1, q_2, \ldots, q_r$. A machine instruction has the form ⟨state, read, write, move, next-state⟩. A program for a Turing machine consists of a finite set of Turing machine instructions.

#### Words and alphabets

Consider the set consisting of the English alphabet $\{a, b, c, d, \ldots, x, y, z\}$. A *finite string or word* over this alphabet is a string of letters; for example, *aaactu* and *zadw* are two strings (words) of letters over the English alphabet.

We will now consider more general alphabets. Let $\Sigma$ be a finite set of symbols. Then $\Sigma^*$ is the set of all words over the alphabet $\Sigma$ (including the empty word $\lambda$). For example, let $\Sigma = \{1, b, c\}$. Then

$$\Sigma^* = \{\lambda, 1, a, b, 1a, a1, bbb, abcabc, \ldots\}.$$

We shall denote the "empty" word with the symbol $\lambda$, and this symbol is not allowed to occur in any alphabet.

#### Turing machine instructions

An instruction has the form $\langle q_i, S_j, S_k, D, q_m\rangle$, where $q_i$ is the current state, $S_j$ and $S_k$ are symbols in $\Sigma$ (the given alphabet), $D$ is either $R$ or $L$ (right or left), and $q_m$ is the next state. If the machine is in state $q_i$, then the instruction $\langle q_i, S_j, S_k, D, q_m\rangle$ tells the machine to look at the square currently under the tape head (the arrow) and do the following (in order):

1. If the square contains symbol $S_j$, then replace it with the symbol $S_k$.
2. If $D = R$, then move the tape head to the next square on the right; if $D = L$, then move the tape head to the next square on the left.
3. Go into state $q_m$.
4. Let $S$ be the current symbol in the square under the tape head.
5. Now execute the instruction that has state $q_m$ and symbol $S$ as its first two components. If there is no such instruction, then halt!

We will not allow the symbol $B$ to be used in any alphabet. The capital letter $B$ will be used to represent a "blank" symbol (that is, an empty square). Thus, an instruction $\langle q_i, S_j, B, D, q_m\rangle$ means to "erase" symbol $S_j$ and the instruction $\langle q_i, B, S_k, D, q_m\rangle$ means that if the square is blank, then write the given symbol $S_k$ into this square. *A Turing machine cannot have two different instructions that have the same first two components.*

Suppose that we are working in the two-letter alphabet $\Sigma = \{a, b\}$ and we want to write a Turing machine that takes a word in this alphabet as input and will append the letter $a$ to each such word (on the right). For example, if the tape has the initial input *abb*, then after running the machine the tape should have the output *abba*. Figure 5.2 illustrates another example. Such a Turing machine is given in Example 5.2.1 below.

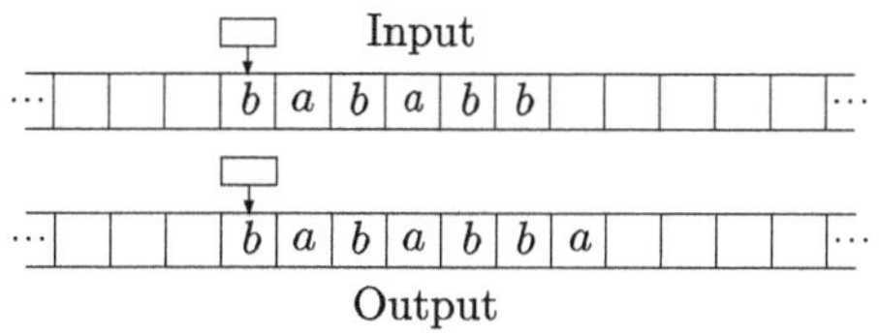


**Figure 5.2:** The Turing input and output.

**Example 5.2.1.** Let $\Sigma = \{a, b\}$ be a two-letter alphabet. We will present a Turing machine $\mathfrak{M}$ which consists of six instructions with just three states $q_1, q_2, q_3$, where $q_1$ is designated as being the initial state. The state $q_3$ will be the halt state, that is, the machine will stop operating. When we start this machine in state $q_1$ it will scan the first letter and then will eventually append the letter $a$ to the word on the right. We show in Figure 5.3 how this machine executes the program on the input word $ab$. The box above the tape head identifies the "current" state of the machine. We have

$$\begin{aligned}&\langle q_1, a, a, R, q_1\rangle\\&\langle q_1, b, b, R, q_1\rangle\\&\langle q_1, B, a, L, q_2\rangle\\&\langle q_2, a, a, L, q_2\rangle\\&\langle q_2, b, b, L, q_2\rangle\\&\langle q_2, B, B, R, q_3\rangle.\end{aligned}\qquad (\mathfrak{M})$$

**Example 5.2.2.** Let $\Sigma = \{a, b\}$ be a two-letter alphabet. Let $f: \Sigma^* \to \Sigma^*$ be the 1-place function defined by $f(w) = wa$, where $w \in \Sigma^*$ is any word in the alphabet $\Sigma$. The machine $\mathfrak{M}$ in Example 5.2.1 computes the total function $f$. So we can say that the function $f$ is Turing computable. Note that $f(\lambda) = a$, where $\lambda$ is the empty word.

Now suppose that $\Sigma$ is a finite alphabet (the blank $B$ does not count as a member of $\Sigma$). Let $\Sigma^*$ be the set of all words over this alphabet (that is, $\Sigma^*$ is the set of all finite strings, including the empty string, consisting of members of $\Sigma$).

**Definition 5.2.3.** Suppose that $f$ is a $k$-place partial function from $(\Sigma^*)^k$ into $\Sigma^*$. We will say that $f$ is *Turing computable* if there exists a Turing machine $\mathfrak{M}$ that, when started in its initial state scanning the first symbol of a $k$-tuple $\vec{w}$ of words (written on the tape, with a blank square between words, and with the rest of the tape blank), behaves as follows:
1. If $f(\vec{w})\downarrow$ (i. e., $\vec{w}$ is in the domain of $f$), then $\mathfrak{M}$ eventually halts and returns the value $f(\vec{w})$, which is the word on the tape whose first letter is under the tape head and whose last letter is followed by a blank square.
2. If $f(\vec{w})\uparrow$ (i. e., $\vec{w}$ is not in the domain of $f$), then $\mathfrak{M}$ never halts.

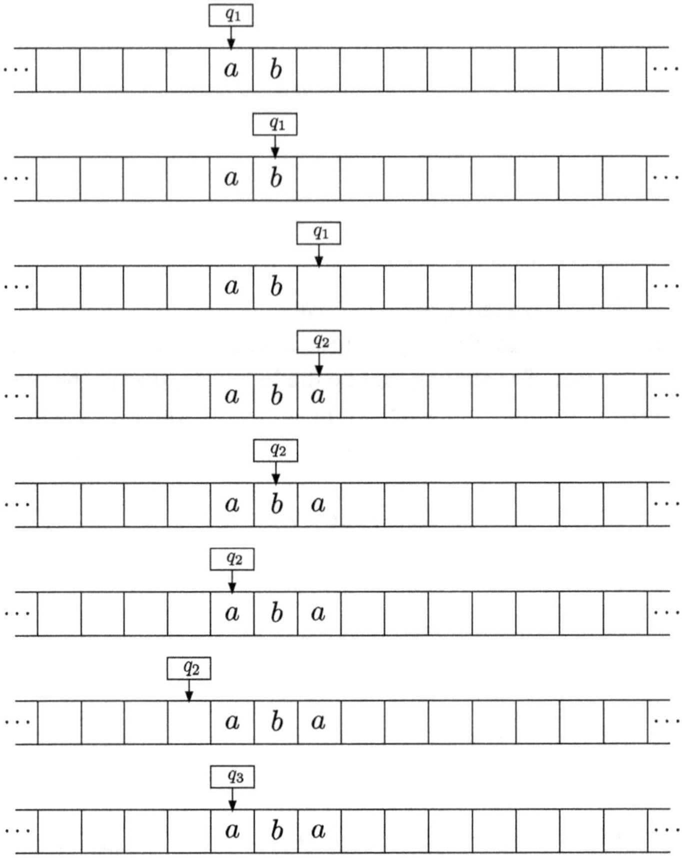

**Figure 5.3:** The machine runs with input *ab* and halts with output *aba*.

Turing developed these ideas prior to the introduction of modern digital computers. After World War II, Turing played an active role in the development of early computers and in the emerging field of artificial intelligence. During the war, he also worked on deciphering the German battlefield code Enigma, which was militarily important work that remained classified until after Turing's death. Turing's remarkable contributions to the war effort have been celebrated in recent years. However, before his death, Turing was persecuted by the English judicial system on account of his sexual orientation.

#### Turing computability of functions on $\mathbb{N}^k$

The definition of Turing computability can be adapted to $k$-place functions on $\mathbb{N}$. One way to do this is to use the single-letter alphabet $\Sigma = \{1\}$ and represent the natural numbers $0, 1, 2, 3, 4, \dots$ as follows: $1, 11, 111, 1111, 11111, \dots$. Thus, the 3-tuple $\langle 3, 0, 5\rangle$ would be represented on a Turing tape as

$$\begin{array}{|c|c|c|c|c|c|c|c|c|c|c|c|c|}\hline 1&1&1&1& &1& &1&1&1&1&1&1\\\hline\end{array}\,.$$

It turns out that every computable function (like the ones in Section 5.1.2) on the natural numbers is Turing computable. For example, the function $f(m,n) = mn$ is Turing computable. Thus, there is a Turing machine such that when given the input $\langle m,n\rangle$, the machine will halt with the output $mn$. For example, if this machine were given the input $\langle 3,4\rangle$ represented on the tape as

$$\begin{array}{|c|c|c|c|c|c|c|c|c|c|}\hline 1&1&1&1& &1&1&1&1&1\\\hline\end{array}\,,$$

then the machine would halt with the output

$$\begin{array}{|c|c|c|c|c|c|c|c|c|c|c|c|c|}\hline 1&1&1&1&1&1&1&1&1&1&1&1&1\\\hline\end{array}\,,$$

which represents 12.

### 5.2.2 Register machines

In theoretical computer science a register machine is an abstract machine that is used in a manner similar to that of a Turing machine. Furthermore, every Turing computable function on the natural numbers can be computed by a register machine and vice versa.

A *register machine* is a conceptual computing device with a finite number of registers, numbered $0, 1, 2, \ldots, K$. Each register is capable of storing a natural number of any magnitude—there is no limit to the size of this number. The operation of the machine is determined by a program. A program is a finite sequence of instructions, drawn from the following list:

- $I\,r$ (where $0 \le r \le K$). "Increment $r$." This instruction results in an increase of the contents of register $r$ by 1. The machine then proceeds to the next instruction in the program (if any).
- $D\,r$ (where $0 \le r \le K$). "Decrement $r$." The effect of this instruction depends on the number in register $r$. If that number is nonzero, it is decreased by 1 and the machine proceeds not to the next instruction, but to the following one. However, if the number in register $r$ is zero, then the machine proceeds to the next instruction. In other words, the machine attempts to decrement register $r$ and if it is successful, then it skips one instruction.
- $J\,q$ (where $q$ is an integer—positive, negative, or zero). "Jump $q$." All registers remain unchanged. The machine takes as its next instruction the $q$-th instruction following this one (if $q \ge 0$) or the $|q|$-th instruction preceding this one (if $q < 0$). The machine halts if there is no such instruction in the program. Thus, the instruction $J\ 0$ results in an infinite loop, by repeating this one instruction over and over again.

This programming language has only these three types of instructions. (Strictly speaking, in these instructions, $r$ and $q$ are numerals, not numbers, that is, an instruction should be a sequence of symbols. If we use base-10 numerals, then the alphabet is $I, D, J, 0, 1, 2, 3, 4, 5, 6, 7, 8, 9, -$. An instruction is a correctly formed word over this alphabet.) A register program will halt when the machine cannot find the "next" instruction.

**Example 5.2.4** (CLEAR 7). Consider the following program, called "CLEAR 7," which will replace the contents of register 7 with the number 0. The comments (on the right) are added to explain the individual steps in the program.

| | | |
|---|---|---|
| D | 7 | Try to decrement register 7. |
| J | 2 | Halt when zero. |
| J | −2 | Go back and repeat. |

The program has three instructions and halts by seeking a fourth instruction. In addition, we can replace 7 with any register number and obtain a program that will clear that particular register, for example, the following program CLEAR 3 will clear register 3.

| | |
|---|---|
| D | 3 |
| J | 2 |
| J | −2 |

**Example 5.2.5** (MOVE $r$ to $s$). Let $r$ and $s$ be different register numbers. Consider the following program, called "MOVE $r$ to $s$," which will move the number in register $r$ into register $s$. This program "calls" on the CLEAR program.

| | | |
|---|---|---|
| CLEAR | $s$ | Ensure that register $s$ is set to zero. |
| D | $r$ | Decrement register $r$. |
| J | 3 | Halt when zero. |
| I | $s$ | Increment register $s$. |
| J | −3 | Go back and repeat. |

The above program leaves a zero in register $r$ and contains seven instructions. The program halts by seeking an eighth instruction.

**Example 5.2.6** (ADD 1 to 2 and 3). Consider the following program, called "ADD 1 to 2 and 3," which will add the number in register 1 to the numbers in register 2 and register 3.

| | | |
|---|---|---|
| D | 1 | Decrement register 1. |
| J | 4 | Halt when zero. |
| I | 2 | Increment register 2. |
| I | 3 | Increment register 3. |
| J | −4 | Go back and repeat. |

This program leaves a zero in register 1. Moreover, the program has five instructions and halts by seeking a sixth instruction. It is clear how to adapt this program to add any register to one or more registers.

**Example 5.2.7** (COPY from $r$ to $s$ (using $t$)). Consider the following program, called "COPY from $r$ to $s$ (using $t$)," which copies the number in register $r$ to register $s$ (leaving register $r$ unchanged). The program "calls" on the programs CLEAR, MOVE, and ADD.

| | | |
|---|---|---|
| CLEAR | $s$ | Set register $s$ to zero. |
| MOVE | $r$ to $t$ | |
| ADD | $t$ to $r$ and $s$ | |

When this program terminates, register $r$ contains the same value as it did when the program began. The program contains 15 instructions and halts by seeking a 16th instruction.

**Example 5.2.8.** Let $R0$, $R1$, $R2$, $R3$ denote registers 0, 1, 2, and 3. Suppose that $x$ and $y$ are in registers 1 and 2. The following program will put the sum $x + y$ in register 0. The comments on the right depict the register contents at each step of the program.

| | | *R0* | *R1* | *R2* | *R3* |
|---|---|---|---|---|---|
| CLEAR | 0 | 0 | $x$ | $y$ | |
| MOVE | 1 to 3 | 0 | 0 | $y$ | $x$ |
| ADD | 3 to 1 and 0 | $x$ | $x$ | $y$ | 0 |
| MOVE | 2 to 3 | $x$ | $x$ | 0 | $y$ |
| ADD | 3 to 2 and 0 | $x + y$ | $x$ | $y$ | 0 |

At the end of this program, registers 1 and 2 contain the same value as they did when the program began. The program has 27 instructions and halts by seeking a 28th instruction.

**Definition 5.2.9.** Suppose that $f\colon \mathbb{N}^n \rightharpoonup \mathbb{N}$ is an $n$-place partial function. Then $f$ is *register-machine computable* if there exists a register program $P$ that, whenever $\vec{x} = \langle x_1, x_2, \ldots, x_n\rangle$ and $x_1$ is in register 1, $x_2$ is in register 2, $\ldots$, $x_n$ is in register $n$, and 0 is in the other registers, behaves as follows:

1. If $f(\vec{x})\downarrow$ (i. e., $\vec{x}$ is in the domain of $f$), then $P$ eventually halts and returns the value $f(\vec{x})$ in register 0. Furthermore, the program halts by seeking a $(p+1)$st instruction when the program $P$ contains $p$ instructions.
2. If $f(\vec{x})\uparrow$ (i. e., $\vec{x}$ is not in the domain of $f$), then $P$ never halts.

**Example 5.2.10.** Let $f\colon \mathbb{N}^2 \to \mathbb{N}$ be defined by $f(x,y) = x+y$. Example 5.2.8 shows that $f$ is register-machine computable.

**Theorem 5.2.11.** *Let $f\colon \mathbb{N}^2 \to \mathbb{N}$ be a partial function. Then $f$ is Turing computable if and only if $f$ is register-machine computable.*

### 5.2.3 Primitive recursiveness and partial search

For a third formalization of the computability concept, we will define a certain class of partial functions on $\mathbb{N}$ to be the smallest class that contains a few simple functions and is also closed under certain constructions that generate more complicated functions. This particular formalization requires no particular type of computing device.

For the *initial functions*, we take the following very simple total functions:

– For each $k \geq 0$, the *zero function* $\mathring{f}\colon \mathbb{N}^k \to \mathbb{N}$ defined by the equation

$$\mathring{f}(x_1, \ldots, x_k) = 0$$

is an *initial* function. The constant 0 is viewed as a 0-place initial function.

– The *successor function* $S\colon \mathbb{N} \to \mathbb{N}$ defined by the equation

$$S(x) = x+1$$

is an *initial* function.

– For all natural numbers $1 \leq i \leq k$, the *projection function* $I_i^k\colon \mathbb{N}^k \to \mathbb{N}$ defined by the equation

$$I_i^k(x_1, \ldots, x_k) = x_i$$

is an *initial* function. The function $I_i^k$ just selects the $i$-th component as its value.

We next identify two ways to generate new functions from the initial functions and those that have already been constructed. Recall that when $f\colon \mathbb{N} \to \mathbb{N}$ and $g\colon \mathbb{N} \to \mathbb{N}$, we can construct the composite function $(f \circ g)\colon \mathbb{N} \to \mathbb{N}$ defined by

$$(f \circ g)(n) = f(g(n))$$

for all $n \in \mathbb{N}$. We will generalize this operation in our next definition.

**Definition 5.2.12.** Let $n \geq 1$ and $k \geq 1$. Suppose that $f: \mathbb{N}^n \to \mathbb{N}$ and $g_i: \mathbb{N}^k \to \mathbb{N}$ for each $i = 1, 2, \ldots, n$. Then we can define the function $h: \mathbb{N}^k \to \mathbb{N}$ by *composition* as follows:

$$h(\vec{x}) = f(g_1(\vec{x}), g_2(\vec{x}), \ldots, g_n(\vec{x})).$$

In Definition 5.2.12, if $f$ and $g_1, g_2, \ldots, g_n$ are partial functions, then $h(\vec{x})$ is defined if and only if $g_1(\vec{x}), g_2(\vec{x}), \ldots, g_n(\vec{x})$ and $f(g_1(\vec{x}), g_2(\vec{x}), \ldots, g_n(\vec{x}))$ are all defined.

**Example 5.2.13.** Let $n = 3$, let $k = 2$, let $f: \mathbb{N}^3 \to \mathbb{N}$ be defined by $f(a, b, c) = ab + c$, and let $g_1(x, y) = 2x$, $g_2(x, y) = 2y$, and $g_3(x, y) = x + y$. Then

$$\begin{aligned} h(x,y) = f(g_1(x,y), g_2(x,y), g_3(x,y)) &= f(2x, 2y, x + y) \\ &= (2x)(2y) + (x + y) = 4xy + x + y. \end{aligned}$$

Another method that we will use to generate functions is called primitive recursion. Recall (Section 1.1.4) that if we have a function $g: \mathbb{N} \to \mathbb{N}$ and an element $a \in \mathbb{N}$, then we can define a function $h: \mathbb{N} \to \mathbb{N}$ by the following recursion:

1. $h(0) = a$,
2. $h(n + 1) = g(h(n))$, for all $n \in \mathbb{N}$.

We will now generalize this definition.

**Definition 5.2.14.** Let $k \geq 1$ and suppose that $f: \mathbb{N}^k \to \mathbb{N}$ and $g: \mathbb{N}^{k+2} \to \mathbb{N}$. Then we can define a function $h: \mathbb{N}^{k+1} \to \mathbb{N}$ by the following *primitive recursion*:

(1) $h(\vec{x}, 0) = f(\vec{x})$,

(2) $h(\vec{x}, n + 1) = g(h(\vec{x}, n), \vec{x}, n)$, for all $n \in \mathbb{N}$.

In Definition 5.2.14, if $f$ and $g$ are partial functions, then $h(\vec{x}, n + 1)$ is defined if and only if $h(\vec{x}, n)$ and $g(h(\vec{x}, n), \vec{x}, n)$ are both defined. Furthermore, the function $h$ is uniquely defined by the above conditions (1) and (2). In fact, Theorem 1.1.27 implies that the function $h$ exists and is unique.

**Example 5.2.15.** Let $k = 2$. Let $f: \mathbb{N}^2 \to \mathbb{N}$ be defined by $f(a, b) = (a+1)b$ and $g: \mathbb{N}^4 \to \mathbb{N}$ be defined by $g(w, x, y, z) = wx + yz$. Then we can define a function $h: \mathbb{N}^3 \to \mathbb{N}$ by the following primitive recursion:

(1) $h(x, y, 0) = f(x, y) = (x + 1)y$,

(2) $h(x, y, n + 1) = g(h(x, y, n), x, y, n)$, for all $n \in \mathbb{N}$.

Since we have the value $h(x, y, 0)$, we can evaluate $h(x, y, 1)$ as follows:

$$h(x, y, 1) = g(h(x, y, 0), x, y, 0) = g((x + 1)y, x, y, 0) = (x + 1)yx + y \cdot 0 = (x^2 + x)y.$$

Now that we have the value $h(x, y, 1)$, we can evaluate $h(x, y, 2)$ as follows:

$$h(x,y,2) = g(h(x,y,1),x,y,1) = g((x^2+x)y,x,y,1) = (x^3+x^2)y+y\cdot 1 = (x^3+x^2+1)y.$$

Continuing in this manner we can evaluate $h(x,y,n)$ for any natural number $n$.

Now that we have the initial functions and two methods of producing functions, we can define the smallest set that contains the initial functions and all the functions that can be generated from the initial functions using composition and primitive recursion.

**Definition 5.2.16.** A function $f\colon \mathbb{N}^k \to \mathbb{N}$ is said to be *primitive recursive* if $f$ can be constructed starting with the zero, successor, and projection functions using composition and primitive recursion.

Thus, if the functions $f\colon \mathbb{N}^n \to \mathbb{N}$ and $g_i\colon \mathbb{N}^k \to \mathbb{N}$ for each $i = 1,2,\ldots,n$ are primitive recursive, then function $h\colon \mathbb{N}^k \to \mathbb{N}$ defined by the composition

$$h(\vec{x}) = f(g_1(\vec{x}), g_2(\vec{x}), \ldots, g_n(\vec{x}))$$

is also primitive recursive. Furthermore, if $f\colon \mathbb{N}^k \to \mathbb{N}$ and $g\colon \mathbb{N}^{k+2} \to \mathbb{N}$ are primitive recursive, the function $h\colon \mathbb{N}^{k+1} \to \mathbb{N}$ that satisfies

(1) $h(\vec{x},0) = f(\vec{x})$,
(2) $h(\vec{x},n+1) = g(h(\vec{x},n),\vec{x},n)$, for all $n \in \mathbb{N}$,

is also primitive recursive.

Again, the collection of all primitive recursive functions is the smallest set that contains the initial functions and is closed under composition and primitive recursion. Thus, the collection of primitive recursive functions is the smallest set $\mathcal{C}$ that contains the zero, successor, and projection functions and whenever a function $f$ is constructed from functions in $\mathcal{C}$ using composition or primitive recursion, $f$ is also in $\mathcal{C}$.

**Example 5.2.17.** Show that the function $h\colon \mathbb{N}^3 \to \mathbb{N}$ that satisfies

(1) $h(x,y,0) = y$,
(2) $h(x,y,n+1) = h(x,y,n) + 2$, for all $n \in \mathbb{N}$,

is primitive recursive.

**Solution.** According to Definition 5.2.14, we need to find primitive recursive functions $f$ and $g$ that satisfy:

(a) $h(x,y,0) = f(x,y)$, where $f$ is a 2-place function,
(b) $h(x,y,n+1) = g(h(x,y,n),x,y,n)$, for all $n \in \mathbb{N}$, where $g$ is a 4-place function.

The function $h$ has two parts of its definition, namely (1) and (2). We will find the functions $f$ and $g$ by using the following tree to show how $h$ is built up from certain primitive recursive functions. The left branch below focuses on part (1) of the definition of $h$ and the right branch addresses part (2) of this definition.

$$h$$

$$\begin{aligned} h(x,y,0) &= y \\ &= I_2^2(x,y) \\ &= f(x,y) \end{aligned}$$

$$\begin{aligned} w &= h(x,y,n) \\ h(x,y,n+1) &= w+2 \\ &= (w+1)+1 \\ &= S(w)+1 \\ &= S(S(w)) \\ &= S(S(I_1^4(w,x,y,n))) \\ &= g(w,x,y,n) \end{aligned}$$

The function $f$ defined by $f(x,y) = I_2^2(x,y)$ is an initial function, so $f$ is primitive recursive. The function $g$ defined by $g(w,x,y,n) = S(S(I_1^4(w,x,y,n)))$ is primitive recursive because $g$ is the result of composing initial functions. Thus, $h$ is a primitive recursive function.

We note that every primitive recursive function is a total function. This is because the initial functions are all total, the composition of total functions is total, and a function obtained by primitive recursion from total functions is also total.

We will show in Section 5.3 that many of the common functions on $\mathbb{N}$ are primitive recursive. For example, we shall show that the operations of addition and multiplication on the natural numbers are primitive recursive.

It seems clear that every primitive recursive function should be regarded as being effectively computable. The initial functions are computable and the composition of computable functions produces a computable function. Whenever $h$ is obtained by primitive recursion from computable functions $f$ and $g$, then it is easy to see how one can effectively find $h(\vec{x}, 99)$, by first finding $h(\vec{x}, 0)$ and then working one's way up as in Example 5.2.15. Therefore, $h$ is computable. However, in order to generate all of the effectively computable functions on the natural numbers, we need to have one more operation. Before we formally identify this new operation, we begin with a motivating discussion.

Let us take a function $g(x,y)$. Let $x \in \mathbb{N}$ and suppose there is at least one value of $y$ which makes $g(x,y) = 0$ and we want to find the least value of $y$ for which $g(x,y) = 0$. There is an effective method for doing this. We know that $y$ is a natural number. We first set $y = 0$ and then compute $g(x,y)$; if we get 0 we stop, because we have found the least $y$ such that $g(x,y) = 0$; but if not, we try the next natural number 1. We try $y = 0, 1, 2, 3\ldots$ until we reach the first value such that $g(x,y) = 0$. Then we define $h(x) = y$ and thereby get a new function. When we know that there is such a value, then this method will terminate in a finite amount of time with the correct answer. Moreover, if $h(x)$ is a function that computes the least $y$ such that $g(x,y) = 0$, then $h$ is computable. We will say that $h$ is produced from $g$ by *minimization*.

However, we do not always know that there is a $y$ where $g(x,y) = 0$. Hence the project of testing $y = 0, 1, 2, 3, \ldots$ may never terminate. If we run the test anyway, which is called *unbounded minimization*, we will get $h$ to be a partial function. For this reason, we will refer to unbounded minimization as *partial search* because such searches may not be successful and thus can produce a partial function.

Since we will be working with partial functions, we will use the notation $h(\vec{x})\downarrow$ to indicate that the function $h$ is defined at $\vec{x}$, and we will write $h(\vec{x})\uparrow$ when $h$ is undefined at $\vec{x}$. Moreover, when we say that $h\colon \mathbb{N}^k \to \mathbb{N}$ is a partial function, we mean that the domain of $h$ may be a proper subset of $\mathbb{N}^k$. We now introduce the $\mu$-operator, which searches for the least natural number that yields a functional value of 0.

**Definition 5.2.18.** Let $g$ be a $(k+1)$-place function on $\mathbb{N}$. We say that the $k$-place function $h$ is obtained (constructed) from $g$ by *partial search* if $h$ satisfies

$$h(\vec{x}) = \mu y(g(\vec{x}, y) = 0),$$

that is, for each $\vec{x} \in \mathbb{N}^k$, we have $h(\vec{x})\downarrow$ and $h(\vec{x}) = y$ if and only if $y$ satisfies the following two conditions:
1. $g(\vec{x}, y)\downarrow$ and $g(\vec{x}, y) = 0$,
2. $g(\vec{x}, s)\downarrow$ and $g(\vec{x}, s) > 0$ for all $s < y$.

When the function $g$ in Definition 5.2.18 is effectively computable, then so is $h$ because we can evaluate $h(\vec{x})$ by investigating the values $g(\vec{x}, 0), g(\vec{x}, 1), \ldots, g(\vec{x}, i)$ (in this order) until we find the first solution $y$ to the equation $g(\vec{x}, y) = 0$. If there is no such $y$, then the search will never end.

Using the initial functions and the operations of composition, primitive recursion, and partial search, we can now define a class of functions that includes the primitive recursive functions.

**Definition 5.2.19.** A function is *partial recursive* if it can be generated by starting with the initial functions and using the operations of composition, primitive recursion, and partial search.

The collection of partial recursive functions is the smallest set that contains the initial functions and is closed under composition, primitive recursion, and partial search. Such functions can be partial functions, because the operation of *partial search* can produce partial functions. However, the expression "partial recursive" is an inseparable phrase, that is, it should be thought of as "partial-recursive."

In Section 5.2.1, we said that a partial function $f\colon \mathbb{N}^k \to \mathbb{N}$ is Turing computable when there is a Turing machine that will evaluate $f$. Thus, the definition of a Turing computable function is very different from the definition of a partial recursive function. Is there a connection between these two dissimilar ways of defining functions? Alan Turing was the first to prove the following striking result.

**Theorem 5.2.20** (Turing). *Let $f\colon \mathbb{N}^k \to \mathbb{N}$ be a partial function. Then $f$ is partial recursive if and only if $f$ is Turing computable.*

There are mathematically formal definitions of a function being Turing computable, register-machine computable, and partial recursive. All three of these computational processes are equivalent, that is, all three approaches define the same class of functions. However, we have not given a formal definition of a function being computable. This is because the notion of being computable is intuitive and consequently cannot be given a precise formal definition. A thesis identified by Alonzo Church and Alan Turing relates the informal idea of being computable to the formal ideas presented in this section. It is an observation that has been verified by very strong evidence.

**Church-Turing Thesis.** Let $f$ be a partial function on the natural numbers. Then the following are equivalent:
- $f$ is computable;
- $f$ is Turing computable;
- $f$ is register-machine computable;
- $f$ is partial recursive.

**Exercises 5.2.**
1. Let $\Sigma = \{a, b\}$ be a two-letter alphabet. Let $f\colon \Sigma^* \to \Sigma^*$ be the 1-place Turing computable function defined by the Turing machine (TM), where $q_1$ is the initial state. Evaluate $f(aabb), f(aba)$, and $f(bbaa)$. We have
$$\begin{aligned}&\langle q_1, a, b, R, q_2\rangle\\&\langle q_1, b, a, R, q_2\rangle\\&\langle q_1, B, a, R, q_2\rangle\\&\langle q_2, a, b, L, q_3\rangle\\&\langle q_2, b, a, R, q_2\rangle\\&\langle q_2, B, a, L, q_3\rangle.\end{aligned}\tag{TM}$$
2. Let $\Sigma = \{a, b\}$ be a two-letter alphabet. Let $f\colon \Sigma^* \to \Sigma^*$ be the 1-place function defined by $f(w) = wbb$ for all $w \in \Sigma^*$. Show that the function $f$ is Turing computable. Then verify that your machine, when given the input *ab*, will produce *abbb* as its output.
3. Let $\Sigma = \{a, b\}$ be a two-letter alphabet. Let $f\colon \Sigma^* \to \Sigma^*$ be the 1-place function that will switch the first letter of every nonempty word in $\Sigma^*$ to the "other" letter. For example, $f(aabb) = babb$ and $f(bbb) = abb$. Show that $f$ is Turing computable. Then verify that your machine, when given the input *aab*, will produce *bab* as its output.
4. Let $\Sigma = \{a, b\}$ be a two-letter alphabet. Let $f\colon \Sigma^* \to \Sigma^*$ be the 1-place function that will take any nonempty word in $\Sigma^*$ and change every occurrence of the letter $a$ in the word to the letter $b$ and will not change any occurrence of letter $b$ in the word. For example, $f(aaba) = bbbb$ and $f(baa) = bbb$. Show that $f$ is Turing computable. Then verify that your machine, when given the input *aab*, will produce *bbb* as its output.
5. Give a register-machine program that computes $f(x, y) = \max\{x - y, 0\}$.
6. Give a register-machine program that computes $f(x, y) = x \cdot y$.

7. Give a register-machine program that computes $f(x,y) = \max\{x,y\}$.
8. Let $h\colon \mathbb{N}^4 \to \mathbb{N}$, $f\colon \mathbb{N} \to \mathbb{N}$, $g\colon \mathbb{N} \to \mathbb{N}$ be primitive recursive functions. Using Definitions 5.2.12 and 5.2.16, show that the function $k(x,y,z) = h(f(x),z,g(y),2)$ is primitive recursive.
9. Show that function $p$ defined by (a) and (b) below is a primitive recursive function:
   (a) $p(m,0) = m$,
   (b) $p(m,n+1) = p(m,n) + 1$.
   Prove that $p(m,n) = m + n$ by induction on $n$.
10. Show that function $h$ defined by (a) and (b) below is a primitive recursive function:
    (a) $h(m,0) = 0$,
    (b) $h(m,n+1) = h(m,n) + m$.
    Prove that $h(m,n) = mn$ by induction on $n$.

**Exercise Notes:** For Exercise 5, $x$ is in register 1 and $y$ is in register 2. Now move register 1 into register 0. Keep decrementing registers 2 and 0 until register 2 contains 0. For Exercise 6, $x$ is in register 1 and $y$ is in register 2. Register 0 initially contains the value 0. Copy register 1 into register 3. Each time you can decrement register 2, add register 3 to register 0 and then copy register 1 into register 3. For Exercise 7, $x$ is in register 1 and $y$ is in register 2. Copy registers 1 and 2 into registers 3 and 4, respectively. Start decrementing registers 1 and 2. The first such register who gets to 0 had the smallest initial value.

## 5.3 Recursive functions

In the previous section, we discovered that the concept of a computable function on the natural numbers has several equivalent definitions (there are more). In this section, we will focus our attention on the class of functions that are recursive. Recursive functions include the primitive recursive functions and are closed under a search operation similar to that of partial search (see Definition 5.2.18).

Recalling Definition 5.2.16, a function $f$ from $\mathbb{N}^k$ to $\mathbb{N}$ is *primitive recursive* if it can be constructed starting with the zero, successor, and projection functions using composition and primitive recursion. Primitive recursion and composition are the key operations that are used to build the primitive recursive functions. Primitive recursive functions are the computable functions that form an important building block on the way to capture all of the computable functions. Most of the functions normally studied in number theory are primitive recursive; for example, addition, division, factorial, exponentiation, and the $n$-th prime are all primitive recursive functions. Let us revisit the initial functions that were introduced in Section 5.2.3 and the operations of composition and primitive recursion.

**Definition 5.3.1.** The *initial functions* are defined as follows:
- For each $k \geq 0$, the *zero function* $\mathring{f}: \mathbb{N}^k \to \mathbb{N}$ is defined by $\mathring{f}(x_1, \ldots, x_k) = 0$. The constant 0 is viewed as a 0-place zero function.
- The *successor function* $S: \mathbb{N} \to \mathbb{N}$ is defined by $S(x) = x + 1$.
- For all $1 \leq i \leq k$, the *projection function* $I_i^k: \mathbb{N}^k \to \mathbb{N}$ is defined by the equation $I_i^k(x_1, \ldots, x_k) = x_i$.

**Definition 5.3.2.** Let $n \geq 1$ and $k \geq 1$. Suppose that $f: \mathbb{N}^n \to \mathbb{N}$ and $g_i: \mathbb{N}^k \to \mathbb{N}$ for each $i = 1, 2, \ldots, n$. We can then form the *composite function* $h: \mathbb{N}^k \to \mathbb{N}$, which is defined by $h(\vec{x}) = f(g_1(\vec{x}), g_2(\vec{x}), \ldots, g_n(\vec{x}))$.

The composition

$$h(\vec{x}) = f(g_1(\vec{x}), g_2(\vec{x}), \ldots, g_n(\vec{x}))$$

can be illustrated as a tree with a vertex having $(n + 1)$ branches:

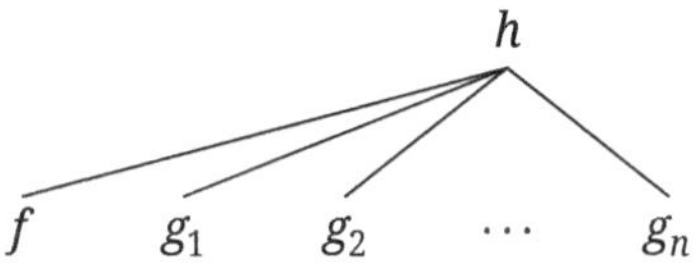


Here, $f$ must be an $n$-place function and $g_1, \ldots, g_n$ must all have the same number of places as the function $h$.

The projection functions can be used to avoid the apparent rigidity in terms of the arity of the functions used in composition. By using compositions with various projection functions, it is possible to pass a subset of the arguments of one function to another function. For example, if $g$ and $h$ are 2-place functions, then the function

$$f(a, b, c) = g(h(c, a), h(a, b))$$

can be obtained by a composition of projection functions, namely,

$$f(a, b, c) = g(h(I_3^3(a, b, c), I_1^3(a, b, c)), h(I_1^3(a, b, c), I_2^3(a, b, c))).$$

**Definition 5.3.3.** Let $k \geq 1$ and suppose that $f: \mathbb{N}^k \to \mathbb{N}$ and $g: \mathbb{N}^{k+2} \to \mathbb{N}$. Then we can define a function $h: \mathbb{N}^{k+1} \to \mathbb{N}$ by the following *primitive recursion*:
(1) $h(\vec{x}, 0) = f(\vec{x})$,
(2) $h(\vec{x}, n + 1) = g(h(\vec{x}, n), \vec{x}, n)$, for all $n \in \mathbb{N}$.

We note that Theorem 1.1.27 implies that the function $h$ in Definition 5.3.3 exists and is unique. The construction of this function $h$ from the functions $f$ and $g$ is illustrated by the following tree, where the left branch illustrates the "base step" (1) and the right branch illustrates the "inductive step" (2).

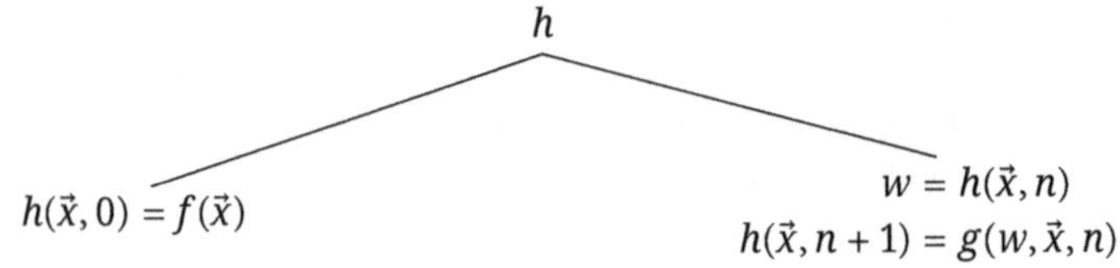


Note that $g$ must have two more places than $f$ and one more place than $h$. For example, if $h$ is a 2-place function, then $g$ must be a 3-place function and $f$ must be a 1-place function.

**Definition 5.3.4.** A function $f$ from $\mathbb{N}^k$ to $\mathbb{N}$ is *primitive recursive* if it can be constructed starting with the zero, successor, and/or projection functions using composition and/or primitive recursion.

Recall that the operation of partial search may not be successful. Consequently, this operation can produce a partial function. The set of recursive functions is obtained by requiring an additional closure condition which is a modification of the partial search operation. This modification, when applicable, will always produce a total function.

**Definition 5.3.5.** Let $g\colon \mathbb{N}^{k+1} \to \mathbb{N}$ be a total function. Suppose that for all $\vec{x} \in \mathbb{N}^k$, there is a $y \in \mathbb{N}$ such that $g(\vec{x}, y) = 0$. We say that the $k$-place function $h$ is obtained (constructed) from $g$ by *total search* if $h$ satisfies

$$h(\vec{x}) = \mu y(g(\vec{x}, y) = 0),$$

that is,

$$h(\vec{x}) = \text{the least } y \text{ such that } g(\vec{x}, y) = 0.$$

**Definition 5.3.6.** A function $f$ from $\mathbb{N}^k$ to $\mathbb{N}$ is *recursive* if it can be constructed starting with the zero, successor, and/or projection functions using composition, primitive recursion, and/or total search.

We observe that every recursive function is a total function. Moreover, every primitive recursive function is a recursive function, and every recursive function is a partial recursive function. We note that there are recursive functions that are not primitive recursive, and there are partial recursive functions that are not recursive. In Chapter 6, we will show that there is a close relationship between deductions in number theory and recursive functions.

The set of recursive functions can be defined by recursion (see Section 1.1.5). Let $B$ be the set that consists of the zero, successor, and projection functions. Let $\mathcal{F}$ be the set of operations that correspond to composition, primitive recursion, and total search. Then, as in Theorem 1.1.24, we can inductively define the sets of functions:

(1) $C_0 = B$,

(2) $C_{n+1} = C_n \cup \mathcal{F}[C_n]$, for all $n \in \mathbb{N}$.

Then $C = \bigcup_{n\in\mathbb{N}} C_n$ is the set of all the recursive functions. Moreover, if a set contains the zero, successor, and projection functions and is closed under composition, primitive recursion, and total search, then the set contains all of the recursive functions.

We will now begin to show that many of the functions on the natural numbers that we use in mathematics are (primitive) recursive. Each function that we construct can be used to construct additional (primitive) recursive functions.

**Proposition 5.3.7.** *The addition function $h(x,y) = x + y$ is primitive recursive.*

*Proof.* Observe that:
- $x + 0 = x$,
- $x + (n + 1) = (x + n) + 1$, for all $n \in \mathbb{N}$.

Thus, letting $h(x,y) = x + y$, we have:
- $h(x,0) = x$,
- $h(x,n + 1) = h(x,n) + 1$, for all $n \in \mathbb{N}$.

We will now show that $h$ is primitive recursive. We must find primitive recursive functions $f$ and $g$ that satisfy Definition 5.3.3:

(1) $h(x,0) = f(x)$, where $f$ is a 1-place function,

(2) $h(x,n + 1) = g(h(x,n),x,n)$, for all $n \in \mathbb{N}$, where $g$ is a 3-place function.

The function $h$ has two parts of its definition, namely items (1) and (2). We will find the two functions $f$ and $g$ by using the following tree to show how $h$ is built up from certain primitive recursive functions. The left branch focuses on part (1) of the definition of $h$ and the right branch addresses part (2) of this definition.

$$h$$

$$\begin{aligned} h(x,0) &= x \\ &= I_1^1(x) \end{aligned} \qquad\qquad \begin{aligned} w &= h(x,n) \\ h(x,n+1) &= w + 1 \\ &= S(w) \\ &= S(I_1^3(w,x,n)) \end{aligned}$$

The desired function $f$ is $f(x) = I_1^1(x)$, which is an initial function and therefore primitive recursive. The desired function $g$ is $g(w,x,n) = S(I_1^3(w,x,n))$, which is primitive recursive because it is the result of composing initial functions. Hence, $h$ is a primitive recursive function. □

The symbol "↦" is read as "maps to." This symbol gives us an easier way to identify a function. The notation $\langle x,y\rangle \mapsto x + y$ indicates that the function is a 2-place function with input $\langle x,y\rangle$ and output $x + y$.

**Proposition 5.3.8.** *Let $k \in \mathbb{N}$. The constant function $\vec{x} \mapsto k$ is primitive recursive.*

Let $k \in \mathbb{N}$. The constant function $\vec{x} \mapsto k$ is primitive recursive because it is the result of composing initial functions. For example, suppose $k = 3$ and let $h(x, y, z, w) = 3$. Observe that

$$h(x, y, z, w) = 3 = S(S(S(0))) = S^3(0) = S^3(\mathring{f}(x, y, z, w)),$$

where $\mathring{f}$ is the 4-place zero function. Since $h$ is the composition of initial functions, it is primitive recursive. For $k \geq 1$ we shall let ${}^k f$ denote the *constant function* ${}^k f(\vec{x}) = k$, for all $\vec{x} \in \mathbb{N}^k$. The constant $i = S^i(0)$ shall be viewed as a 0-place constant function.

**Proposition 5.3.9.** *The multiplication function $\langle x, y \rangle \mapsto x \times y$ is primitive recursive.*

*Proof.* We note that:
(1) $x \times 0 = 0$,
(2) $x \times (n + 1) = (x \times n) + x$, for all $n \in \mathbb{N}$.

We will find the functions $f$ and $g$ by using the following tree to show how $h$ is built up from certain primitive recursive functions. The following left branch focuses on part (1) of the definition of $h$ and the right branch addresses part (2) of this definition.

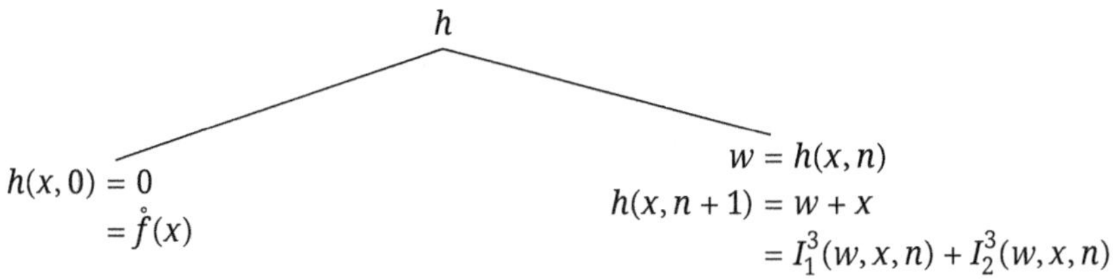


The desired function $f$ is $f(x) = \mathring{f}(x)$, the zero function which is primitive recursive. The desired function $g$ is $g(w, x, n) = I_1^3(w, x, n) + I_2^3(w, x, n)$, which is primitive recursive because addition is primitive recursive. Therefore, $h$ is primitive recursive. □

Since the function $h(x, y) = x \times y$ is primitive recursive, we see that

$$h(I_1^2(x, y), I_1^2(x, y)) = x \times x = x^2,$$
$$h({}^3 f(x, y), I_1^2(x, y)) = 3 \times x = 3x$$

are primitive recursive. Propositions 5.3.7, 5.3.8, and 5.3.9 imply that any polynomial function with coefficients from $\mathbb{N}$ is primitive recursive. Thus, $p(z, y, z) = 3xy^8 + z^2$ is primitive recursive.

**Remark 5.3.10.** Given any (primitive) recursive function $f$, we can define another function by replacing any of the variables in $f$ with constants or by interchanging and/or repeating the variables of $f$. The new function will then be (primitive) recursive. This follows by composing $f$ with the appropriate projection and constant functions.

**Proposition 5.3.11.** *The exponentiation function $\langle x,y\rangle \mapsto x^y$ is primitive recursive.*

*Proof.* Since
- $x^0 = 1$,
- $x^{n+1} = x^n \times x$ for all $n \in \mathbb{N}$,

one can now show, in a manner similar to that in the proof of Proposition 5.3.9, that exponentiation is primitive recursive. □

The exponentiation function $\langle x,y\rangle \mapsto y^x$ is also primitive recursive. This is established by composing certain projection functions with the primitive recursive function in Proposition 5.3.11; namely, let $h(x,y) = x^y$. Define the function $g$ by

$$g(x,y) = h(I_2^2(x,y), I_1^2(x,y)) = h(y,x) = y^x.$$

Since $g$ is constructed by composing primitive recursive functions, it follows that $g$ is primitive recursive. Since $h(3,4) \neq g(3,4)$, $x^y$ and $y^x$ are different functions.

**Proposition 5.3.12.** *The factorial function $x \mapsto x!$ is primitive recursive.*

*Proof.* We have:
- $0! = 1$,
- $(n+1)! = n! \times (n+1)$, for all $n \in \mathbb{N}$.

Thus, letting $h(x) = x!$, we see that:
- $h(0) = 1$,
- $h(n+1) = h(n) \times (n+1)$, for all $n \in \mathbb{N}$.

The following tree shows that $h$ is primitive recursive.

$$h$$

$$h(0) = 1$$

$$\begin{aligned} w &= h(n) \\ h(n+1) &= w \times (n+1) \\ &= w \times S(n) \\ &= I_1^2(w,n) \times S(I_2^2(w,n)) \end{aligned}$$

Here, 1 is a 0-place primitive recursive function and $g(w,n) = I_1^2(w,n) \times S(I_2^2(w,n))$ is a 2-place primitive recursive function. □

**Proposition 5.3.13.** *The predecessor function pred$(x) = x - 1$ (where pred$(0) = 0$) is primitive recursive.*

*Proof.* We have:
- $\text{pred}(0) = 0$,
- $\text{pred}(n+1) = n$, for all $n \in \mathbb{N}$.

Thus, letting $h(x) = \text{pred}(x)$, we see that:
- $h(0) = 0$,
- $h(n+1) = n$, for all $n \in \mathbb{N}$.

The following tree shows that $h$ is primitive recursive.

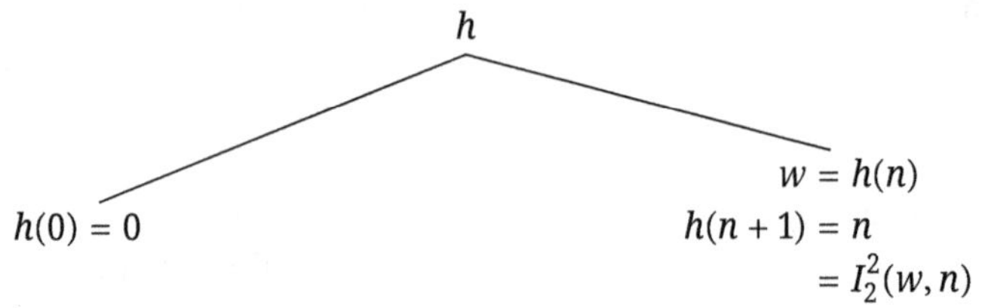


Here, 0 is a 0-place primitive recursive function and $g(w,n) = I_2^2(w,n)$ is a 2-place primitive recursive function. □

**Proposition 5.3.14.** *The proper subtraction function $x \dot{-} y$ is primitive recursive, where $x \dot{-} y$ is defined by $x \dot{-} y = \max\{x - y, 0\}$.*

*Proof.* To see this, observe that:
- $x \dot{-} 0 = x$,
- $x \dot{-} (n+1) = \text{pred}(x \dot{-} n)$, for all $n \in \mathbb{N}$.

Thus, letting $h(x,y) = x \dot{-} y$, we see that:
- $h(x,0) = x$,
- $h(x,n+1) = \text{pred}(h(x,n))$, for all $n \in \mathbb{N}$.

The following tree shows that $h$ is primitive recursive.

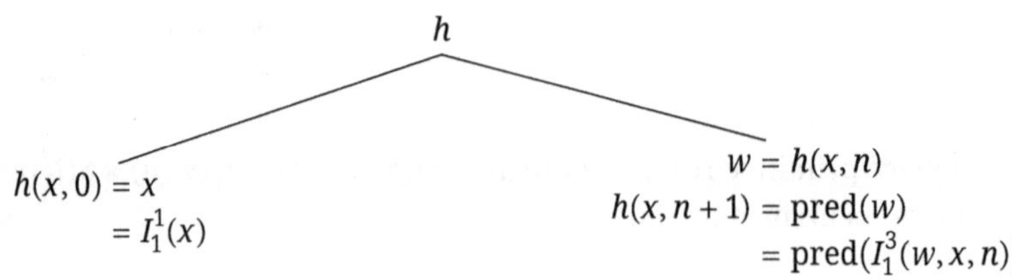

Here, $I_1^1$ is a 1-place projection function and $g(w, x, n) = \text{pred}(I_1^3(w, x, n))$ is a 3-place primitive recursive function. □

We are now ready to show that bounded sums and products are recursive. First, we review summation and product notation. Let $f\colon \mathbb{N} \to \mathbb{N}$ and $n \in \mathbb{N}$. The summation notation $\sum_{t<n} f(t)$ and the product notation $\prod_{t<n} f(t)$ are defined as follows:

$$\sum_{t<n} f(t) = f(0) + f(1) + f(2) + \cdots + f(n-1),$$
$$\prod_{t<n} f(t) = f(0) \times f(1) \times f(2) \times \cdots \times f(n-1),$$

where $\sum_{t<0} f(t) = 0$ and $\prod_{t<0} f(t) = 1$. Observe that

$$\sum_{t<(n+1)} f(t) = \Big(\sum_{t<n} f(t)\Big) + f(n) \quad \text{and} \quad \prod_{t<(n+1)} f(t) = \Big(\prod_{t<n} f(t)\Big) \times f(n).$$

**Proposition 5.3.15.** *Suppose that $f\colon \mathbb{N}^{k+1} \to \mathbb{N}$ is (primitive) recursive. Then the summation function $s$ and the product function $p$ defined by*

$$s(\vec{x}, y) = \sum_{t<y} f(\vec{x}, t) \quad \textit{and} \quad p(\vec{x}, y) = \prod_{t<y} f(\vec{x}, t)$$

*are (primitive) recursive.*

*Proof.* Since addition and multiplication are primitive recursive, the functions $s$ and $p$ are defined, respectively, by primitive recursion as follows:
(1) $s(\vec{x}, 0) = 0$,
(2) $s(\vec{x}, n+1) = s(\vec{x}, n) + f(\vec{x}, n)$, for all $n \in \mathbb{N}$;
(1) $p(\vec{x}, 0) = 1$
(2) $p(\vec{x}, n+1) = p(\vec{x}, n) \times f(\vec{x}, n)$, for all $n \in \mathbb{N}$. □

**Proposition 5.3.16.** *Define the function $z$ by*

$$z(x) = \begin{cases} 1, & \textit{if } x = 0, \\ 0, & \textit{if } x > 0. \end{cases}$$

*Then $z$ is primitive recursive.*

*Proof.* We have $z(x) = 1 \dot{-} x$, and $1 \dot{-} x$ is primitive recursive (see Remark 5.3.10). To see why $z(x) = 1 \dot{-} x$, observe that:
- $z(x) = 1$ if and only if $x = 0$ if and only if $1 \dot{-} x = 1$,
- $z(x) = 0$ if and only if $x > 0$ if and only if $1 \dot{-} x = 0$.

Another way to show that $z$ is primitive recursive is to observe that:

(1) $z(0) = 1$,
(2) $z(n+1) = 0$, for all $n \in \mathbb{N}$.

Thus, one can define $z$ by primitive recursion as illustrated in the following tree.

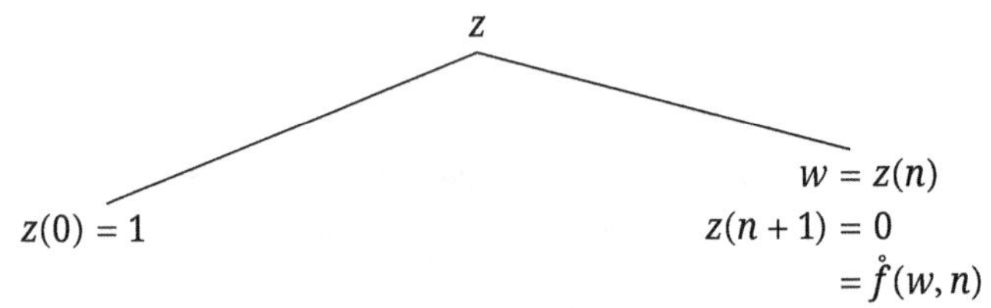


Here, 1 is a 0-place constant function and $\mathring{f}(w,n)$ is the 2-place zero function. □

**Proposition 5.3.17.** *Define the function h by*

$$h(x,y) = \begin{cases} 1, & \text{if } x \leq y, \\ 0, & \text{if } x > y. \end{cases} \tag{5.4}$$

*Then h is primitive recursive.*

*Proof.* Clearly the function $z(x \mathbin{\dot{-}} y)$ is the composition of primitive recursive functions. To see that $h(x,y) = z(x \mathbin{\dot{-}} y)$ satisfies (5.4), note that:
- $h(x,y) = 1$ if and only if $x \leq y$ if and only if $x \mathbin{\dot{-}} y = 0$ if and only if $z(x \mathbin{\dot{-}} y) = 1$,
- $h(x,y) = 0$ if and only if $x > y$ if and only if $x \mathbin{\dot{-}} y > 0$ if and only if $z(x \mathbin{\dot{-}} y) = 0$.

Thus, the function defined by (5.4) is primitive recursive. □

We have been building recursive functions. We can thus build a recursive relation by using its characteristic function.

**Definition 5.3.18.** Let $k \geq 1$. We say that a $k$-place relation $R$ on $\mathbb{N}$ is (primitive) recursive if its characteristic function is (primitive) recursive.

In other words, a $k$-place relation $R$ on $\mathbb{N}$ is (primitive) recursive if the function

$$C_R(x_1, x_2, \ldots, x_k) = \begin{cases} 1, & \text{if } \langle x_1, x_2, \ldots, x_k \rangle \in R, \\ 0, & \text{if } \langle x_1, x_2, \ldots, x_k \rangle \notin R \end{cases}$$

is (primitive) recursive. Thus, Proposition 5.3.17 shows that the 2-place relation

$$\{\langle x,y \rangle : x \leq y\}$$

is primitive recursive. Proposition 5.3.16 shows that the set $\{0\}$ is primitive recursive.

Given an $n$-place (primitive) recursive relation $R$ and $n$ $k$-place (primitive) recursive functions $g_1, g_2, \dots, g_n$, we can define the $k$-place relation $Q$ by

$$Q = \{\vec{x} : \langle g_1(\vec{x}), g_2(\vec{x}), \dots, g_n(\vec{x})\rangle \in R\}.$$

Since the characteristic function for $Q$ is equal to the composition

$$C_Q(\vec{x}) = C_R(g_1(\vec{x}), g_2(\vec{x}), \dots, g_n(\vec{x})),$$

it follows that the relation $Q$ is also (primitive) recursive. This observation shall be referred to as the *substitution rule.*

For any $n$-place relation $R$ on $\mathbb{N}$, we shall write $R(\vec{x})$ to mean $\vec{x} \in R$. Thus, the *substitution rule* states that if $R$ is (primitive) recursive and $g_1, g_2, \dots, g_n$ are $k$-place (primitive) recursive functions, then a relation $Q$ that satisfies

$$Q(\vec{x}) \quad \text{if and only if} \quad R(g_1(\vec{x}), g_2(\vec{x}), \dots, g_n(\vec{x})) \tag{5.5}$$

is also (primitive) recursive.

**Proposition 5.3.19.** *Show that the relation* $\{\langle x, y\rangle : x \geq y\}$ *is primitive recursive.*

*Proof.* Let $Q = \{\langle x, y\rangle : x \geq y\}$. We know by Proposition 5.3.17 that $\leq$ is primitive recursive. Recall that $x \geq y$ if and only if $y \leq x$. Since

$$Q(x, y) \quad \text{if and only if} \quad I_2^2(x, y) \leq I_1^2(x, y),$$

the substitution rule implies that the relation $Q$ is primitive recursive. □

**Example 5.3.20.** Show that the relation $\{\langle x, y\rangle : x \leq y + 1\}$ is primitive recursive.

**Solution.** Let $Q$ be the relation $\{\langle x, y\rangle : x \leq y + 1\}$. We know that $\leq$ is primitive recursive by Proposition 5.3.17. Since the successor function $S(y) = y + 1$ is primitive recursive and

$$\begin{aligned} Q(x, y) \quad &\text{iff} \quad x \leq S(y) \\ &\text{iff} \quad I_1^2(x, y) \leq S(I_2^2(x, y)), \end{aligned}$$

the substitution rule implies that the relation $\{\langle x, y\rangle : x \leq y + 1\}$ is primitive recursive.

From two $k$-place relations $R$ and $Q$ on $\mathbb{N}$, we define the following three new relations:
- $\overline{R} = \{\vec{x} \in \mathbb{N}^k : \text{not } R(\vec{x})\}$,
- $R \cap Q = \{\vec{x} \in \mathbb{N}^k : R(\vec{x}) \text{ and } Q(\vec{x})\}$,
- $R \cup Q = \{\vec{x} \in \mathbb{N}^k : R(\vec{x}) \text{ or } Q(\vec{x})\}$.

The relation $\overline{R}$ is called the *complement* of $R$ and satisfies $\overline{R}(\vec{x})$ if and only if $\vec{x} \notin R$. The relation $R \cap Q$ is the *intersection* of $R$ and $Q$, and the relation $R \cup Q$ is the *union* of the relations $R$ and $Q$.

**Theorem 5.3.21.** *Suppose that R and Q are k-place (primitive) recursive relations. Then the following relations are also (primitive) recursive:*
1. $\overline{R} = \{\vec{x} \in \mathbb{N}^k : \textit{not } R(\vec{x})\}$,
2. $R \cap Q = \{\vec{x} \in \mathbb{N}^k : R(\vec{x}) \textit{ and } Q(\vec{x})\}$,
3. $R \cup Q = \{\vec{x} \in \mathbb{N}^k : R(\vec{x}) \textit{ or } Q(\vec{x})\}$.

*Proof.* Let $C_R$ and $C_Q$ be the respective characteristic functions of the relations $R$ and $Q$. Assuming that $C_R$ and $C_Q$ are (primitive) recursive, we shall show that the relations $\overline{R}$, $R \cap Q$, and $R \cup Q$ are also (primitive) recursive.
1. First we show that $C_{\overline{R}}(\vec{x}) = z(C_R(\vec{x}))$. Observe that
$$\begin{aligned} C_{\overline{R}}(\vec{x}) = 1 \quad &\text{iff} \quad C_R(\vec{x}) = 0 \\ &\text{iff} \quad z(C_R(\vec{x})) = 1. \end{aligned}$$
Furthermore,
$$\begin{aligned} C_{\overline{R}}(\vec{x}) = 0 \quad &\text{iff} \quad C_R(\vec{x}) = 1 \\ &\text{iff} \quad z(C_R(\vec{x})) = 0. \end{aligned}$$
Thus, $C_{\overline{R}}(\vec{x}) = z(C_R(\vec{x}))$, which is a composition of (primitive) recursive functions. Hence, $\overline{R}$ is a (primitive) recursive relation.
2. One can easily check that $C_{R\cap Q}(\vec{x}) = C_R(\vec{x}) \times C_Q(\vec{x})$, which is a composition of (primitive) recursive functions. Hence, $R \cap Q$ is a (primitive) recursive relation.
3. First we show that $C_{R\cup Q}(\vec{x}) = 1 \dot{-} z(C_R(\vec{x}) + C_Q(\vec{x}))$. Note that
$$\begin{aligned} C_{R\cup Q}(\vec{x}) = 1 \quad &\text{iff} \quad C_R(\vec{x}) + C_Q(\vec{x}) \geq 1 \\ &\text{iff} \quad z(C_R(\vec{x}) + C_Q(\vec{x})) = 0 \\ &\text{iff} \quad 1 \dot{-} z(C_R(\vec{x}) + C_Q(\vec{x})) = 1. \end{aligned}$$
Moreover,
$$\begin{aligned} C_{R\cup Q}(\vec{x}) = 0 \quad &\text{iff} \quad C_R(\vec{x}) + C_Q(\vec{x}) = 0 \\ &\text{iff} \quad z(C_R(\vec{x}) + C_Q(\vec{x})) = 1 \\ &\text{iff} \quad 1 \dot{-} z(C_R(\vec{x}) + C_Q(\vec{x})) = 0. \end{aligned}$$
Thus, $C_{R\cup Q}(\vec{x}) = 1 \dot{-} z(C_R(\vec{x}) + C_Q(\vec{x}))$, a composition of (primitive) recursive functions. Therefore, $R \cup Q$ is a (primitive) recursive relation. □

By a pertinent application of the projection functions, Theorem 5.3.21 implies that the conjunction and disjunction of any two (primitive) recursive relations are also (primitive) recursive. For example, if $R(x,y)$ and $S(y,z)$ are recursive, then the relation $P(x,y,z)$ defined by "$R(x,y)$ and $S(y,z)$" is also recursive.

From Theorem 5.3.21(2)(3), using a proof by induction, one can now show that any finite intersection or union of recursive sets is also recursive.

**Corollary 5.3.22.** *Let $n \in \mathbb{N}$ and suppose that $R_1, R_2, \ldots, R_n$ are k-place (primitive) recursive relations. Then*

$$\bigcap_{1\le i\le n} R_i \quad \textit{and} \quad \bigcup_{1\le i\le n} R_i$$

*are (primitive) recursive.*

We now apply Theorem 5.3.21 to show that the relations $<$, $>$, and $=$ are primitive recursive.

**Proposition 5.3.23.** *The following relations are primitive recursive:*
1. $\{\langle x,y\rangle : x < y\}$,
2. $\{\langle x,y\rangle : x > y\}$,
3. $\{\langle x,y\rangle : x = y\}$.

*Proof.* We show that the relations are primitive recursive as follows:
1. Theorem 5.3.21(1) implies that $\{\langle x,y\rangle : x < y\}$ is primitive recursive, since $<$ is the complement of the relation $\geq$, which is primitive recursive by Proposition 5.3.19.
2. Theorem 5.3.21(1) implies that $\{\langle x,y\rangle : x > y\}$ is primitive recursive, as $>$ is the complement of the relation $\leq$, which is primitive recursive by Proposition 5.3.17.
3. Clearly, $E = \{\langle x,y\rangle : x = y\}$ is the intersection of $\leq$ and $\geq$. Propositions 5.3.17 and 5.3.19 and Theorem 5.3.21(2) imply that $E$ is primitive recursive. □

We can now prove that any finite set of natural numbers is primitive recursive.

**Proposition 5.3.24.** *Let $A = \{n_1, n_2, \ldots, n_k\}$ be a finite set of natural numbers. Then A is primitive recursive.*

*Proof.* Let $n_i \in A$, where $1 \leq i \leq k$. We first show that the singleton $\{n_i\}$ is primitive recursive. Let ${}^{n_i}f$ be the constant function defined by ${}^{n_i}f(x) = n_i$, for all $x \in \mathbb{N}$. The function ${}^{n_i}f$ is primitive recursive by Proposition 5.3.8. Clearly, $x \in \{n_i\}$ if and only if $x = n_i$. It follows from the substitution rule that $\{n_i\}$ is primitive recursive. To formally establish this, let $I_1^1$ be the 1-place projection function. By Proposition 5.3.23, the relation $x = y$ is primitive recursive. As

$$x \in \{n_i\} \quad \text{iff} \quad I_1^1(x) = {}^{n_i}f(x),$$

the substitution rule (see (5.5)) implies that $\{n_i\}$ is primitive recursive. Hence, since $A = \bigcup_{1\le i\le k}\{n_i\}$, Corollary 5.3.22(2) implies that $A$ is primitive recursive. □

We now show that the graph of a recursive function is itself recursive. The converse also holds (see Exercise 14).

**Proposition 5.3.25.** *Let $f\colon \mathbb{N}^n \to \mathbb{N}$ be a (primitive) recursive function. Then the relation $G = \{\langle \vec{x}, y\rangle : f(\vec{x}) = y\}$, the graph of $f$, is (primitive) recursive.*

*Proof.* Since the relation $=$ is primitive recursive and

$$G(\vec{x}, y) \quad \text{if and only if} \quad f(\vec{x}) = y,$$

it follows from the substitution rule that $G$ is a (primitive) recursive relation. More specifically, since $\vec{x}$ denotes the $n$-tuple $x_1, x_2, \ldots, x_n$, we see that $I_i^{n+1}(\vec{x}, y) = x_i$ for each $1 \le i \le n$ and $I_{n+1}^{n+1}(\vec{x}, y) = y$. For every $\vec{x} \in \mathbb{N}^n$ and $y \in \mathbb{N}$, define

$$g_1(\vec{x}, y) = f(I_1^{n+1}(\vec{x}, y), \ldots, I_n^{n+1}(\vec{x}, y)) \quad \text{and} \quad g_2(\vec{x}, y) = I_{n+1}^{n+1}(\vec{x}, y).$$

Then $g_1$ and $g_2$ are (primitive) recursive, where $g_1(\vec{x}, y) = f(\vec{x})$ and $g_2(\vec{x}, y) = y$. Thus,

$$G(\vec{x}, y) \quad \text{if and only if} \quad g_1(\vec{x}, y) = g_2(\vec{x}, y).$$

As the relation $=$ is primitive recursive, the substitution rule (see (5.5)) implies that the relation $G$ is (primitive) recursive. □

The next theorem gives a condition for which one can define a (primitive) recursive function by cases.

**Theorem 5.3.26.** *Let $Q$ be a $k$-place (primitive) recursive relation. If $f$ and $g$ are $k$-place (primitive) recursive functions, then the $k$-place function $h$ defined by*

$$h(\vec{x}) = \begin{cases} f(\vec{x}), & \text{if } Q(\vec{x}), \\ g(\vec{x}), & \text{if not } Q(\vec{x}) \end{cases}$$

*is also (primitive) recursive.*

*Proof.* Note that

$$h(\vec{x}) = (f(\vec{x}) \times C_Q(\vec{x})) + (g(\vec{x}) \times C_{\overline{Q}}(\vec{x})),$$

which is (primitive) recursive by Propositions 5.3.9 and 5.3.7 and Theorem 5.3.21. □

Theorem 5.3.26 can be extended to more than just two exclusive cases. For example, suppose that $f_1, f_2, f_3, f_4$ are $k$-place (primitive) recursive functions and suppose that $R$ and $Q$ are $k$-place (primitive) recursive relations. Then Theorem 5.3.21 implies that the function $h\colon \mathbb{N}^k \to \mathbb{N}$ defined by

$$h(\vec{x}) = \begin{cases} f_1(\vec{x}), & \text{if } Q(\vec{x}) \text{ and } R(\vec{x}), \\ f_2(\vec{x}), & \text{if } Q(\vec{x}) \text{ and not } R(\vec{x}), \\ f_3(\vec{x}), & \text{if } R(\vec{x}) \text{ and not } Q(\vec{x}), \\ f_4(\vec{x}), & \text{if not } Q(\vec{x}) \text{ and not } R(\vec{x}) \end{cases}$$

is also (primitive) recursive. In the above definition of $h$, no two of the four cases occur at the same time, that is, the cases are *exclusive*. Furthermore, for each $\vec{x}$ exactly one of these cases holds.

As another example, let $f_1, f_2, f_3, f_4$ be $k$-place (primitive) recursive functions and let $Q_1, Q_2, Q_3$ be $k$-place (primitive) recursive relations. Suppose that the relations are exclusive, that is, for any $\vec{x}$, no two of the relations $Q_1(\vec{x})$, $Q_2(\vec{x})$, and $Q_3(\vec{x})$ hold at the same time. Then the function $h\colon \mathbb{N}^k \to \mathbb{N}$ defined by

$$h(\vec{x}) = \begin{cases} f_1(\vec{x}), & \text{if } Q_1(\vec{x}), \\ f_2(\vec{x}), & \text{if } Q_2(\vec{x}), \\ f_3(\vec{x}), & \text{if } Q_3(\vec{x}), \\ f_4(\vec{x}), & \text{if none of the above hold} \end{cases}$$

is (primitive) recursive.

Bounded number quantifiers are very useful when one wants to put some restriction on the numbers being quantified. To say that all natural numbers $x < 9$ satisfy the property $P(x)$, we shall write $(\forall x < 9)P(x)$. Similarly, to say that some natural number $x < 4$ satisfies $P(x)$, we can write $(\exists x < 4)P(x)$.

**Definition 5.3.27** (Bounded number quantifiers). When $a$ is a specific number, we write $(\forall x < a)P(x)$ to mean that *for every natural number $x < a$, $P(x)$ is true.* We also write $(\exists x < a)P(x)$ to assert that *for some natural number $x < a$, $P(x)$ is true.*

In our proof of the next theorem we will be using the 1-place function pos defined by

$$\operatorname{pos}(x) = \begin{cases} 1, & \text{if } x > 0, \\ 0, & \text{if } x = 0. \end{cases} \tag{5.6}$$

In Exercise 3, one is asked to prove that pos is primitive recursive.

**Theorem 5.3.28.** *If $Q$ is a $(k+1)$-place (primitive) recursive relation, then the two following $(k+1)$-place relations are also (primitive) recursive:*
1. $R = \{\langle \vec{x}, y\rangle : (\forall t < y)Q(\vec{x}, t)\}$,
2. $P = \{\langle \vec{x}, y\rangle : (\exists t < y)Q(\vec{x}, t)\}$.

*Proof.* Let $Q$ be a $(k+1)$-place (primitive) recursive relation. Thus, the characteristic function $C_Q$ is (primitive) recursive. The characteristic function of the relation $R$ is

$$C_R(\vec{x}, y) = \prod_{t<y} C_Q(\vec{x}, t),$$

which is (primitive) recursive by Proposition 5.3.15. Furthermore, the characteristic function of the relation $P$ is

$$C_P(\vec{x}, y) = \text{pos}\Big(\sum_{t<y} C_Q(\vec{x}, t)\Big),$$

which is also (primitive) recursive by Exercise 3 and Proposition 5.3.15. □

Theorem 5.3.28 implies that whenever $Q$ is a $(k+1)$-place (primitive) recursive relation and we define a $(k+1)$-place relation $R$ by

$$R(\vec{x}, y) \quad \text{if and only if} \quad (\forall t < y)Q(\vec{x}, t),$$

then $R$ is (primitive) recursive. Similarly, if we define a $(k+1)$-place relation $P$ by

$$P(\vec{x}, y) \quad \text{if and only if} \quad (\exists t < y)Q(\vec{x}, t),$$

we can conclude that $P$ is (primitive) recursive.

**Example 5.3.29.** Show that the 2-place relation $R = \{\langle x, y\rangle : (\exists q < y+1)(x \times q = y)\}$ is primitive recursive.

**Solution.** Observe that the 3-place relation $Q$ defined by

$$Q(x, q, y) \quad \text{if and only if} \quad x \times q = y$$

is primitive recursive by Proposition 5.3.25. Define the 3-place relation $P$ by

$$P(x, y, z) \quad \text{if and only if} \quad (\exists q < z)(x \times q = y).$$

Theorem 5.3.28(2) implies that $P$ is primitive recursive. Now let $g_1(x, y) = I_1^2(x, y)$, $g_2(x, y) = I_2^2(x, y)$, and $g_3(x, y) = S(I_2^2(x, y))$. Clearly, $g_1$, $g_2$, and $g_3$ are primitive recursive functions. By the substitution rule (see page 181) the relation

$$P(g_1(x, y), g_2(x, y), g_3(x, y))$$

is primitive recursive. Observe that

$$\begin{aligned} P(g_1(x, y), g_2(x, y), g_3(x, y)) \quad &\text{iff} \quad P(x, y, y+1) \\ &\text{iff} \quad (\exists q < y+1)(x \times q = y) \\ &\text{iff} \quad R(x, y). \end{aligned}$$

Therefore, the relation $R$ is primitive recursive.

**Proposition 5.3.30.** *The divisibility relation $x \mid y$ is primitive recursive.*

*Proof.* Since $x \mid y$ if and only if $x \times q = y$ for some $q$, we must show that the relation

$$\{\langle x,y\rangle : x \times q = y \text{ for some } q\}$$

is primitive recursive. To verify this, observe that

$$\begin{aligned} x \mid y \quad &\text{iff} \quad \exists q(x \times q = y) \\ &\text{iff} \quad (\exists q \le y)(x \times q = y) \\ &\text{iff} \quad (\exists q < y+1)(x \times q = y). \end{aligned}$$

Example 5.3.29 now implies that the divisibility relation $x \mid y$ is primitive recursive. □

**Remark 5.3.31.** It now follows that if we define a relation using bounded quantifiers, constants, (primitive) recursive relations, (primitive) recursive functions, and the three logical connectives "and", "or", and "not", then the relation is (primitive) recursive.

**Proposition 5.3.32.** *The set* $\{2,3,5,7,\dots\}$ *of prime numbers is primitive recursive.*

*Proof.* We show that the 1-place relation $\{x \in \mathbb{N} : x \text{ is a prime}\}$ is primitive recursive. Observe that

$$x \text{ is a prime} \quad \text{iff} \quad 1 < x \text{ and } (\forall a < x)(\forall b < x)(a \times b \ne x),$$

where the relations $<$ and $\ne$ are primitive recursive, and the function $\times$ is also primitive recursive. Therefore, the set of primes forms a primitive recursive set. □

### 5.3.1 Bounded search

The total search operator (see Definition 5.3.5), also called the $\mu$-operator, provides a method for defining a function whose value is the least number that satisfies a particular condition. We will now define a search operation that will only perform a search when there is an upper bound on the number of the searches allowed, that is, the search will terminate after looking at finitely many cases.

**Definition 5.3.33.** Let $R$ be a $(k+1)$-place relation on $\mathbb{N}$. For each $\vec{x} \in \mathbb{N}^k$, define the number $(\mu t < y)R(\vec{x},t)$ by

$$(\mu t < y)R(\vec{x},t) = \begin{cases} \text{the least } t \text{ that satisfies } t < y \text{ and } R(\vec{x},t), \\ y, \quad \text{if no such } t \text{ exists.} \end{cases}$$

Thus, for any $(k+1)$-place relation $R$ on $\mathbb{N}$, we can define the following total $(k+1)$-place function $f\colon \mathbb{N}^{k+1} \to \mathbb{N}$ by

$$f(\vec{x},y) = (\mu t < y)R(\vec{x},t).$$

The function $f$ is said to be defined by *bounded minimization* or *bounded search*. Observe that $f(\vec{x},y) = y$ if and only if there is no $t < y$ that satisfies $R(\vec{x},t)$.

For example, consider the 2-place relation "$t$ is a prime and $t > x$." So we can define the 2-place function $f\colon \mathbb{N}^2 \to \mathbb{N}$ by

$$f(x,y) = (\mu t < y)[t \text{ is a prime and } t > x].$$

To illustrate how to evaluate this function, we obtain

$$\begin{aligned} f(4,9) &= (\mu t < 9)[t \text{ is a prime and } t > 4] = 5, \\ f(4,20) &= (\mu t < 20)[t \text{ is a prime and } t > 4] = 5, \\ f(9,4) &= (\mu t < 4)[t \text{ is a prime and } t > 9] = 4, \\ f(9,0) &= (\mu t < 0)[t \text{ is a prime and } t > 9] = 0. \end{aligned}$$

**Theorem 5.3.34.** *If $R$ is a (primitive) recursive relation, then the function*

$$f(\vec{x},y) = (\mu t < y)R(\vec{x},t)$$

*is (primitive) recursive.*

*Proof.* We will show that $f$ can be defined by primitive recursion. Observe that:

(1) $f(\vec{x},0) = 0,$

(2) $f(\vec{x},n+1) = \begin{cases} f(\vec{x},n), & \text{if } f(\vec{x},n) < n, \\ n, & \text{if } f(\vec{x},n) = n \text{ and } R(\vec{x},n), \\ n+1, & \text{if } f(\vec{x},n) = n \text{ and not } R(\vec{x},n), \end{cases}$ for all $n \in \mathbb{N}$.

Thus, we get the following tree, where $g$ is (primitive) recursive (see Proposition 5.3.23, Remark 5.3.31, and Theorem 5.3.26).

$$f$$

$$f(\vec{x},0) = 0 = \mathring{f}(\vec{x})$$

$$w = f(\vec{x},n)$$

$$f(\vec{x},n+1) = \begin{cases} w, & \text{if } w < n, \\ n, & \text{if } w = n \,\&\, R(\vec{x},n), \\ n+1, & \text{if } w = n \,\&\, \neg R(\vec{x},n) \end{cases} = g(w,\vec{x},n)$$

□

Euclid proved that there are infinitely many prime numbers. So the function $h\colon \mathbb{N} \to \mathbb{N}$ defined in our next theorem is a total function.

**Theorem 5.3.35.** *Let $h\colon \mathbb{N} \to \mathbb{N}$ be defined by*

$$h(x) = \text{the smallest prime number that is strictly larger than } x. \tag{5.7}$$

*Then h is primitive recursive.*

*Proof.* First we prove that for every natural number $x$ there is a prime number $p$ such that $x < p \le x! + 1$. Note that every prime number $q \le x$ evenly divides $x!$. We know that every natural number greater than 1 is divisible by a prime. Since $x! + 1 > 1$, the natural number $x! + 1$ is divisible by a prime $p$. Because $p$ evenly divides $x! + 1$, it follows that $p$ does not evenly divide $x!$. Therefore, $x < p \le x! + 1$. It now follows that

$$h(x) = (\mu t < (x! + 2))[t \text{ is a prime and } t > x]. \tag{5.8}$$

The predicate

$$t \text{ is a prime and } t > x$$

is primitive recursive by Propositions 5.3.32 and 5.3.23 and Theorem 5.3.21(2). Also, the function $x \mapsto x!$ is primitive recursive by Proposition 5.3.12. So $f\colon \mathbb{N} \to \mathbb{N}$ defined by $f(x) = x! + 2$ is primitive recursive. Theorem 5.3.34 implies that the function

$$\langle x, y\rangle \mapsto (\mu t < y)[t \text{ is a prime and } t > x]$$

is primitive recursive. Therefore, by composition, the function $h$ satisfying (5.8) is also primitive recursive. □

For each $x \in \mathbb{N}$, let $p_x$ be the $(x + 1)$-st prime number. Thus,

$$p_0 = 2, \quad p_1 = 3, \quad p_2 = 5, \quad p_3 = 7, \quad p_4 = 11, \ \ldots, \ p_{25} = 101, \ \ldots.$$

One can easily prove, by induction, that $p_x > x + 1$ for all $x \in \mathbb{N}$.

**Proposition 5.3.36.** *The function $x \mapsto p_x$ is primitive recursive.*

*Proof.* Let $g$ be defined by $g(x) = p_x$. Now let $h$ be the primitive recursive function in Theorem 5.3.35. Then $g$ can be defined by primitive recursion as follows:
(1) $g(0) = 2$,
(2) $g(n + 1) = h(g(n))$, for all $n \in \mathbb{N}$.

Thus, the function $g$ is primitive recursive. □

An important feature of the natural numbers is that one can code a finite sequence of natural numbers by a single natural number. The fundamental theorem of arithmetic (see Theorem 1.1.29 on page 11) states that every natural number $x$ has a unique prime factorization. This theorem allows us to encode any finite tuple of natural numbers by a single natural number using the following bracket notation:

$$\begin{aligned}[\,] &= 1,\\ [x,y] &= 2^{x+1}3^{y+1},\\ [x,y,z] &= 2^{x+1}3^{y+1}5^{z+1},\\ &\vdots\\ [x_0,x_1,\ldots,x_k] &= 2^{x_0+1}3^{x_1+1}5^{x_2+1}\cdots p_k^{x_k+1},\end{aligned} \tag{5.9}$$

where $[\,]$ encodes the "empty" tuple and $p_k$ denotes the $(k+1)$-st prime. For example, $[2,1] = 72$ and $[2,1,0] = 360$. Let $S$ be the set of all finite sequences (tuples) of natural numbers. Then the function $h\colon S \to \mathbb{N}$ defined by $h(\langle x_0,x_1,\ldots,x_k\rangle) = [x_0,x_1,\ldots,x_k]$ is one-to-one; however, $h$ is not onto since $10 = 2\times 5$ is not a value of this function.

**Theorem 5.3.37.** *Let $k$ be a natural number. The function $h\colon \mathbb{N}^{k+1} \to \mathbb{N}$ defined by*

$$h(x_0,x_1,\ldots,x_k) = [x_0,x_1,\ldots,x_k] \tag{5.10}$$

*is primitive recursive.*

*Proof.* Let $k \in \mathbb{N}$. As $h(x_0,x_1,\ldots,x_k) = 2^{x_0+1}3^{x_1+1}5^{x_2+1}\cdots p_k^{x_k+1}$, it follows from Proposition 5.3.15 that $h$ is primitive recursive because exponentiation is primitive recursive (see Propositions 5.3.11). □

Let $s$ be a natural number which codes a sequence, that is, $s = [a_0,a_1,a_2,\ldots,a_k]$ for some finite $(k+1)$-tuple of natural numbers $a_0,a_1,a_2,\ldots,a_k$. We will next show that there is a primitive recursive "decoding" function $\langle s,i\rangle \mapsto (s)_i$ such that $(s)_i = a_i$ for each $i \le k$. For example, since $[2,1] = 72$, we have $(72)_0 = 2$ and $(72)_1 = 1$.

**Proposition 5.3.38.** *A primitive recursive decoding function $\langle s,i\rangle \mapsto (s)_i$ exists.*

*Proof.* For a prime number $q$, the largest exponent $e$ such that $q^e \mid s$ is also the least natural number $k$ such that $q^{k+1} \nmid s$, that is, $e = \mu k(q^{k+1} \nmid s)$. Since $q \ge 2$ and $q^e \mid s$, we have $e < q^e \le s$. Hence, $e = (\mu k < s)(q^{k+1} \nmid s)$. Thus, the exponent of $q$ in the prime factorization of $s$ is equal to

$$(\mu k < s)(q^{k+1} \nmid s).$$

When $q = p_i$, the $(i+1)$-st prime, define

$$(s)_i^* = (\mu k < s)(p_i^{k+1} \nmid s),$$

which is the largest exponent of $p_i$ in the prime factorization of $s$. For our decoding function we need one less than the exponent of the prime $p_i$ in the prime factorization of $s$. Thus, we have

$$(s)_i = (s)_i^* \mathbin{\dot{-}} 1 = (\mu k < s)(p_i^{k+1} \nmid s) \mathbin{\dot{-}} 1,$$

which is primitive recursive since the far right hand side of the above equations involves the composition of primitive recursive functions (see Theorems 5.3.34 and 5.3.21(1) and Propositions 5.3.36 and 5.3.14). □

Note that $(s)_i$ is defined for all $s, i \in \mathbb{N}$, even when $s$ does not code a sequence.

**Definition 5.3.39.** A natural number $s$ is called a *sequence number* if $s = [\,]$ or $s = [a_0, a_1, a_2, \ldots, a_k]$ for some $(k+1)$-tuple of natural numbers $a_0, a_1, a_2, \ldots, a_k$.

For example, 1 and 108 are sequence numbers because $1 = [\,]$ and $[1,2] = 2^2 3^3 = 108$, but 10 is not a sequence number.

**Proposition 5.3.40.** *The set of sequence numbers is primitive recursive.*

*Proof.* See Exercise 11. □

One can prove (by induction) that $n + 1 < p_n$ for all natural numbers $n$. Let $x \in \mathbb{N}$ and let $k$ be the smallest such that $p_k \nmid x$. So $x \geq 1$. If $k = 0$, then $k < x$. If $k > 0$, then $p_{k-1} \mid x$. Thus, $p_{k-1} \leq x$ and hence, $k = (k-1) + 1 < p_{k-1} \leq x$, that is, $k < x$. Therefore, if $k$ is the smallest such that $p_k \nmid x$, then $k < x$. This justifies the upper bound of $x$ on the $\mu$-operator in the following proposition.

**Proposition 5.3.41.** *The function* $lh(x) = (\mu k < x)(p_k \nmid x)$ *is primitive recursive.*

*Proof.* This holds because $\mathrm{lh}(x) = (\mu k < x)(p_k \nmid x)$ and the right hand side of this equation is primitive recursive. □

The function $\mathrm{lh}(x)$ is called the *length function* and gives the *length* of a sequence coded by $x$; for example, since $360 = 2^3 \times 3^2 \times 5 = [2, 1, 0]$ and $p_3 = 7$ is the least prime that does not divide 360, we have $\mathrm{lh}(360) = 3$. In general, if $x = [a_0, a_1, a_2, \ldots, a_k]$, then $\mathrm{lh}(x) = k + 1$. In addition, $(x)_{\mathrm{lh}(x) \dot{-} 1} = a_k$ is the *last* component of the sequence. We note that $\mathrm{lh}(0) = 0$ by Definition 5.3.33.

Let $x = [a_0, a_1, a_2, \ldots, a_k]$ be a sequence number and let $y \leq k + 1$. Consider the function $\langle x, y \rangle \mapsto x \upharpoonright y$ defined by

$$x \upharpoonright y = [a_0, a_1, a_2, \ldots, a_{y-1}].$$

We say that $x \upharpoonright y$ is the *restriction* of $x$ to $y$ and gives us the code for the sequence consisting of the first $y$ components of the sequence coded by $x$.

**Proposition 5.3.42.** *The function* $\langle x, y \rangle \mapsto x \upharpoonright y$ *is primitive recursive.*

*Proof.* This holds because

$$x \upharpoonright y = \prod_{t<y} p_t^{(x)_t^*} \tag{5.11}$$

and the right hand side of (5.11) is a composition of primitive recursive functions. □

For example, if $x = [a_0, a_1, \ldots, a_k]$ and $y \leq k+1$, then

$$\begin{aligned} x \upharpoonright y = [a_0, a_1, \ldots, a_{y-1}] &= p_0^{a_0+1} p_1^{a_1+1} \cdots p_{y-1}^{a_{y-1}+1} \\ &= p_0^{(x)_0^*} p_1^{(x)_1^*} \cdots p_{y-1}^{(x)_{y-1}^*} = \prod_{t<y} p_t^{(x)_t^*}. \end{aligned}$$

Furthermore, if $k = \mathrm{lh}(x) \dot{-} 1$, then $x \upharpoonright k = [a_0, a_1, a_2, \ldots, a_{k-1}]$ results in the deletion of the last component in the sequence coded by $x$.

Let $x = [a_0, a_1, a_2, \ldots, a_k]$ and $y = [b_0, b_1, b_2, \ldots, b_\ell]$ be two sequence numbers. Consider the function $\langle x, y\rangle \mapsto x * y$ defined by

$$x * y = [a_0, a_1, a_2, \ldots, a_k, b_0, b_1, b_2, \ldots, b_\ell].$$

We say that $x * y$ is the *concatenation* of $x$ to $y$.

**Proposition 5.3.43.** *The concatenation function $\langle x, y\rangle \mapsto x * y$ is primitive recursive.*

*Proof.* We must show that there is a 2-place primitive recursive function such that if $x$ and $y$ are sequence numbers, then $\langle x, y\rangle \mapsto x * y$. This is done by defining

$$x * y = x \times \prod_{t<\mathrm{lh}(y)} p_{t+\mathrm{lh}(x)}^{(y)_t^*},$$

which is a composition of primitive recursive functions. □

Since $[2, 1] = 2^3 \times 3^2 = 72$,

$$72 * 72 = [2, 1] * [2, 1] = [2, 1, 2, 1] = 72 \times 5^3 \times 7^2 = 441,000.$$

Moreover, if $x$ is a sequence number and $i \in \mathbb{N}$, then $x * [i]$ is the sequence number of the sequence obtained by adjoining $i$ at the end of the sequence coded by $x$. We note that $x * y$ is defined for all $x$ and $y$ in $\mathbb{N}$, even if $x$ and $y$ are not sequence numbers.

We can define the operation of concatenation of more than two sequence numbers. Suppose that $x_0, x_1, \ldots x_{k-1}$ are sequence numbers. Then we can concatenate all of the sequences coded by $x_0, x_1, \ldots x_{k-1}$, denoted by $\mathop{\Huge *}_{t<k} x_t$, as follows:

$$\mathop{\Huge *}_{t<k} x_t = x_0 * x_1 * \cdots * x_{k-1},$$

where on the right hand side of the above equation the operation $*$ is associative.

**Proposition 5.3.44.** *Let $f$ be a (primitive) recursive $(k+1)$-place function. Then the function $\langle \vec{x}, y\rangle \mapsto \mathop{\ast}_{t<y} f(\vec{x}, t)$ is (primitive) recursive.*

*Proof.* Note that the function $\langle \vec{x}, y\rangle \mapsto \mathop{\ast}_{t<y} f(\vec{x}, t)$ can be defined by primitive recursion as follows:

(1) $\mathop{\ast}_{t<0} f(\vec{x}, t) = 1$,

(2) $\mathop{\ast}_{t<(n+1)} f(\vec{x}, t) = (\mathop{\ast}_{t<n} f(\vec{x}, t)) * f(x, n)$, for all $n \in \mathbb{N}$.

Therefore, the function $\langle \vec{x}, y\rangle \mapsto \mathop{\ast}_{t<y} f(\vec{x}, t)$ is (primitive) recursive. □

For any $(k+1)$-place function $f$, we define a new $(k+1)$-place function $\overline{f}$ by

$$\overline{f}(\vec{x}, y) = [f(\vec{x}, 0), f(\vec{x}, 1), \ldots, f(\vec{x}, y-1)] = \prod_{t<y} p_t^{f(\vec{x},t)+1}.$$

So the function $\overline{f}$ encodes the first $y$ values $f(\vec{x}, 0), f(\vec{x}, 1), \ldots, f(\vec{x}, y-1)$ of $f$ as a single natural number. For example, $\overline{f}(\vec{x}, 0) = [\,] = 1$ and $\overline{f}(\vec{x}, 2) = [f(\vec{x}, 0), f(\vec{x}, 1)]$. Clearly, $\overline{f}(\vec{x}, y)$ is always a sequence number of length $y$.

**Proposition 5.3.45.** *Let $f$ be a (primitive) recursive $(k+1)$-place function. Then the $(k+1)$-place function $\overline{f}$ is (primitive) recursive.*

*Proof.* Since $\overline{f}(\vec{x}, y) = \prod_{t<y} p_t^{f(\vec{x},t)+1}$, we see that the right hand side of this equation is a composition of (primitive) recursive functions. □

Given any $(k+2)$-place function $g$, there exists a unique $(k+1)$-place function $h$ that satisfies the equation

$$h(\vec{x}, y) = g(\overline{h}(\vec{x}, y), \vec{x}, y),$$

where we must first know the value of

$$\overline{h}(\vec{x}, y) = [h(\vec{x}, 0), h(\vec{x}, 1), \ldots, h(\vec{x}, y-1)]$$

before we can evaluate $h(\vec{x}, y)$. For example,

$$\begin{aligned}
h(\vec{x}, 0) &= g([\,], \vec{x}, 0),\\
h(\vec{x}, 1) &= g([h(\vec{x}, 0)], \vec{x}, 1),\\
h(\vec{x}, 2) &= g([h(\vec{x}, 0), h(\vec{x}, 1)], \vec{x}, 2), \qquad (\blacktriangle)\\
&\ \ \vdots\\
h(\vec{x}, n) &= g([h(\vec{x}, 0), h(\vec{x}, 1), \ldots, h(\vec{x}, n-1)], \vec{x}, n),\\
h(\vec{x}, n+1) &= g([h(\vec{x}, 0), h(\vec{x}, 1), \ldots, h(\vec{x}, n-1), h(\vec{x}, n)], \vec{x}, n+1).
\end{aligned}$$

The definition of a function by primitive recursion allows one to define the value of a function $h(\vec{x}, n + 1)$ in terms of its preceding value $h(\vec{x}, n)$ (see Definition 5.2.14). The above (▲) clearly illustrates how to define the value of a function in terms of all its preceding values, namely, by coding the sequence of preceding values as a sequence number. In computability theory, a course-of-values recursion is a technique for defining number-theoretic functions by recursion.

**Proposition 5.3.46** (Course of values recursion). *Let $g$ be a (primitive) recursive $(k + 2)$-place function. Then the $(k + 1)$-place function $h$ that satisfies*

$$h(\vec{x}, n) = g(\overline{h}(\vec{x}, n), \vec{x}, n), \textit{ for all } \vec{x} \textit{ and } n,$$

*is (primitive) recursive.*

*Proof.* To prove that $h$ is (primitive) recursive, we first show (paradoxically) that $\overline{h}$ is (primitive) recursive. Note that $\overline{h}$ satisfies the primitive recursion:
(1) $\overline{h}(\vec{x}, 0) = 1$,
(2) $\overline{h}(\vec{x}, n + 1) = \overline{h}(\vec{x}, n) * [g(\overline{h}(\vec{x}, n)), \vec{x}, n)]$, for all $n \in \mathbb{N}$.

Thus, $\overline{h}$ is (primitive) recursive. Since $(\overline{h}(\vec{x}, n+1))_n = h(\vec{x}, n)$, we conclude that $h$ is (primitive) recursive. From (2), we also conclude that

$$(\overline{h}(\vec{x}, n + 1))_n = (\overline{h}(\vec{x}, n) * [g(\overline{h}(\vec{x}, n)), \vec{x}, n)])_n.$$

Thus, $h(\vec{x}, n) = g(\overline{h}(\vec{x}, y), \vec{x}, n)$. □

We end this section by showing how the total search operation can be used to construct a recursive function.

**Proposition 5.3.47.** *Let $R \subseteq \mathbb{N}^{k+1}$ be a recursive relation such that for all $\vec{x} \in \mathbb{N}^k$ there is a $p \in \mathbb{N}$ such that $R(\vec{x}, p)$. Then the function $f\colon \mathbb{N}^k \to \mathbb{N}$ defined by*

$$f(\vec{x}) = \textit{the least } p \in \mathbb{N} \textit{ such that } R(\vec{x}, p) \tag{5.12}$$

*is recursive and $R(\vec{x}, f(\vec{x}))$ for all $\vec{x} \in \mathbb{N}^k$.*

*Proof.* Since for all $\vec{x} \in \mathbb{N}^k$ there exists a $p \in \mathbb{N}$ such that $R(\vec{x}, p)$, the function $f\colon \mathbb{N}^k \to \mathbb{N}$ defined by (5.12) satisfies $f(\vec{x}) = \mu p(1 \dot{-} C_R(\vec{x}, p) = 0)$. Therefore, $f$ is recursive and $R(\vec{x}, f(\vec{x}))$ for all $\vec{x} \in \mathbb{N}^k$. □

**Exercises 5.3.**

1. Show that the function $f\colon \mathbb{N} \to \mathbb{N}$ defined by $f(x) = x^2$ is primitive recursive.
2. Show that the function $f\colon \mathbb{N} \to \mathbb{N}$ defined by

$$f(n) = \begin{cases} 1, & \text{if } n \text{ is even,} \\ 0, & \text{if } n \text{ is odd} \end{cases}$$

is primitive recursive.

*3. Show that the function pos defined by (5.6) is primitive recursive.

4. Let $f: \mathbb{N} \to \mathbb{N}$ and the 2-place relation $Q$ be (primitive) recursive. Show, as in Example 5.3.29, that $R = \{\langle x, y\rangle : (\exists q < f(y))Q(x, q)\}$ is (primitive) recursive.

5. Show that the set $\{i^2 : i \in \mathbb{N}\}$ is primitive recursive.

6. Show that the set $\{2^i : i \in \mathbb{N}\}$ is primitive recursive.

7. Using definition by cases, show that the following functions $f: \mathbb{N}^2 \to \mathbb{N}$ are primitive recursive:
   (a) $f(x, y) = |x - y|$,
   (b) $f(x, y) = \max\{x, y\}$,
   (c) $f(x, y) = \min\{x, y\}$.

8. Find a nontotal 2-place function $g$ such that the function $h$ defined by

$$h(x) = \mu y(g(x, y) = 0),$$

that is,

$$h(x) = \text{the least } y \text{ such that } g(x, y) = 0,$$

is a total function.

9. For each $i \in \mathbb{N}$, by Proposition 5.3.38, the function $t \mapsto (t)_i$ is primitive recursive. Evaluate $(24)_0$, $(24)_1$, $(45)_0$, $(45)_1$, $(45)_2$, $(23)_0$, and $(23)_1$.

10. Define $m: \mathbb{N}^{k+1} \to \mathbb{N}$ by $m(\vec{x}, n) = \max\{f(\vec{x}, i) : 0 \leq i \leq n\}$, where $f: \mathbb{N}^{k+1} \to \mathbb{N}$ is (primitive) recursive. Show that $m$ is (primitive) recursive.

*11. Prove Proposition 5.3.40.

12. For each $n \in \mathbb{N}$, let $A_n$ be a recursive subset of $\mathbb{N}$. Show that $\bigcap_{i=0}^{n} A_i$ and $\bigcup_{i=0}^{n} A_i$ are recursive, for all $n \in \mathbb{N}$.

13. Let $g: \mathbb{N} \to \mathbb{N}$ be (primitive) recursive. Let $g^{(0)}(i) = i$ and whenever $p \geq 1$, let

$$g^{(p)} = \underbrace{g \circ g \circ \cdots \circ g}_{p\text{-times}}.$$

Define $f: \mathbb{N}^2 \to \mathbb{N}$ by $f(i, n) = g^{(n)}(i)$. Show that $f$ is (primitive) recursive.

*14. Let $f: \mathbb{N}^k \to \mathbb{N}$ be a total function. Show that if the graph of $f$ is recursive, then $f$ is recursive.

## 5.4 Recursively enumerable sets and relations

Some sets of natural numbers are recursive and some are *almost* recursive. Recall that a recursive nonempty set $A \subseteq \mathbb{N}$ is one whose characteristic function $C_A$ is recursive. Using $C_A$, one can effectively enumerate the elements of $A$. Let $k_0 \in A$ and define the function $f: \mathbb{N} \to \mathbb{N}$ by

$$f(n) = \begin{cases} n, & \text{if } C_A(n) = 1, \\ k_0, & \text{if } C_A(n) = 0. \end{cases}$$

Clearly, $f$ is a recursive function and the range of $f$ is equal to $A$. Thus, we can list all the values of $f$ as follows: $f(0), f(1), f(2), \ldots$. This listing also effectively enumerates all of the elements of $A$. Theorem 5.3.21 implies that $\mathbb{N} \setminus A$ is recursive. Thus, in a similar manner, we can effectively enumerate the elements of $\mathbb{N} \setminus A$ (if nonempty). This motivates our next definition.

**Definition 5.4.1.** A set $A \subseteq \mathbb{N}$ is *recursively enumerable* if and only if $A = \varnothing$ or there is a recursive function $f : \mathbb{N} \to \mathbb{N}$ such that $A$ is the range of $f$.

The function $f$ in Definition 5.4.1 is not required to be one-to-one. However, Exercise 7 shows that every infinite recursively enumerable subset of $\mathbb{N}$ is the range of a one-to-one recursive function.

A recursively enumerable set of natural numbers can be viewed as being "almost recursive." A recursively enumerable set is also said to be *semi-recursive*. There is an alternative interpretation of a set being recursively enumerable.

**Proposition 5.4.2.** *A set $A \subseteq \mathbb{N}$ is recursively enumerable if and only if there is a 2-place recursive relation $R$ such that $A = \{n \in \mathbb{N} : \exists p R(n, p)\}$.*

*Proof.* Let $A \subseteq \mathbb{N}$.

($\Rightarrow$). Assume that $A$ is recursively enumerable. Let $f : \mathbb{N} \to \mathbb{N}$ be a recursive function such that $A$ equals the range of $f$. Consider the relation

$$R = \{\langle n, p\rangle : n = f(p)\}.$$

Since the relation $=$ is primitive recursive, the substitution rule (see (5.5) on page 181) implies that the relation $R$ is recursive. Clearly, $A = \{n \in \mathbb{N} : \exists p R(n, p)\}$.

($\Leftarrow$). Assume that $A = \{n \in \mathbb{N} : \exists p R(n, p)\}$ for some recursive relation $R$. If $A = \varnothing$, then we are done. So let $k \in A$. By Proposition 5.3.38, the "decoding" operation $(s)_i$ is defined for all $s, i \in \mathbb{N}$. Let $f : \mathbb{N} \to \mathbb{N}$ be defined by

$$f(n) = \begin{cases} (n)_0, & \text{if } R((n)_0, (n)_1), \\ k, & \text{if not } R((n)_0, (n)_1). \end{cases}$$

Then $f$ is recursive (see Proposition 5.3.38) and the range of $f$ is $A$. □

Proposition 5.4.2 allows us to broaden the definition of recursively enumerable.

**Definition 5.4.3.** A relation $P \subseteq \mathbb{N}^k$ is *recursively enumerable* if and only if there is a $(k + 1)$-place recursive relation $R$ such that

$$P(\vec{x}) \quad \text{iff} \quad \exists p R(\vec{x}, p)$$

for all $\vec{x} \in \mathbb{N}^k$.

Our goal now is to establish a surprising result: Every total partial recursive function is, in fact, a recursive function (see Theorem 5.4.12).

**Theorem 5.4.4** (Selection theorem). *Let $S \subseteq \mathbb{N}^{k+1}$ be recursively enumerable and suppose that for all $\vec{x} \in \mathbb{N}^k$ there exists an $i \in \mathbb{N}$ such that $S(\vec{x}, i)$. Then there is a recursive function $f: \mathbb{N}^k \to \mathbb{N}$ such that $S(\vec{x}, f(\vec{x}))$ for all $\vec{x} \in \mathbb{N}^k$.*

*Proof.* Given that $S$ is recursively enumerable, there is a recursive relation $R$ such that

$$S(\vec{x}, i) \quad \text{iff} \quad \exists j R(\vec{x}, i, j).$$

Suppose that for all $\vec{x} \in \mathbb{N}^k$ there exists an $i \in \mathbb{N}$ such that $S(\vec{x}, i)$. Thus, for all $\vec{x} \in \mathbb{N}^k$ there are $i, j \in \mathbb{N}$ such that $R(\vec{x}, i, j)$. Define $h: \mathbb{N}^k \to \mathbb{N}$ by

$$h(\vec{x}) = \text{the least } p \in \mathbb{N} \text{ such that } R(\vec{x}, (p)_0, (p)_1).$$

Note that $h(\vec{x}) = \mu p(1 \dot{-} C_R(\vec{x}, (p)_0, (p)_1) = 0)$ and this is obtained from the recursive function $g(\vec{x}, p) = 1 \dot{-} C_R(\vec{x}, (p)_0, (p)_1)$ by a *total* search (Definition 5.3.5). So $h$ is recursive. Define $f: \mathbb{N}^k \to \mathbb{N}$ by $f(\vec{x}) = (h(\vec{x}))_0$. Thus, $f$ is recursive and $S(\vec{x}, f(\vec{x}))$ for all $\vec{x} \in \mathbb{N}^k$. □

We now show that the set of recursively enumerable relations is closed under recursive substitutions, conjunction, and disjunction.

**Lemma 5.4.5.** *Suppose that $P$ is an $n$-place recursively enumerable relation and that $g_1, g_2, \ldots, g_n$ are $k$-place recursive functions. Then the $k$-place relation $Q$ defined by*

$$Q(\vec{x}) \quad \textit{iff} \quad P(g_1(\vec{x}), g_2(\vec{x}), \ldots, g_n(\vec{x}))$$

*is recursively enumerable.*

*Proof.* Given that $P$ is recursively enumerable, there is an $(n+1)$-place recursive relation $R$ such that $P(x_1, \ldots, x_n)$ iff $\exists i R(x_1, \ldots, x_n, i)$. Thus, for all $\vec{x} \in \mathbb{N}^k$, we have

$$Q(\vec{x}) \quad \text{iff} \quad P(g_1(\vec{x}), \ldots, g_n(\vec{x})) \quad \text{iff} \quad \exists i R(g_1(\vec{x}), \ldots, g_n(\vec{x}), i).$$

By the substitution rule, the relation $R(g_1(\vec{x}), \ldots, g_n(\vec{x}), i)$ is recursive. Therefore, $Q$ is recursively enumerable. □

**Lemma 5.4.6.** *Let $R \subseteq \mathbb{N}^k$ and $Q \subseteq \mathbb{N}^k$.*
1. *If $R$ is a recursive relation, then $R$ is recursively enumerable.*
2. *If $R$ and $Q$ are recursively enumerable relations, then the following relations are also recursively enumerable:*

(a) $R \cap Q = \{\vec{x} \in \mathbb{N}^k : R(\vec{x}) \text{ and } Q(\vec{x})\}$,
(b) $R \cup Q = \{\vec{x} \in \mathbb{N}^k : R(\vec{x}) \text{ or } Q(\vec{x})\}$.

*Proof.* Suppose that $R \subseteq \mathbb{N}^k$ and $Q \subseteq \mathbb{N}^k$.

1. Let $R$ be a recursive relation. The relation $S \subseteq \mathbb{N}^{k+1}$ defined by $S(\vec{x}, p)$ iff $R(\vec{x})$ is recursive, as $S$ is the result of a composition of $R$ with the projection functions $I_i^{k+1}(x_1, \ldots, x_k, p) = x_i$, where $1 \leq i \leq k$. Clearly, $R(\vec{x})$ iff $\exists p S(\vec{x}, p)$. So $R$ is recursively enumerable.
2. Assume that $R$ and $Q$ are recursively enumerable. Therefore, there are recursive relations $S \subseteq \mathbb{N}^{k+1}$ and $U \subseteq \mathbb{N}^{k+1}$ such that

$$(R(\vec{x}) \quad \text{iff} \quad \exists p S(\vec{x}, p)) \quad \text{and} \quad (Q(\vec{x}) \quad \text{iff} \quad \exists p U(\vec{x}, p)).$$

Thus,

$$\begin{aligned} (R(\vec{x}) \text{ and } S(\vec{x})) \quad &\text{iff} \quad \exists p(S(\vec{x}, (p)_0) \text{ and } U(\vec{x}, (p)_1)), \\ (R(\vec{x}) \text{ or } S(\vec{x})) \quad &\text{iff} \quad \exists p(S(\vec{x}, p) \text{ or } U(\vec{x}, p)). \end{aligned}$$

Therefore, Theorem 5.3.21 and the substitution rule imply that the conjunction "$R(\vec{x})$ and $S(\vec{x})$" and disjunction "$R(\vec{x})$ or $S(\vec{x})$" are recursively enumerable. □

By a suitable use of the projection functions, part 2 of Lemma 5.4.6 implies that the conjunction and disjunction of any two recursively enumerable relations are recursively enumerable. For example, if $R(x, y)$ and $S(y, z)$ are recursively enumerable, then the relation $P(x, y, z)$ defined by "$R(x, y)$ and $S(y, z)$" is recursively enumerable. Moreover, by applying a proof by induction, part 2 of Lemma 5.4.6 implies that the conjunction or disjunction of any finite number of recursively enumerable relations is also recursively enumerable.

**Corollary 5.4.7.** *Let $n \in \mathbb{N}$ and suppose that $R_1, R_2, \ldots, R_n$ are k-place recursively enumerable relations. Then*

1. $\bigwedge_{i=1}^n R_i(\vec{x}) = R_1(\vec{x}) \wedge R_2(\vec{x}) \wedge \cdots \wedge R_n(\vec{x})$ *is recursively enumerable,*
2. $\bigvee_{i=1}^n R_i(\vec{x}) = R_1(\vec{x}) \vee R_2(\vec{x}) \vee \cdots \vee R_n(\vec{x})$ *is recursively enumerable.*

We will now show that the set of recursively enumerable relations is closed under the bounded number quantifiers and the existential quantifier.

**Lemma 5.4.8.** *Let $Q \subseteq \mathbb{N}^{k+1}$ be recursively enumerable and let $n \in \mathbb{N}$. Then:*

1. *The relation $(\forall i < n)Q(\vec{x}, i)$ is recursively enumerable.*
2. *The relation $(\exists i < n)Q(\vec{x}, i)$ is recursively enumerable.*
3. *The relation $\exists i Q(\vec{x}, i)$ is recursively enumerable.*

*Proof.* Let $Q \subseteq \mathbb{N}^{k+1}$ be recursively enumerable. Thus, there is a recursive relation $R$ such that

$$Q(\vec{x}, i) \quad \text{iff} \quad \exists j R(\vec{x}, i, j).$$

The desired conclusion follows from the following equivalences:

$$(\forall i < n) Q(\vec{x}, i) \quad \text{iff} \quad (\forall i < n) \exists j R(\vec{x}, i, j) \quad \text{iff} \quad \exists j (\forall i < n) R(\vec{x}, i, (j)_i), \tag{5.13}$$

$$(\exists i < n) Q(\vec{x}, i) \quad \text{iff} \quad (\exists i < n) \exists j R(\vec{x}, i, j) \quad \text{iff} \quad \exists j (\exists i < n) R(\vec{x}, i, j), \tag{5.14}$$

$$\exists i Q(\vec{x}, i) \quad \text{iff} \quad \exists i \exists j R(\vec{x}, i, j) \quad \text{iff} \quad \exists k R(\vec{x}, (k)_0, (k)_1). \tag{5.15}$$

To confirm (5.13), given that for each $i < n$ there is a $j_i \in \mathbb{N}$ such that $R(\vec{x}, i, j_i)$, let $j$ be the sequence number $j = [j_0, j_1, \ldots, j_{n-1}]$. Then $(\forall i < n) R(\vec{x}, i, (j)_i)$. Thus, $\exists j (\forall i < n) R(\vec{x}, i, (j)_i)$. The converse holds similarly. Theorem 5.3.28 implies that the right hand sides of (5.13) and (5.14) are recursively enumerable, and the substitution rule implies that the right hand side of (5.15) is recursively enumerable. □

The set of partial recursive functions can be defined by induction (see Section 1.1.5). Let $B$ be the set consisting only of the zero, successor, and projection functions. Let $\mathcal{F}$ be the set of functional operations that correspond to composition, primitive recursion, and partial search. Then, as in Theorem 1.1.24, we inductively define the following sets of functions:

(1) $C_0 = B$,
(2) $C_{n+1} = C_n \cup \mathcal{F}[C_n]$, for all $n \in \mathbb{N}$.

Then $C = \bigcup_{n \in \mathbb{N}} C_n$ is the set of all the partial recursive functions. Moreover, whenever a set contains the zero, successor, and projection functions and it is closed under composition, primitive recursion, and partial search, then the set contains all of the partial recursive functions (see Exercise 5(b) on page 15). We will apply this observation in the proof of our next theorem. First we give a definition.

**Definition 5.4.9.** Let $h$ be a $k$-place partial recursive function. Then the *graph* of $h$ is the $(k+1)$-place relation

$$G_h = \{(\vec{x}, y) : h(\vec{x}) = y\}.$$

The next theorem shows that there exists a close connection between partial recursive functions and recursively enumerable relations.

**Theorem 5.4.10.** *For every partial recursive function $h$, the graph of $h$ is recursively enumerable.*

*Proof.* Let $S$ be the set of all partial recursive functions whose graph is recursively enumerable. We shall prove by induction that $S$ contains all partial recursive functions.

*Base step:* We must show that every initial function is in $S$. To do this, let $h$ be an initial function. Since $h$ is a recursive function, $G_h$ is recursive by Proposition 5.3.25. Lemma 5.4.6(1) now implies that the relation $G_h$ is recursively enumerable.

*Inductive step:* We must show that $S$ is closed under (1) composition, (2) primitive recursion, and (3) partial search. Let $f, g, g_1, g_2, \ldots, g_n$ be partial recursive functions and assume that all of these functions are in $S$, that is, assume the induction hypothesis

$$G_f, G_g, G_{g_1}, \ldots, G_{g_n} \text{ are recursively enumerable.} \tag{IH}$$

We now establish closure under the above identified operations (1), (2), and (3).

(1) Assume that $f: \mathbb{N}^n \to \mathbb{N}$ and $g_i: \mathbb{N}^k \to \mathbb{N}$ for each $i = 1, 2, \ldots, n$. We must show that function $h: \mathbb{N}^k \to \mathbb{N}$ defined by the composition

$$h(\vec{x}) = f(g_1(\vec{x}), g_2(\vec{x}), \ldots, g_n(\vec{x}))$$

is also in $S$. Since

$$h(\vec{x}) = y \quad \text{iff} \quad \exists v_1 \cdots \exists v_n \left( \bigwedge_{i=1}^{n} g_i(\vec{x}) = v_i \text{ and } f(v_1, v_2, \ldots, v_n) = y \right),$$

(IH), Corollary 5.4.7, and Exercise 11 imply that $G_h$ is recursively enumerable.

(2) Assume that $f: \mathbb{N}^k \to \mathbb{N}$ and $g: \mathbb{N}^{k+2} \to \mathbb{N}$. We must show that the function $h: \mathbb{N}^{k+1} \to \mathbb{N}$ defined by the primitive recursion
(a) $h(\vec{x}, 0) = f(\vec{x})$,
(b) $h(\vec{x}, n+1) = g(h(\vec{x}, n), \vec{x}, n)$, for all $n \in \mathbb{N}$,
is also in $S$. Because

$$h(\vec{x}, n) = y \quad \text{iff} \quad \exists p (f(\vec{x}) = (p)_0 \text{ and } (\forall i < n)(g((p)_i, \vec{x}, i) = (p)_{i+1}) \text{ and } (p)_n = y),$$

Lemmas 5.4.5, 5.4.6, and 5.4.8(1)(2) and (IH) imply that $G_h$ is recursively enumerable.

(3) Assume that $g$ is a $(k+1)$-place function. We must show that the $k$-place function $h$ defined by the partial search

$$h(\vec{x}) = \mu y (g(\vec{x}, y) = 0)$$

is in $S$. Since

$$h(\vec{x}) = y \quad \text{iff} \quad g(\vec{x}, y) = 0 \text{ and } (\forall i < y) \exists v (g(\vec{x}, s) = v \text{ and } v > 0),$$

Lemmas 5.4.6 and 5.4.8(1)(2) and (IH) imply that $G_h$ is recursively enumerable. □

Theorem 5.4.10 and a modification of the proof of Theorem 5.4.4 (see Exercise 5) now imply the following equivalence.

**Theorem 5.4.11.** *Let $h: \mathbb{N}^k \to \mathbb{N}$ be a partial function. Then $h$ is partial recursive if and only if $G_h$ is recursively enumerable.*

Theorems 5.4.10 and 5.4.4 also imply that every total partial recursive function is recursive.

**Theorem 5.4.12.** *Let $h\colon \mathbb{N}^k \to \mathbb{N}$ be a total partial recursive function. Then $h$ is a recursive function.*

*Proof.* Let $h$ be a total partial recursive function. Theorem 5.4.10 implies the relation

$$G_h(\vec{x}, y) \quad \text{iff} \quad h(x) = y \tag{5.16}$$

is recursively enumerable. Since $h$ is total, we know that for all $\vec{x} \in \mathbb{N}^k$, there is a $y$ such that $G_h(\vec{x}, y)$. The selection theorem (Theorem 5.4.4) implies that there is a recursive function $f\colon \mathbb{N}^k \to \mathbb{N}$ such that $G_h(\vec{x}, f(\vec{x}))$ for all $\vec{x} \in \mathbb{N}^k$. Therefore, by (5.16), we conclude that $h(\vec{x}) = f(\vec{x})$ for all $\vec{x} \in \mathbb{N}^k$, that is, $h$ is a recursive function. □

### 5.4.1 Decidability revisited

In Section 5.1.1 we presented a definition of a decidable (semi-decidable) set. However, this definition was given in terms of the intuitive concept of an "effective procedure." Because of Theorems 5.2.11 and 5.2.20 and the Church–Turing thesis, we can now give a mathematically precise definition of a set being decidable (semi-decidable).

**Definition 5.4.13.** Let $S \subseteq \mathbb{N}^k$. Then $S$ is *decidable* if its characteristic function

$$C_S(\vec{x}) = \begin{cases} 1, & \text{if } \vec{x} \in S, \\ 0, & \text{if } \vec{x} \notin S \end{cases}$$

is partial recursive. Moreover, $S$ is *semi-decidable* if its semi-characteristic function

$$c_S(\vec{x}) = \begin{cases} 1, & \text{if } \vec{x} \in S \\ \uparrow, & \text{if } \vec{x} \notin S \end{cases}$$

is partial recursive.

In light of Theorem 5.4.12, the above definition of a set being decidable can be made stronger. Our next result confirms this by showing that a relation is decidable if and only if its characteristic function is recursive.

**Theorem 5.4.14.** *Let $S \subseteq \mathbb{N}^k$. Then $S$ is decidable if and only if $S$ is recursive.*

*Proof.* Assume that $S$ is decidable. Thus, the characteristic function $C_S$ is total and partial recursive. Theorem 5.4.12 implies that $C_S$ is recursive, so $S$ is recursive. Conversely, suppose that $S$ is recursive. Then $C_S$ is recursive and hence, it is partial recursive. Therefore, $S$ is decidable. □

**Theorem 5.4.15.** *Let $S \subseteq \mathbb{N}^k$. Then $S$ is semi-decidable if and only if $S$ is recursively enumerable.*

*Proof.* Assume that $S$ is semi-decidable. So $c_S$ is partial recursive. By Theorem 5.4.10, the graph $G_{c_S}(\vec{x}, y)$ of $c_S$ is recursively enumerable. Since

$$\vec{x} \in S \quad \text{iff} \quad \exists y G_{C_S}(\vec{x}, y),$$

Lemma 5.4.8(3) implies that $S$ is recursively enumerable.

For the converse, suppose that $S$ is recursively enumerable. Definition 5.4.3 implies that there is a recursive $(k+1)$-place relation $R$ such that

$$S(\vec{x}) \quad \text{iff} \quad \exists p R(\vec{x}, p).$$

Since $R$ is recursive, the characteristic function $C_R$ is recursive. Let $h$ be the partial recursive function

$$h(\vec{x}) = \mu p(1 \mathbin{\dot{-}} C_R(\vec{x}, p) = 0),$$

where $h$ is the result of applying *partial* search (see Definition 5.2.18) to the recursive function $g(\vec{x}, p) = 1 \mathbin{\dot{-}} C_R(\vec{x}, p)$. Then $C_R(\vec{x}, h(\vec{x}))$ is the semi-characteristic function of $S$. Therefore, $S$ is semi-decidable. □

We now show that a relation is recursive if and only if the relation and its complement are recursively enumerable.

**Theorem 5.4.16.** *Let $P \subseteq \mathbb{N}^k$. Then $P$ is recursive if and only if $P$ and $\mathbb{N}^k \setminus P$ are recursively enumerable.*

*Proof.* Let $P \subseteq \mathbb{N}^k$. Then $P$ and $\mathbb{N}^k \setminus P$ are recursive. Lemma 5.4.6(1) implies that both of these sets are recursively enumerable. Now assume that $P$ and $\mathbb{N}^k \setminus P$ are recursively enumerable. Thus, there are recursive $(k+1)$-relations $R$ and $S$ such that

$$\vec{x} \in P \quad \text{iff} \quad \exists i R(\vec{x}, i), \tag{5.17}$$

$$\vec{x} \notin P \quad \text{iff} \quad \exists i S(\vec{x}, i), \tag{5.18}$$

for all $\vec{x} \in \mathbb{N}^k$. Hence, for all $\vec{x} \in \mathbb{N}^k$, there exists an $i$ such that $R(\vec{x}, i)$ or $S(\vec{x}, i)$. By Lemma 5.4.6(2b), the relation "$R(\vec{x}, i)$ or $S(\vec{x}, i)$" is recursive. So, by total search, the function $h\colon \mathbb{N}^k \to \mathbb{N}$ defined by

$$h(\vec{x}) = \mu i(R(\vec{x}, i) \quad \text{or} \quad S(\vec{x}, i))$$

is recursive. Since $\vec{x} \in P$ iff $R(\vec{x}, h(\vec{x}))$, it follows that $P$ is recursive. □

In Section 5.1.1 we presented an intuitive argument that was designed only to confirm Theorem 5.1.17. Theorems 5.4.14, 5.4.15, and 5.4.16 now provide a mathematically rigorous proof of Theorem 5.1.17, which is restated below.

**Theorem 5.4.17** (Kleene). *Let $S \subseteq \mathbb{N}^k$. Then $S$ is decidable if and only if $S$ and its complement $\mathbb{N}^k \setminus S$ are semi-decidable.*

**Exercises 5.4.**

1. Let $f: \mathbb{N} \to \mathbb{N}$ be recursive. Show that the semi-characteristic function of the range of $f$, $\{f(x) : x \in \mathbb{N}\}$, is partial recursive.
2. Let $f: \mathbb{N} \to \mathbb{N}$ be a recursive bijection. Show that $f^{-1}$ is recursive.
3. Let $A \subseteq \mathbb{N}$ be the nonempty range of a partial recursive function. Show that $A$ is recursively enumerable.
4. Let $A \subseteq \mathbb{N}$ be recursively enumerable. Show that $A$ is the domain of a partial recursive function.

*5. Let $h: \mathbb{N}^k \to \mathbb{N}$ be a partial function. Show that if $G_h$ is recursively enumerable, then $h$ is partial recursive.

*6. Let $f: \mathbb{N} \to \mathbb{N}$ be a recursive function. Show that if $f$ is strictly increasing, then its range $\{f(x) : x \in \mathbb{N}\}$ is recursive.

*7. Let $f: \mathbb{N} \to \mathbb{N}$ be recursive, where $\operatorname{ran}(f) = \{f(x) : x \in \mathbb{N}\}$ is infinite.
   (a) For all $n \in \mathbb{N}$, show that there is an $i > n$ such that $(\forall j \leq n)(f(j) \neq f(i))$.
   (b) For all $n \in \mathbb{N}$, let $i$ be the least such that $i > n$ and $(\forall j \leq n)(f(j) \neq f(i))$. Show that $(\forall j < i)(f(j) \neq f(i))$.
   (c) Define $g: \mathbb{N} \to \mathbb{N}$ by

$$g(n) = \text{the least } i \in \mathbb{N} \text{ such that } i > n \text{ and } (\forall j < i)(f(j) \neq f(i)).$$

   Show that $g$ is recursive. Clearly, $n < g(n)$ for all $n \in \mathbb{N}$.
   (d) Define $h: \mathbb{N} \to \mathbb{N}$ by $h(0) = 0$ and $h(n+1) = g(h(n))$. Thus, $h$ is recursive. Show that $h(n) < h(n+1)$ for all $n \in \mathbb{N}$. Hence, $h$ is one-to-one and strictly increasing.
   (e) Show that $f(h(n)) \neq f(j)$ for all $j < h(n)$, for each $n \in \mathbb{N}$.
   (f) Show that $v: \mathbb{N} \to \mathbb{N}$ defined by $v(n) = f(h(n))$ is one-to-one.
   (g) Let $y \in \operatorname{ran}(f)$ and let $i \in \mathbb{N}$ be the least such that $f(i) = y$. Therefore, $(\forall j < i)(f(j) \neq f(i))$. Show that $i = h(n)$ for some $n$. Now conclude that $\operatorname{ran}(v) = \operatorname{ran}(f)$.

8. Let $A \subseteq \mathbb{N}$ be an infinite recursive set. Show that $A$ is the range of a strictly increasing recursive function.
9. Let $A \subseteq \mathbb{N}$ be an infinite recursively enumerable set. Show that there is an infinite recursive set $B$ such that $B \subseteq A$.
10. Prove Corollary 5.4.7.

*11. Let $Q \subseteq \mathbb{N}^{k+n}$ be recursively enumerable. Prove that the relation

$$\exists i_1 \exists i_2 \ldots \exists i_n R(\vec{x}, i_1, i_2, \ldots, i_n),$$

where $\vec{x} \in \mathbb{N}^k$, is recursively enumerable.

**Exercise Notes:** For Exercise 7(c), see Proposition 5.3.47. For Exercise 7(f), if not, then by (d) there exists an $n$ such that $h(n) < i < h(n+1)$. Exercise 7 shows that every infinite recursively enumerable subset of $\mathbb{N}$ is the range of a one-to-one recursive function. For Exercise 9, by Definition 5.4.1 there is a recursive function $f : \mathbb{N} \to \mathbb{N}$ such that $A$ is the range of $f$. Use $f$ to construct a strictly increasing recursive function $h$ such that the range of $h$ is a subset of $A$; Exercise 6 (above) implies that the range of $h$ is recursive.

# 6 Undecidability and incompleteness

## 6.1 Introduction

Let $\mathcal{L}$ be a language that is commonly used in mathematics. Let us say that a set of $\mathcal{L}$-sentences is decidable if there is an effective procedure to decide whether or not a sentence belongs to the set. In particular, a finite set of sentences is decidable. Now let $\mathfrak{A}$ be an $\mathcal{L}$-structure.

$$\begin{aligned}&\text{Can we find a decidable set } \Gamma \text{ of } \mathcal{L}\text{-sentences such that } \mathfrak{A} \vDash \Gamma \text{ and}\\ &\mathfrak{A} \vDash \varphi \text{ if and only if } \Gamma \vdash \varphi, \text{ for each } \mathcal{L}\text{-sentence } \varphi?\end{aligned} \tag{Q1}$$

That is, is there a reasonable set of axioms that can be used to deduce the statements that are true in the structure $\mathfrak{A}$? This is a question that is often posed in mathematics. If one is going to make deductions from a set of axioms, then one must be able to clearly identify the axioms, that is, the set of axioms must be decidable.

Let us direct question (Q1) to some mathematical languages and structures. Let $\mathcal{L} = \{\dot{<}, \dot{=}\}$ and let $\Psi$ be the axioms for dense linear orders without endpoints (see page 143). Recall that $\mathcal{Q} \vDash \Psi$. Corollary 4.3.19 shows that $\mathcal{Q} \vDash \varphi$ if and only if $\Psi \vdash \varphi$, for all $\mathcal{L}$-sentences $\varphi$. Thus, in this case, there is a positive answer to question (Q1). Moreover, let $\theta$ be an $\mathcal{L}$-sentence. Corollary 4.3.19 implies that either $\Psi \vdash \theta$ or $\Psi \vdash \neg\theta$, but not both. Since $\Psi$ is finite and the set of logical axioms is decidable, we can effectively enumerate all of the deductions from $\Psi$ and eventually find a deduction of either $\theta$ or $\neg\theta$. So we can effectively answer the question: Does $\Psi \vdash \theta$?

Now let us consider the language of elementary number theory (see Section 4.3.1) $\mathcal{L} = \{\dot{<}, \dot{0}, \dot{S}, \dot{+}, \dot{\times}, \dot{E}, \dot{=}\}$ and the standard model of arithmetic (see page 137)

$$\mathcal{N} = \langle \mathbb{N}; 0, S, <, +, \times, E\rangle.$$

Let us rephrase question (Q1) with respect to $\mathcal{L}$ and $\mathcal{N}$:

$$\begin{aligned}&\text{Is there a decidable set of } \mathcal{L}\text{-sentences } \Gamma \text{ such that } \mathcal{N} \vDash \Gamma \text{ and}\\ &\mathcal{N} \vDash \varphi \text{ if and only if } \Gamma \vdash \varphi, \text{ for each } \mathcal{L}\text{-sentence } \varphi?\end{aligned} \tag{Q2}$$

Our development in Sections 5.3 and 5.4 on recursive functions and decidability will allow us to address question (Q2). As we will see, mathematical logic and computability are intimately connected.

**Remark.** The above question (Q2) inspired much of the early growth of mathematical logic. At the International Congress of Mathematicians, a meeting held in Paris in 1900, David Hilbert challenged mathematicians to identify a set of axioms for number theory that would positively address question (Q2). David Hilbert (1862–1943), whose name is attached to the concept of a Hilbert space, was one of the most influential and comprehensive mathematicians of the late nineteenth and early twentieth centuries.

https://doi.org/10.1515/9783110782073-006

## 6.2 Basic axioms for number theory

In our quest to address the preceding question (Q2), we will first identify a finite set of axioms for number theory which will help us to answer this question. These axioms are first-order sentences in the language $\mathcal{L} = \{\dot{<}, \dot{0}, \dot{S}, \dot{+}, \dot{\times}, \dot{E}, \dot{=}\}$ and hold in $\mathcal{N}$, the standard model of arithmetic. We will use $x \dot{\leq} y$ as an abbreviation for $(x \dot{<} y \vee x \dot{=} y)$ and use $x \dot{\not<} y$, $x \dot{\neq} y$ to respectively abbreviate $\neg(x \dot{<} y)$, $\neg(x \dot{=} y)$.

**Basic Axioms 6.2.1.** Let $\mathcal{L} = \{\dot{<}, \dot{0}, \dot{S}, \dot{+}, \dot{\times}, \dot{E}, \dot{=}\}$. Let $\Omega$ consist of the following 11 axioms, where $x, y$ are variables:
(A1) $\forall x(\dot{S}x \dot{\neq} \dot{0})$,
(A2) $\forall x \forall y(\dot{S}x \dot{=} \dot{S}y \rightarrow x \dot{=} y)$,
(A3) $\forall x \forall y(x \dot{<} \dot{S}y \leftrightarrow x \dot{\leq} y)$,
(A4) $\forall x(x \dot{\not<} \dot{0})$,
(A5) $\forall x \forall y(x \dot{<} y \vee x \dot{=} y \vee y \dot{<} x)$,
(A6) $\forall x(x \dot{+} \dot{0} \dot{=} x)$,
(A7) $\forall x \forall y(x \dot{+} \dot{S}y \dot{=} \dot{S}(x \dot{+} y))$,
(A8) $\forall x(x \dot{\times} \dot{0} \dot{=} \dot{0})$,
(A9) $\forall x \forall y(x \dot{\times} \dot{S}y \dot{=} (x \dot{\times} y) \dot{+} x)$,
(A10) $\forall x(\dot{E}x\dot{0} \dot{=} \dot{S}\dot{0})$,
(A11) $\forall x \forall y(\dot{E}x\dot{S}y \dot{=} \dot{E}xy \dot{\times} x)$.

These 11 sentences shall be called the $\Omega$-*axioms* (Omega axioms).

Axioms (A1)–(A5) concern the successor operation and the less than relation. The other six axioms, (A6)–(A11), relate to the steps used to generate the primitive recursive operations of addition, multiplication, and exponentiation (see the proofs of Propositions 5.3.7, 5.3.9, and 5.3.11, respectively). Clearly, $\mathcal{N} \models \Omega$. Thus, $\Omega$ is consistent by Corollary 4.1.11. Let $\Sigma$ be a set of $\mathcal{L}$-sentences such that $\Omega \subseteq \Sigma$. If $\mathcal{N} \models \Sigma$, then for any $\mathcal{L}$-sentence $\varphi$ we know that $\Sigma \vdash \varphi$ implies $\mathcal{N} \models \varphi$, by the soundness theorem. Can we find a decidable set $\Sigma$ so that the converse implication holds?

We will now show that some basic sentences that are true in $\mathcal{N}$ are deducible from the $\Omega$-axioms. For any natural number $n$, to simplify the notation, we will let $\overline{n}$ denote the $\mathcal{L}$-term $\dot{S}^n\dot{0}$, that is,

$$\overline{n} = \dot{S}^n\dot{0} = \overbrace{\dot{S}\dot{S}\cdots\dot{S}}^{n \text{ times}}\dot{0}. \tag{6.1}$$

So $\overline{n+1} = \dot{S}\overline{n}$, $\overline{1} = \dot{S}\dot{0}$, and $\overline{0} = \dot{0}$. The terms $\overline{n}$ are called *numerals*, and we have $\overline{n}^{\mathcal{N}} = n$, where $\overline{n}^{\mathcal{N}}$ is the term $\overline{n}$ interpreted in $\mathcal{N}$. Our next lemma shows that when $n > 0$, one can deduce from $\Omega$ that the only objects less than $\overline{n}$ are $\dot{0}, \overline{1}, \overline{2}, \ldots, \overline{n-1}$. So in any model of $\Omega$, the interpretation of the terms $\dot{0}, \overline{1}, \overline{2}, \ldots$ are ordered like the standard natural numbers.

**Lemma 6.2.2.** *For every natural number $n > 0$, we have*

$$\Omega \vdash x \dot{<} \overline{n} \leftrightarrow (x \doteq \dot{0} \vee x \doteq \overline{1} \vee \cdots \vee x \doteq \overline{n-1}).$$

*Proof.* We shall use natural number induction on $n$.

*Base step:* Let $n = 1$. We must show that $\Omega \vdash (x \dot{<} \overline{1} \leftrightarrow x \doteq \dot{0})$. Since $\overline{1} = \dot{S}\dot{0}$, we have $\Omega \vdash x \dot{<} \overline{1} \leftrightarrow x \dot{\leq} \dot{0}$ by axiom (A3). Moreover, by axiom (A4), we see that $\Omega \vdash x \dot{\leq} \dot{0} \leftrightarrow x \doteq \dot{0}$. Therefore, $\Omega \vdash (x \dot{<} \overline{1} \leftrightarrow x \doteq \dot{0})$.

*Inductive step:* Let $n \in \mathbb{N}$ be such that $n \geq 1$. Assume the induction hypothesis

$$\Omega \vdash x \dot{<} \overline{n} \leftrightarrow (x \doteq \dot{0} \vee x \doteq \overline{1} \vee \cdots \vee x \doteq \overline{n-1}). \tag{IH}$$

From the induction hypothesis we conclude that

$$\Omega \vdash (x \dot{<} \overline{n} \vee x \doteq \overline{n}) \leftrightarrow (x \doteq \dot{0} \vee x \doteq \overline{1} \vee \cdots \vee x \doteq \overline{n-1} \vee x \doteq \overline{n}).$$

Since $x \dot{\leq} \overline{n}$ is an abbreviation for $(x \dot{<} \overline{n} \vee x \doteq \overline{n})$, axiom (A3) implies

$$\Omega \vdash x \dot{<} \overline{n+1} \leftrightarrow (x \doteq \dot{0} \vee x \doteq \overline{1} \vee \cdots \vee x \doteq \overline{n}).$$

□

Thus, any nonstandard "number" $x$ in a model of $\Omega$ satisfies $\overline{n} \dot{<} x$ for all $n \in \mathbb{N}$.

We now introduce two more abbreviations. For any $\mathcal{L}$-formula $\varphi(x)$ and $n > 0$, we let $(\forall x \dot{<} \overline{n})\varphi(x)$ abbreviate the wff $\forall x(x \dot{<} \overline{n} \to \varphi(x))$ and we let $(\exists x \dot{<} \overline{n})\varphi(x)$ abbreviate the wff $\exists x(x \dot{<} \overline{n} \wedge \varphi(x))$. The quantifiers $(\forall x \dot{<} \overline{n})$ and $(\exists x \dot{<} \overline{n})$ are called *bounded quantifiers*.

Let $\varphi_1, \varphi_2, \varphi_3$ be $\mathcal{L}$-wffs. To show that $\Omega \vdash \varphi_1 \leftrightarrow \varphi_2$ and $\Omega \vdash \varphi_2 \leftrightarrow \varphi_3$, it can be more illuminating to derive these consecutive results by using the vertical list

$$\begin{aligned} \Omega \vdash \varphi_1 &\leftrightarrow \varphi_2 \\ &\leftrightarrow \varphi_3 \end{aligned}$$

and thereby conclude that $\Omega \vdash \varphi_1 \leftrightarrow \varphi_3$. This will be done in the proof of Lemma 6.2.3.

**Lemma 6.2.3.** *Let $\varphi(x)$ be an $\mathcal{L}$-formula. For any natural number $n > 0$,*

(1) $\Omega \vdash (\forall x \dot{<} \overline{n})\varphi(x) \leftrightarrow (\varphi(\dot{0}) \wedge \varphi(\overline{1}) \wedge \cdots \wedge \varphi(\overline{n-1}))$,

(2) $\Omega \vdash (\exists x \dot{<} \overline{n})\varphi(x) \leftrightarrow (\varphi(\dot{0}) \vee \varphi(\overline{1}) \vee \cdots \vee \varphi(\overline{n-1}))$.

*Proof.* We now prove (1). We first observe that

$$\begin{aligned} \Omega \vdash (\forall x \dot{<} \overline{n})\varphi(x) &\leftrightarrow \forall x(x \dot{<} \overline{n} \to \varphi(x)) && \text{abbreviation,} \\ &\leftrightarrow \forall x((x \doteq \dot{0} \vee x \doteq \overline{1} \vee \cdots \vee x \doteq \overline{n-1}) \to \varphi(x)) && \text{by Lemma 6.2.2,} \\ &\leftrightarrow (\varphi(\dot{0}) \wedge \varphi(\overline{1}) \wedge \cdots \wedge \varphi(\overline{n-1})) && \text{logical equivalence.} \end{aligned}$$

As $\forall x((x \doteq \dot{0} \vee x \doteq \overline{1} \vee \cdots \vee x \doteq \overline{n-1}) \rightarrow \varphi(x))$ and $(\varphi(\dot{0}) \wedge \varphi(\overline{1}) \wedge \cdots \wedge \varphi(\overline{n-1}))$ are logically equivalent, the completeness theorem justifies the above derivation. Therefore, $\Omega \vdash (\forall x \dot{<} \overline{n})\varphi(x) \leftrightarrow (\varphi(\dot{0}) \wedge \varphi(\overline{1}) \wedge \cdots \wedge \varphi(\overline{n-1}))$. A similar argument proves (2). □

The following lemma will be used to show that for every $\mathcal{L}$-term $t$ which contains no variables, there is a natural number $n$ such that $\Omega \vdash t \doteq \overline{n}$.

**Lemma 6.2.4.** *For all natural numbers m and n,*
(1) $\Omega \vdash \overline{m} \dot{+} \overline{n} \doteq \overline{m+n}$,
(2) $\Omega \vdash \overline{m} \dot{\times} \overline{n} \doteq \overline{m \times n}$,
(3) $\Omega \vdash \dot{E}\,\overline{m}\,\overline{n} \doteq \overline{m^n}$.

*Proof.* Let $m \in \mathbb{N}$. We will prove (1) by induction on $n$.

*Base step:* Let $n = 0$. We must show that $\Omega \vdash \overline{m} \dot{+} \overline{0} \doteq \overline{m+0}$. By axiom (A6), we have $\Omega \vdash \overline{m} \dot{+} \overline{0} \doteq \overline{m}$. Therefore, $\Omega \vdash \overline{m} \dot{+} \overline{0} \doteq \overline{m+0}$, as $m = m + 0$.

*Inductive step:* Let $n \in \mathbb{N}$ be arbitrary. Assume the induction hypothesis

$$\Omega \vdash \overline{m} \dot{+} \overline{n} \doteq \overline{m+n}. \tag{IH}$$

We need to prove that $\Omega \vdash \overline{m} \dot{+} \overline{n+1} \doteq \overline{m+n+1}$. Since $\overline{n+1} = \dot{S}\overline{n}$, axiom (A7) implies that $\Omega \vdash \overline{m} \dot{+} \overline{n+1} \doteq \dot{S}(\overline{m} \dot{+} \overline{n})$. From the induction hypothesis (IH), we conclude that $\Omega \vdash \overline{m} \dot{+} \overline{n+1} \doteq \dot{S}(\overline{m+n})$. As $\dot{S}(\overline{m+n}) = \overline{m+n+1}$, we see that

$$\Omega \vdash \overline{m} \dot{+} \overline{n+1} \doteq \overline{m+n+1}.$$

The proofs of (2) and (3) follow in a similar manner using axioms (A8)–(A9) and (A10)–(A11), respectively. □

We now show that $\Omega$-axioms "agree" with $\mathcal{N}$ about equality.

**Lemma 6.2.5.** *For all natural numbers m and n,*
(1) *if* $m = n$, *then* $\Omega \vdash \overline{m} \doteq \overline{n}$;
(2) *if* $m \neq n$, *then* $\Omega \vdash \overline{m} \not\doteq \overline{n}$.

*Proof.* For (1), if $m = n$, then the terms $\overline{m}$ and $\overline{n}$ are identical, that is, $\overline{m} = \overline{n}$. Hence, $\Omega \vdash \overline{m} \doteq \overline{n}$. The proof of (2) is by the following induction on $n$.

*Base step:* Let $n = 0$ and assume that $m \neq 0$. We must show that $\Omega \vdash \overline{m} \not\doteq \overline{0}$. Since $m \neq 0$, there is an $x \in \mathbb{N}$ such that $m = x + 1$. Thus, $\overline{m} = \dot{S}\,\overline{x}$. Axiom (A1) implies that $\Omega \vdash \dot{S}\,\overline{x} \not\doteq \dot{0}$. Thus, $\Omega \vdash \overline{m} \not\doteq \overline{0}$.

*Inductive step:* Let $n \in \mathbb{N}$ be arbitrary. Assume the induction hypothesis

$$\text{for all } m \in \mathbb{N}, \text{ if } m \neq n, \text{ then } \Omega \vdash \overline{m} \not\doteq \overline{n}. \tag{IH}$$

Let $m \in \mathbb{N}$ be such that $m \neq n+1$. We prove that $\Omega \vdash \overline{m} \mathrel{\dot{\neq}} \overline{n+1}$. If $m = 0$, then $\overline{m} = \dot{0}$ and $\Omega \vdash \dot{S}\overline{n} \mathrel{\dot{\neq}} \dot{0}$ by axiom (A1). Thus, $\Omega \vdash \overline{m} \mathrel{\dot{\neq}} \overline{n+1}$. If $m > 0$, then $m = x+1$ for some $x \in \mathbb{N}$. So $\overline{m} = \dot{S}\overline{x}$ and $x \neq n$. The induction hypothesis (IH) implies that $\Omega \vdash \overline{x} \mathrel{\dot{\neq}} \overline{n}$. Axiom (A2) now implies that $\Omega \vdash \dot{S}\overline{x} \mathrel{\dot{\neq}} \dot{S}\overline{n}$, that is, $\Omega \vdash \overline{m} \mathrel{\dot{\neq}} \overline{n+1}$. □

**Lemma 6.2.6.** *For each variable-free $\mathcal{L}$-term $t$, there is a unique $n \in \mathbb{N}$ such that $\Omega \vdash t \doteq \overline{n}$.*

*Proof.* Uniqueness follows from Lemma 6.2.5(2). We use induction on terms to prove that for all terms $t^*$,

$$\text{if } t^* \text{ is variable-free, then } \Omega \vdash t^* \doteq \overline{n}, \text{ for some natural number } n. \tag{6.2}$$

*Base step:* $t^* = \dot{0}$. Since $\overline{0} = \dot{0}$, we clearly have $\Omega \vdash \dot{0} \doteq \overline{0}$.

*Inductive step:* Let $t$ and $\tau$ be variable-free $\mathcal{L}$-terms. Assume the induction hypothesis

$$\Omega \vdash t \doteq \overline{k} \quad \text{and} \quad \Omega \vdash \tau \doteq \overline{m}, \tag{IH}$$

for $k, m \in \mathbb{N}$. We must prove that (6.2) holds for each of the variable-free terms

$$\dot{S}t,\ t \mathbin{\dot{+}} \tau,\ t \mathbin{\dot{\times}} \tau,\ \dot{E}t\tau.$$

Since $\Omega \vdash t \doteq \overline{k}$, we have $\Omega \vdash \dot{S}t \doteq \dot{S}\overline{k}$. Thus, as $\dot{S}\overline{k} = \overline{k+1}$, $\Omega \vdash \dot{S}t \doteq \overline{k+1}$. The induction hypothesis (IH) implies that $\Omega \vdash t \mathbin{\dot{+}} \tau \doteq \overline{k} \mathbin{\dot{+}} \overline{m}$. So, by Lemma 6.2.4(1), $\Omega \vdash t \mathbin{\dot{+}} \tau \doteq \overline{k+m}$. The argument for the terms $t \mathbin{\dot{\times}} \tau$ and $\dot{E}t\tau$ is similar. □

The $\Omega$-axioms also "agree" with $\mathcal{N}$ about inequality.

**Lemma 6.2.7.** *For all natural numbers $n$ and $m$,*
(1) *if $m < n$, then $\Omega \vdash \overline{m} \mathrel{\dot{<}} \overline{n}$,*
(2) *if $m \not< n$, then $\Omega \vdash \overline{m} \mathrel{\dot{\not<}} \overline{n}$.*

*Proof.* The proofs of both conditionals (1) and (2) are by induction on $n$.
(1) Let $m \in \mathbb{N}$. We prove the above item (1) by the following induction on $n$.

*Base step:* Let $n = 0$. In this case, the conditional (1) is vacuously true.

*Inductive step:* Let $n \in \mathbb{N}$ be arbitrary. Assume the induction hypothesis

$$\text{if } m < n, \quad \text{then } \Omega \vdash \overline{m} \mathrel{\dot{<}} \overline{n}. \tag{IH}$$

Assume that $m < n+1$. So either $m < n$ or $m = n$. By the induction hypothesis and Lemma 6.2.5(1), we have either $\Omega \vdash \overline{m} \mathrel{\dot{<}} \overline{n}$ or $\Omega \vdash \overline{m} \doteq \overline{n}$. Hence,

$$\Omega \vdash (\overline{m} \mathrel{\dot{<}} \overline{n} \lor \overline{m} \doteq \overline{n}).$$

So $\Omega \vdash \overline{m} \mathrel{\dot{\leq}} \overline{n}$. Axiom (A3) implies that $\Omega \vdash \overline{m} \mathrel{\dot{<}} \dot{S}\overline{n}$. Therefore, $\Omega \vdash \overline{m} \mathrel{\dot{<}} \overline{n+1}$.

(2) Let $m \in \mathbb{N}$. We now prove the above item (2) by induction.

*Base step:* Let $n = 0$. Axiom (A4) implies that $\Omega \vdash \overline{m} \dot{\nless} \dot{0}$. So (2) holds for $n = 0$.

*Inductive step:* Let $n \in \mathbb{N}$ be arbitrary. Assume the induction hypothesis

$$\text{if } m \nless n, \quad \text{then } \Omega \vdash \overline{m} \dot{\nless} \overline{n}. \tag{IH}$$

Assume that $m \nless n + 1$. Thus, $m \nless n$ and $m \neq n$. By the induction hypothesis and Lemma 6.2.5(2), we have $\Omega \vdash \overline{m} \dot{\nless} \overline{n}$ and $\Omega \vdash \overline{m} \dot{\neq} \overline{n}$. Hence,

$$\Omega \vdash (\overline{m} \dot{\nless} \overline{n} \wedge \overline{m} \dot{\neq} \overline{n}).$$

So $\Omega \vdash \neg(\overline{m} \dot{<} \overline{n} \vee \overline{m} \doteq \overline{n})$ (De Morgan's law), that is, $\Omega \vdash \neg(\overline{m} \dot{\leq} \overline{n})$. Axiom (A3) implies that $\Omega \vdash \neg(\overline{m} \dot{<} \dot{S}\overline{n})$, that is, $\Omega \vdash \overline{m} \dot{\nless} \overline{n+1}$. □

Lemmas 6.2.5–6.2.7 imply a partial answer to question (Q2) posed on page 205.

**Theorem 6.2.8.** *For every quantifier-free sentence $\varphi$,*
(1) *if $\mathcal{N} \models \varphi$, then $\Omega \vdash \varphi$,*
(2) *if $\mathcal{N} \models \neg\varphi$, then $\Omega \vdash \neg\varphi$.*

*Proof.* The proof is by induction on the construction of the quantifier-free sentences as outlined in Exercise 2. □

An existential sentence has the form $\exists x_1 \cdots \exists x_n \theta$, where $\theta$ is quantifier-free. Our next result is an extension of Theorem 6.2.8.

**Corollary 6.2.9.** *Let $\varphi$ be an existential sentence. If $\mathcal{N} \models \varphi$, then $\Omega \vdash \varphi$.*

*Proof.* Let $\varphi$ be an existential sentence of the form $\exists x \alpha$, where $\alpha$ is quantifier-free. Suppose that $\mathcal{N} \models \exists x \alpha$. Thus, for some natural number $n$, we have $\mathcal{N} \models \alpha_{\overline{n}}^{x}$. Since $\alpha_{\overline{n}}^{x}$ is a quantifier-free sentence, Theorem 6.2.8 implies that $\Omega \vdash \alpha_{\overline{n}}^{x}$. Exercise 10 on page 117 implies that $\Omega \vdash \alpha_{\overline{n}}^{x} \to \exists x \alpha$. Thus, $\Omega \vdash \exists x \alpha$. One can now extend this argument, by induction, to existential sentences with more than one such quantifier. □

Corollary 6.2.9 does not hold, in general, for the negation of an existential sentence.

**Exercises 6.1.**

1. Let $m, n \in \mathbb{N}$. Show that if $\Omega \nvdash \overline{m} \dot{<} \overline{n}$, then $\Omega \vdash (\overline{m} \doteq \overline{n} \vee \overline{n} \dot{<} \overline{m})$.

*2. Let $t$ and $\tau$ be variable-free terms in the language $\mathcal{L} = \{\dot{<}, \dot{0}, \dot{S}, \dot{+}, \dot{\times}, \dot{E}, \doteq\}$. Prove Theorem 6.2.8 by induction as follows:
Base step:
(a) Show that if $\mathcal{N} \models t \doteq \tau$, then $\Omega \vdash t \doteq \tau$.
(b) Show that if $\mathcal{N} \models t \dot{\neq} \tau$, then $\Omega \vdash t \dot{\neq} \tau$.
(c) Show that if $\mathcal{N} \models t \dot{<} \tau$, then $\Omega \vdash t \dot{<} \tau$.
(d) Show that if $\mathcal{N} \models t \dot{\nless} \tau$, then $\Omega \vdash t \dot{\nless} \tau$.

Inductive step: Let $\varphi$ and $\psi$ be quantifier-free sentences. Assume the following:

(a) If $\mathcal{N} \models \varphi$, then $\Omega \vdash \varphi$.

(b) If $\mathcal{N} \models \neg\varphi$, then $\Omega \vdash \neg\varphi$.

(c) If $\mathcal{N} \models \psi$, then $\Omega \vdash \psi$.

(d) If $\mathcal{N} \models \neg\psi$, then $\Omega \vdash \neg\psi$.

Let $\widetilde{\varphi} \in \{\varphi, \neg\varphi\}$ and $\overline{\psi} \in \{\psi, \neg\psi\}$. Show the following:

- If $\mathcal{N} \models (\widetilde{\varphi} \to \overline{\psi})$, then $\Omega \vdash (\widetilde{\varphi} \to \overline{\psi})$,
- If $\mathcal{N} \models \neg(\widetilde{\varphi} \to \overline{\psi})$, then $\Omega \vdash \neg(\widetilde{\varphi} \to \overline{\psi})$.

*3. Show that for all natural numbers $n$, $\Omega \vdash \dot{S}\overline{n} \doteq \overline{Sn}$.

4. Show that there is a formula $\varphi(x, y)$ such that for all $m, n \in \mathbb{N}$, $m \mid n$ if and only if $\Omega \vdash \varphi(\overline{m}, \overline{n})$.

5. Prove Proposition 6.2.4(2).

6. Prove Proposition 6.2.4(3).

7. Let $\varphi(x)$ be an $\mathcal{L}$-formula. Let $n \in \mathbb{N}$. Suppose that for all $k < n + 1$, $\Omega \vdash \varphi(\overline{k})$. Show that $\Omega \vdash (\forall x \mathrel{\dot<} \overline{n+1})\varphi(x)$.

8. Show that $\Omega \vdash \overline{m} \mathbin{\dot+} \overline{n} \doteq \overline{n} \mathbin{\dot+} \overline{m}$, for all $m, n \in \mathbb{N}$.

9. Show that $\Omega \vdash \overline{n} \mathrel{\dot<} \overline{n} \mathbin{\dot+} \overline{1}$, for all $n \in \mathbb{N}$.

10. Let $\varphi(x)$ be an $\mathcal{L}$-formula. Show that if $\Omega \vdash \varphi(\overline{4})$, then $\Omega \vdash (\exists x \mathrel{\dot<} \overline{44})\varphi(x)$.

11. Let $\varphi(x)$ be an $\mathcal{L}$-formula. Show that if $\Omega \vdash \neg\varphi(\overline{2})$, then $\Omega \vdash \neg(\forall x \mathrel{\dot<} \overline{4})\varphi(x)$.

## 6.3 Representable relations and functions

Again, let $\mathcal{L} = \{\dot<, \dot0, \dot S, \dot+, \dot\times, \dot E, \doteq\}$ be the language for elementary number theory and let $\Omega$ be the basic axioms of number theory. In this section, we present a concept called representability, which shows that recursive functions and relations can be represented by formulas that can be deduced from the $\Omega$-axioms. In Section 6.2, we have already shown that the relation $<$ and the operations $+$ and $\times$, when applied to specific natural numbers, can be interpreted as $\mathcal{L}$-sentences that are deducible from $\Omega$. In particular, Lemma 6.2.7 shows that if $m < n$, then $\Omega \vdash \overline{m} \mathrel{\dot<} \overline{n}$, where $\overline{m} \mathrel{\dot<} \overline{n}$ is an atomic sentence. Moreover, Propositions 5.3.23, 5.3.7, and 5.3.9 show that $<$, $+$, and $\times$ are all recursive. In this section, we will show that these results can be extended to include all recursive relations and functions.

We will use the notation $\theta(x_1, x_2, \ldots, x_k)$ to indicate that the free variables in the wff $\theta$ are among the distinct variables $x_1, x_2, \ldots, x_k$. Moreover, the notation $\theta(\overline{n}_1, \overline{n}_2, \ldots, \overline{n}_k)$ will be used to denote the result of substituting the free variables in $\theta$ with the corresponding terms $\overline{n}_1, \overline{n}_2, \ldots, \overline{n}_k$ where we let $\overline{n}_i = \dot S^{n_i}\dot0$ for $i = 1, \ldots, k$, that is, we let

$$\theta(\overline{n}_1, \overline{n}_2, \ldots, \overline{n}_k) = \theta^{x_1 \cdots x_k}_{\overline{n}_1 \cdots \overline{n}_k}.$$

Similarly, for a term $\tau$, the notation $\tau(x_1, x_2, \ldots, x_k)$ will be used to indicate that the variables which appear in $\tau$ are among the distinct variables $x_1, x_2, \ldots, x_k$. We will also let

$$\tau(\overline{n}_1, \overline{n}_2, \ldots, \overline{n}_k) = \tau^{x_1 \cdots x_k}_{\overline{n}_1 \cdots \overline{n}_k},$$

that is, $\tau(\overline{n}_1, \overline{n}_2, \ldots, \overline{n}_k)$ is the result of replacing the variables $x_1, x_2, \ldots, x_k$ in $\tau$ with the respective terms $\overline{n}_1, \overline{n}_2, \ldots, \overline{n}_k$ (see Exercise 11 on page 65).

**Definition 6.3.1.** A relation $R \subseteq \mathbb{N}^k$ is *representable* if there exists an $\mathcal{L}$-formula $\varphi(x_1, x_2, \ldots, x_k)$ such that for all $n_1, n_2, \ldots, n_k \in \mathbb{N}$,
(1) if $\langle n_1, n_2, \ldots, n_k \rangle \in R$, then $\Omega \vdash \varphi(\overline{n}_1, \overline{n}_2, \ldots, \overline{n}_k)$,
(2) if $\langle n_1, n_2, \ldots, n_k \rangle \notin R$, then $\Omega \vdash \neg\varphi(\overline{n}_1, \overline{n}_2, \ldots, \overline{n}_k)$.

When (1) and (2) hold, we shall say that $\varphi$ *represents* the relation $R$.

Lemma 6.2.7 shows that the formula $x_1 \dot{<} x_2$ represents the relation $<$. Similarly, Lemma 6.2.5 implies that the relation $=$ is representable.

The representability concept bears resemblance to that of definability over $\mathcal{N}$. However, definability over $\mathcal{N}$ involves truth in the structure $\mathcal{N}$. Representability, on the other hand, involves deductions from the $\Omega$-axioms. Nevertheless, in our next lemma we establish a connection between representability and definability over $\mathcal{N}$.

**Definition 6.3.2.** Let $\varphi(x_1, x_2, \ldots, x_k)$ be an $\mathcal{L}$-well-formed formula ($\mathcal{L}$-wff) where all of the free variables in $\varphi$ are among $x_1, x_2, \ldots, x_k$. We say that $\varphi$ is *numeralwise determined* if for all natural numbers $n_1, n_2, \ldots, n_k$, we have either

$$\begin{aligned} &\Omega \vdash \varphi(\overline{n}_1, \overline{n}_2, \ldots, \overline{n}_k), \quad \text{or} \\ &\Omega \vdash \neg\varphi(\overline{n}_1, \overline{n}_2, \ldots, \overline{n}_k). \end{aligned} \tag{6.3}$$

The following lemma shows that a relation is representable if the relation is definable over $\mathcal{N}$ by means of a formula that is numeralwise determined. First observe that if $\varphi(x_1, \ldots, x_k)$ is numeralwise determined and $\mathcal{N} \models \varphi(\overline{n}_1, \ldots, \overline{n}_k)$, then it follows that $\Omega \vdash \varphi(\overline{n}_1, \ldots, \overline{n}_k)$. Otherwise, Definition 6.3.2 would imply that $\Omega \vdash \neg\varphi(\overline{n}_1, \ldots, \overline{n}_k)$ and hence, $\mathcal{N} \models \neg\varphi(\overline{n}_1, \ldots, \overline{n}_k)$ as $\mathcal{N} \models \Omega$.

**Lemma 6.3.3.** *A formula $\varphi$ represents a relation $R$ if and only if $\varphi$ is numeralwise determined and $\varphi$ defines $R$ over $\mathcal{N}$.*

*Proof.* Recall that (▲) $\mathcal{N} \models \Omega$. If $\varphi$ represents $R$, then by Definition 6.3.1(1)(2) we see that $\varphi$ is numeralwise determined and, by the soundness theorem, $\varphi$ defines $R$ over $\mathcal{N}$. Conversely, if (a) $\varphi$ is numeralwise determined and (b) $\varphi$ defines $R$ over $\mathcal{N}$, then

$$\begin{aligned} \langle n_1, n_2, \ldots, n_k \rangle \in R &\Rightarrow \mathcal{N} \models \varphi(\overline{n}_1, \overline{n}_2, \ldots, \overline{n}_k) && \text{by (b),} \\ &\Rightarrow \Omega \vdash \varphi(\overline{n}_1, \overline{n}_2, \ldots, \overline{n}_k) && \text{by (a), (6.3), and (▲).} \end{aligned}$$

Similarly, $\langle n_1, n_2, \ldots, n_k \rangle \notin R$ implies $\Omega \vdash \neg\varphi(\overline{n}_1, \overline{n}_2, \ldots, \overline{n}_k)$. So $\varphi$ represents $R$. ☐

So if a formula $\varphi$ is numeralwise determined, then $\varphi$ identifies a representable relation. Our next result shows how one can take advantage of Lemma 6.3.3 to build new representable relations from existing ones. Recall that $(\forall x \dot{<} y)\varphi$ and $(\exists x \dot{<} y)\varphi$ are the abbreviated forms of $\forall x(x \dot{<} y \to \varphi)$ and $\exists x(x \dot{<} y \wedge \varphi)$, respectively.

**Theorem 6.3.4.** *Let $\varphi$ and $\psi$ be $\mathcal{L}$-wffs.*
(1) *If $\varphi$ is an atomic formula, then $\varphi$ is numeralwise determined.*
(2) *If $\varphi$ and $\psi$ are numeralwise determined, then so are $\neg\varphi$ and $\varphi \to \psi$.*
(3) *If $\varphi$ is numeralwise determined, then so are $(\forall x \dot{<} y)\varphi$ and $(\exists x \dot{<} y)\varphi$.*

*Proof.* Let $\varphi$ and $\psi$ be wffs in the language $\mathcal{L}$.
(1) Let $\varphi$ be the atomic formula $t \dot{<} \tau$, where $t$ and $\tau$ are both terms. Let $n_1, n_2, \ldots, n_k$ be natural numbers. Theorem 6.2.8 implies that either

$$\Omega \vdash t(\overline{n}_1, \overline{n}_2, \ldots, \overline{n}_k) \dot{<} \tau(\overline{n}_1, \overline{n}_2, \ldots, \overline{n}_k), \quad \text{or}$$
$$\Omega \vdash t(\overline{n}_1, \overline{n}_2, \ldots, \overline{n}_k) \dot{\not<} \tau(\overline{n}_1, \overline{n}_2, \ldots, \overline{n}_k).$$

Similarly, if $\varphi$ is the atomic formula $t \doteq \tau$, then Theorem 6.2.8 implies that either

$$\Omega \vdash t(\overline{n}_1, \overline{n}_2, \ldots, \overline{n}_k) \doteq \tau(\overline{n}_1, \overline{n}_2, \ldots, \overline{n}_k), \quad \text{or}$$
$$\Omega \vdash t(\overline{n}_1, \overline{n}_2, \ldots, \overline{n}_k) \not\doteq \tau(\overline{n}_1, \overline{n}_2, \ldots, \overline{n}_k).$$

(2) See Exercise 4.
(3) Let $\varphi(x, y, x_1, \ldots, x_k)$ be numeralwise determined. Let $n, n_1, n_2, \ldots, n_k$ be natural numbers and let $\widetilde{\varphi}(x)$ denote the wff $\varphi(x, \overline{n}, \overline{n}_1, \overline{n}_2, \ldots, \overline{n}_k)$. If $n = 0$, then one can show that

$$\Omega \vdash (\forall x \dot{<} \dot{0})\widetilde{\varphi}(x) \quad \text{and} \quad \Omega \vdash \neg(\exists x \dot{<} \dot{0})\widetilde{\varphi}(x).$$

So let $n > 0$. We first show that either

$$\Omega \vdash (\forall x \dot{<} \overline{n})\widetilde{\varphi}(x), \quad \text{or}$$
$$\Omega \vdash \neg(\forall x \dot{<} \overline{n})\widetilde{\varphi}(x).$$

By Lemma 6.2.2, there are two cases two consider.

Case 1: For every $m < n$, we have $\Omega \vdash \widetilde{\varphi}(\overline{m})$. In this case, Lemma 6.2.3(1) implies that $\Omega \vdash (\forall x \dot{<} \overline{n})\widetilde{\varphi}(x)$.

Case 2: For some $m < n$, we have $\Omega \nvdash \widetilde{\varphi}(\overline{m})$. Since $\varphi$ is numeralwise determined, we conclude that $\Omega \vdash \neg\widetilde{\varphi}(\overline{m})$. Lemma 6.2.3(1) now implies that $\Omega \vdash \neg(\forall x \dot{<} \overline{n})\widetilde{\varphi}(x)$. Using Lemma 6.2.3(2), one can similarly prove that $(\exists x \dot{<} y)\varphi$ is also numeralwise determined. □

Theorem 6.3.4 is a useful tool for showing that many relations are representable. Let $\varphi(v)$ be the formula

$$\dot{S}\dot{0} \dot{<} v \land (\forall x \dot{<} v)(\forall y \dot{<} v)(x \dot{\times} y \not\doteq v).$$

Theorem 6.3.4 implies that $\varphi(v)$ is numeralwise determined. Since $\varphi(v)$ defines the set of primes in $\mathcal{N}$, Lemma 6.3.3 shows that the set of primes is representable.

**Definition 6.3.5.** A function $f\colon \mathbb{N}^k \to \mathbb{N}$ is said to be *functionally representable* if there is an $\mathcal{L}$-formula $\varphi(x_1, x_2, \ldots, x_k, v)$ such that for all $n_1, n_2, \ldots, n_k \in \mathbb{N}$,

$$\Omega \vdash \forall v(\varphi(\overline{n}_1, \overline{n}_2, \ldots, \overline{n}_k, v) \leftrightarrow v \doteq \overline{f(n_1, n_2, \ldots, n_k)}). \tag{6.4}$$

When (6.4) holds, we will say that $\varphi$ *functionally represents* $f$.

**Remark 6.3.6.** Let $f$ satisfy (6.4). Thus,

$$\Omega \vdash \forall v(v \doteq \overline{f(n_1, n_2, \ldots, n_k)} \to \varphi(\overline{n}_1, \overline{n}_2, \ldots, \overline{n}_k, v)), \tag{6.5}$$

$$\Omega \vdash \forall v(\varphi(\overline{n}_1, \overline{n}_2, \ldots, \overline{n}_k, v) \to v \doteq \overline{f(n_1, n_2, \ldots, n_k)}). \tag{6.6}$$

The above (6.5) implies that

$$\Omega \vdash \varphi(\overline{n}_1, \overline{n}_2, \ldots, \overline{n}_k, \overline{f(n_1, n_2, \ldots, n_k)}),$$

and therefore $\Omega \vdash \exists v \varphi(\overline{n}_1, \overline{n}_2, \ldots, \overline{n}_k, v)$. Thus, (6.5) and (6.6) imply that

$$\Omega \vdash \exists! v \varphi(\overline{n}_1, \overline{n}_2, \ldots, \overline{n}_k, v),$$

where $\exists!$ is the uniqueness quantifier (see page 22).

The substitution property of equality asserts that if two quantities are equal, then one can replace one with the other. Theorem 3.3.54 shows that this substitution property is formally deducible. Thus, Lemma 6.2.4(1) and Theorem 3.3.54 imply that

$$\Omega \vdash \forall v(\overline{m} \dot{+} \overline{n} \doteq v \leftrightarrow v \doteq \overline{m + n})$$

for all $m, n \in \mathbb{N}$. Let $\varphi(x_1, x_2, v)$ be the $\mathcal{L}$-formula $x_1 \dot{+} x_2 \doteq v$. Thus, the particular formula $\varphi(x_1, x_2, v)$ is an equation and functionally represents $+$, that is, it represents the function $f\colon \mathbb{N}^2 \to \mathbb{N}$ defined by $f(x_1, x_2) = x_1 + x_2$. Furthermore, Lemma 6.2.4 and Exercise 3 on page 211 imply that the three equational formulas $x_1 \dot{\times} x_2 \doteq v$, $\dot{E}x_1x_2 \doteq v$, and $\dot{S}x_1 \doteq v$ functionally represent the operations $\times$, $E$, and $S$, respectively. These four examples motivate a generalization. Again, $\tau(x_1, x_2, \ldots, x_k)$ denotes a term whose variables, if any, are among $x_1, x_2, \ldots, x_k$.

**Lemma 6.3.7.** *Let $\tau(x_1, x_2, \ldots, x_k)$ be an $\mathcal{L}$-term and define $f_\tau: \mathbb{N}^k \to \mathbb{N}$ by*

$$f_\tau(n_1, n_2, \ldots, n_k) = \text{the unique } n \text{ such that } \Omega \vdash (\tau(\overline{n}_1, \overline{n}_2, \ldots, \overline{n}_k) \doteq \overline{n}).$$

*Then the equational formula $\tau(x_1, x_2, \ldots, x_k) \doteq v$ functionally represents $f_\tau$.*

*Proof.* Let $n_1, n_2, \ldots, n_k \in \mathbb{N}$. Lemma 6.2.6 implies that there exists a unique $n \in \mathbb{N}$ such that $\Omega \vdash \tau(\overline{n}_1, \overline{n}_2, \ldots, \overline{n}_k) = \overline{n}$. Thus, by definition of $f_\tau$, $\overline{n} = \overline{f_\tau(n_1, n_2, \ldots, n_k)}$. Therefore, $\Omega \vdash \tau(\overline{n}_1, \overline{n}_2, \ldots, \overline{n}_k) \doteq \overline{f_\tau(n_1, n_2, \ldots, n_k)}$, and this implies that

$$\Omega \vdash \forall v(\tau(\overline{n}_1, \overline{n}_2, \ldots, \overline{n}_k) \doteq v \leftrightarrow v \doteq \overline{f_\tau(n_1, n_2, \ldots, n_k)}).$$

Hence, the equation $\tau(x_1, x_2, \ldots, x_k) \doteq v$ functionally represents $f_\tau$. □

**Corollary 6.3.8.** *Let $f: \mathbb{N}^k \to \mathbb{N}$ and let $\tau$ be a term so that for all $n_1, n_2, \ldots, n_k, n \in \mathbb{N}$,*

$$f(n_1, n_2, \ldots, n_k) = n \quad \textit{iff} \quad \mathcal{N} \models \tau(\overline{n}_1, \overline{n}_2, \ldots, \overline{n}_k) \doteq \overline{n}. \tag{6.7}$$

*Then the equational formula $\tau(x_1, x_2, \ldots, x_k) \doteq v$ functionally represents $f$.*

*Proof.* For all $n_1, n_2, \ldots, n_k, n \in \mathbb{N}$, Theorems 6.2.8 and 4.1.5 imply that

$$\mathcal{N} \models \tau(\overline{n}_1, \overline{n}_2, \ldots, \overline{n}_k) \doteq \overline{n} \quad \text{iff} \quad \Omega \vdash \tau(\overline{n}_1, \overline{n}_2, \ldots, \overline{n}_k) \doteq \overline{n}.$$

Thus, from (6.7), we conclude that

$$f(n_1, n_2, \ldots, n_k) = \text{the unique } n \text{ such that } \Omega \vdash (\tau(\overline{n}_1, \overline{n}_2, \ldots, \overline{n}_k) \doteq \overline{n}).$$

So, by Lemma 6.3.7, $f = f_\tau$ and $\tau(x_1, x_2, \ldots, x_k) \doteq v$ functionally represents $f$. □

We now illustrate how to apply Corollary 6.3.8. Let $f: \mathbb{N}^2 \to \mathbb{N}$ and $\tau$ be defined by

$$f(x_1, x_2) = S(x_1) \times x_2 \quad \text{and} \quad \tau(x_1, x_2) = \dot{S}x_1 \mathbin{\dot{\times}} x_2.$$

Since $f$ and $\tau$ satisfy condition (6.7) of Corollary 6.3.8, we conclude that the formula $\tau(x_1, x_2) \doteq v$ functionally represents $f$. For a second example, let $f: \mathbb{N}^2 \to \mathbb{N}$ be defined by $f(x_1, x_2) = 0$ and let $\tau = \dot{0}$. Clearly, $f$ and $\tau$ satisfy condition (6.7). Therefore, by Corollary 6.3.8, the equational formula $\dot{0} \doteq v$ functionally represents $f$.

**Corollary 6.3.9.** *The initial and constant functions are functionally representable.*

*Proof.* Corollary 6.3.8 implies the following:
1. The *zero function* $\mathring{f}: \mathbb{N}^k \to \mathbb{N}$, as defined by $\mathring{f}(x_1, \ldots, x_k) = 0$, is functionally represented by the equational formula $\dot{0} \doteq v$.
2. For $m \in \mathbb{N}$, the *constant function* ${}^m f: \mathbb{N}^k \to \mathbb{N}$, defined by ${}^m f(x_1, \ldots, x_k) = m$, is functionally represented by the equational formula $\overline{m} \doteq v$.
3. The *successor function* $S(x)$ is represented by the equational formula $\dot{S}x \doteq v$.

4. For all natural numbers $1 \leq i \leq k$, the *projection function* $I_i^k \colon \mathbb{N}^k \to \mathbb{N}$, defined by $I_i^k(x_1, \ldots, x_k) = x_i$, is functionally represented by the formula $x_i \doteq v$. □

Corollary 6.3.8 therefore implies that many functions are represented by an equational formula. However, this is not true for all representable functions.

There is another natural notion that concerns the representability of a function.

**Definition 6.3.10.** A function $f \colon \mathbb{N}^k \to \mathbb{N}$ is said to be *graph representable* if the graph of $f$

$$G_f = \{\langle n_1, n_2, \ldots, n_k, n\rangle : f(n_1, n_2, \ldots, n_k) = n\}$$

is representable as a relation, that is, there is a formula $\varphi(x_1, \ldots, x_k, x)$ such that for all $n_1, n_2, \ldots, n_k, n \in \mathbb{N}$,

(1) if $f(n_1, n_2, \ldots, n_k) = n$, then $\Omega \vdash \varphi(\overline{n}_1, \overline{n}_2, \ldots, \overline{n}_k, \overline{n})$,
(2) if $f(n_1, n_2, \ldots, n_k) \neq n$, then $\Omega \vdash \neg\varphi(\overline{n}_1, \overline{n}_2, \ldots, \overline{n}_k, \overline{n})$.

It turns out that a function is functionally representable if and only if the function is graph representable. However, the proof of this equivalence is a bit subtle. The next lemma will be used to prove Theorem 6.3.12 below, which establishes this equivalence.

**Lemma 6.3.11.** *Let $f \colon \mathbb{N}^k \to \mathbb{N}$ and let $n_1, n_2, \ldots, n_k \in \mathbb{N}$. Let $\varphi(x_1, \ldots, x_k, x)$ be such that (1) and (2) of Definition* 6.3.10 *hold for all $n \in \mathbb{N}$. If $f(n_1, n_2, \ldots, n_k) = n$, then*

$$\Omega \vdash \varphi(\overline{n}_1, \overline{n}_2, \ldots, \overline{n}_k, \overline{n}) \wedge (\forall y \mathrel{\dot{<}} \overline{n})\neg\varphi(\overline{n}_1, \overline{n}_2, \ldots, \overline{n}_k, y). \tag{6.8}$$

*Proof.* Let $n_1, n_2, \ldots, n_k \in \mathbb{N}$ and $f$ be as stated and let $f(n_1, n_2, \ldots, n_k) = n$. Therefore, $\Omega \vdash \varphi(\overline{n}_1, \overline{n}_2, \ldots, \overline{n}_k, \overline{n})$ by Definition 6.3.10(1). Let $m < n$. Thus, $f(n_1, n_2, \ldots, n_k) \neq m$. Definition 6.3.10(2) implies that $\Omega \vdash \neg\varphi(\overline{n}_1, \overline{n}_2, \ldots, \overline{n}_k, \overline{m})$. Lemma 6.2.3 now implies that $\Omega \vdash (\forall y \mathrel{\dot{<}} \overline{n})\neg\varphi(\overline{n}_1, \overline{n}_2, \ldots, \overline{n}_k, y)$. Hence, (6.8) holds. □

**Theorem 6.3.12.** *Let $f \colon \mathbb{N}^k \to \mathbb{N}$. Then $f$ is functionally representable if and only if $f$ is graph representable.*

*Proof.* Let $f \colon \mathbb{N}^k \to \mathbb{N}$.

($\Rightarrow$). Assume that $f$ is functionally representable. By Definition 6.3.5, there exists an $\mathcal{L}$-formula $\varphi(x_1, \ldots, x_k, v)$ such that for all $n_1, n_2, \ldots, n_k \in \mathbb{N}$,

$$\Omega \vdash \forall v(\varphi(\overline{n}_1, \overline{n}_2, \ldots, \overline{n}_k, v) \leftrightarrow v \doteq \overline{f(n_1, n_2, \ldots, n_k)}).$$

By Logical axiom 3.3.17(2), we conclude that

$$\Omega \vdash \varphi(\overline{n}_1, \overline{n}_2, \ldots, \overline{n}_k, \overline{n}) \leftrightarrow \overline{n} \doteq \overline{f(n_1, n_2, \ldots, n_k)} \tag{6.9}$$

for all $n_1, n_2, \dots, n_k, n \in \mathbb{N}$. Lemma 6.2.5 and (6.9) now easily imply (1) and (2) of Definition 6.3.10. Therefore, $f$ is graph representable.

($\Leftarrow$). Assume that $f$ is graph representable. Let $\varphi(x_1, \dots, x_k, x)$ be a formula that satisfies (1) and (2) of Definition 6.3.10 for all $n_1, n_2, \dots, n_k, n \in \mathbb{N}$. We need to find a formula $\theta(x_1, \dots, x_k, v)$ such that for all $n_1, n_2, \dots, n_k \in \mathbb{N}$,

$$\Omega \vdash \forall v(\theta(\overline{n}_1, \overline{n}_2, \dots, \overline{n}_k, v) \leftrightarrow v \doteq \overline{f(n_1, n_2, \dots, n_k)}).$$

Let $n_1, n_2, \dots, n_k \in \mathbb{N}$ be arbitrary and let $\widetilde{\varphi}(v)$ denote the $\mathcal{L}$-wff $\varphi(\overline{n}_1, \overline{n}_2, \dots, \overline{n}_k, v)$. Now consider the formula

$$\widetilde{\varphi}(v) \wedge (\forall y \dot{<} v)\neg\widetilde{\varphi}(y),$$

which appears in (6.8) of Lemma 6.3.11 (where $v$ replaces $\overline{n}$). We will now show that

$$\Omega \vdash \forall v((\widetilde{\varphi}(v) \wedge (\forall y \dot{<} v)\neg\widetilde{\varphi}(y)) \leftrightarrow v \doteq \overline{f(n_1, n_2, \dots, n_k)}). \tag{6.10}$$

Let $f(n_1, n_2, \dots, n_k) = n$. Since $\overline{n}$ and $\overline{f(n_1, n_2, \dots, n_k)}$ are exactly the same term, we can replace $\overline{f(n_1, n_2, \dots, n_k)}$ in (6.10) with $\overline{n}$. By Theorem 3.3.29 and the biconditional law, it is now enough to show that

$$\Omega \vdash (\widetilde{\varphi}(v) \wedge (\forall y \dot{<} v)\neg\widetilde{\varphi}(y)) \to v \doteq \overline{n}, \tag{6.11}$$

$$\Omega \vdash v \doteq \overline{n} \to (\widetilde{\varphi}(v) \wedge (\forall y \dot{<} v)\neg\widetilde{\varphi}(y)). \tag{6.12}$$

To establish (6.11), Theorem 3.3.33 implies that we just need to prove that

$$\Omega \cup \{\widetilde{\varphi}(v) \wedge (\forall y \dot{<} v)\neg\widetilde{\varphi}(y)\} \vdash v \doteq \overline{n}. \tag{6.13}$$

To do this, we will first show that

$$\Omega \cup \{\widetilde{\varphi}(v) \wedge (\forall y \dot{<} v)\neg\widetilde{\varphi}(y)\} \vdash v \not\dot{<} \overline{n}, \tag{6.14}$$

$$\Omega \cup \{\widetilde{\varphi}(v) \wedge (\forall y \dot{<} v)\neg\widetilde{\varphi}(y)\} \vdash \overline{n} \not\dot{<} v. \tag{6.15}$$

Recall that $(\forall y \dot{<} v)\neg\widetilde{\varphi}(y)$ is the abbreviated form of $\forall y(y \dot{<} v \to \neg\widetilde{\varphi}(y))$. To verify (6.14), we shall use Corollary 3.3.37. Let

$$\Gamma = \Omega \cup \{\widetilde{\varphi}(v) \wedge (\forall y \dot{<} v)\neg\widetilde{\varphi}(y), v \dot{<} \overline{n}\}.$$

We will show that $\Gamma$ is inconsistent. By Lemma 6.3.11, we have

$$\Omega \vdash (\forall y \dot{<} \overline{n})\neg\widetilde{\varphi}(y).$$

Therefore, $\Gamma \vdash (\forall y \dot{<} \overline{n})\neg\widetilde{\varphi}(y)$. Since $v \dot{<} \overline{n}$ is in $\Gamma$, we conclude that (▲) $\Gamma \vdash \neg\widetilde{\varphi}(v)$. However, as $\widetilde{\varphi}(v)$ is in $\Gamma$, we also have $\Gamma \vdash \widetilde{\varphi}(v)$. This together with (▲) shows that $\Gamma$ is inconsistent. Therefore, (6.14) holds by Corollary 3.3.37.

We now prove (6.15). By Lemma 6.3.11, we have (▼) $\Omega \vdash \widetilde{\varphi}(\overline{n})$. Let

$$\Gamma^* = \Omega \cup \{\widetilde{\varphi}(v) \wedge (\forall y \dot{<} v)\neg\widetilde{\varphi}(y), \overline{n} \dot{<} v\}.$$

So $\Gamma^* \vdash \neg\widetilde{\varphi}(\overline{n})$ and $\Gamma^*$ is inconsistent by (▼). Hence, (6.15) is confirmed. Axiom (A5), together with (6.14) and (6.15), establishes (6.13), that is,

$$\Omega \cup \{\widetilde{\varphi}(v) \wedge (\forall y \dot{<} v)\neg\widetilde{\varphi}(y)\} \vdash v \doteq \overline{n}.$$

To prove (6.12), we only need to show that

$$\Omega \cup \{v \doteq \overline{n}\} \vdash \widetilde{\varphi}(v) \wedge (\forall y \dot{<} v)\neg\widetilde{\varphi}(y).$$

Lemma 6.3.11 implies that $\Omega \vdash \widetilde{\varphi}(\overline{n}) \wedge (\forall y \dot{<} \overline{n})\neg\widetilde{\varphi}(y)$. Hence, by Exercise 7 on page 135, $\Omega \cup \{v \doteq \overline{n}\} \vdash \widetilde{\varphi}(v) \wedge (\forall y \dot{<} v)\neg\widetilde{\varphi}(y)$. Therefore, $f$ is functionally representable. □

By Theorem 6.3.12, we can now say that a function $f$ is *representable* to mean that $f$ is functionally representable and/or graph representable. The next theorem summarizes some of the observations made in this section.

**Theorem 6.3.13.** *The relations $<$ and $=$ and the functions $+$, $\times$, $E$, and $S$ defined on the natural numbers are all representable.*

We end this section by proving that all representable relations give rise to representable functions.

**Theorem 6.3.14.** *Let $R \subseteq \mathbb{N}^k$ be a relation. Then $R$ is representable if and only if its characteristic function $C_R$ is representable.*

*Proof.* Let $R \subseteq \mathbb{N}^k$ be a relation.

($\Rightarrow$). Assume the relation $R$ is representable. By Definition 6.3.1, let $\varphi(x_1, \ldots, x_k)$ represent $R$. Thus, (▲) $\varphi$ is numeralwise determined and (▼) $\varphi$ defines $R$ over $\mathcal{N}$, by Lemma 6.3.3. We will show that $C_R$ is graph representable. Let $\psi$ be the $\mathcal{L}$-wff

$$(\varphi(x_1, \ldots, x_k) \wedge v \doteq \overline{1}) \vee (\neg\varphi(x_1, \ldots, x_k) \wedge v \doteq \dot{0}). \tag{6.16}$$

Clearly, by (▼), the formula $\psi$ defines the graph of $C_R$ over $\mathcal{N}$. Moreover, by (▲), $\psi$ is numeralwise determined (see Exercises 1 and 1). Therefore, by Lemma 6.3.3, $C_R$ is graph representable.

($\Leftarrow$). Assume $C_R$ is representable. So let $\theta(x_1, \ldots, x_k, v)$ be a formula that verifies that $C_R$ is graph representable. The formula $\theta(x_1, \ldots, x_k, \overline{1})$ thus represents $R$. □

### 6.3.1 Recursive relations and functions are representable

We have shown that the initial functions are representative (see Corollary 6.3.9). So, to prove that every recursive function is representable, we need to prove that the set of all representable functions is closed under composition, total search, and primitive recursion.

**Theorem 6.3.15.** *The composition of representable functions is representable.*

*Proof.* Let $g_1, \ldots, g_j$ be functionally representable functions from $\mathbb{N}^k$ to $\mathbb{N}$ and let $f : \mathbb{N}^j \to \mathbb{N}$ be functionally representable. We will show that the composition

$$f(g_1(x_1, \ldots, x_k), \ldots, g_j(x_1, \ldots, x_k))$$

is functionally representable. Let $\theta_1(x_1, \ldots, x_k, v_1), \ldots, \theta_j(x_1, \ldots, x_k, v_j)$ functionally represent $g_1, \ldots g_j$, respectively. Thus, for each $i$, where $1 \leq i \leq j$, we have

$$\Omega \vdash \forall v_i(\theta_i(\overline{n}_1, \ldots, \overline{n}_k, v_i) \leftrightarrow v_i \doteq \overline{g_i(n_1, \ldots, n_k)}), \quad \text{for all } n_1, \ldots, n_k \in \mathbb{N}. \tag{6.17}$$

Also let $\psi(x_1, \ldots, x_j, v)$ functionally represent $f$, that is,

$$\Omega \vdash \forall v(\psi(\overline{n}_1, \ldots, \overline{n}_j, v) \leftrightarrow v \doteq \overline{f(n_1, \ldots, n_j)}), \quad \text{for all } n_1, \ldots, n_j \in \mathbb{N}. \tag{6.18}$$

Remark 6.3.6 shows that (6.17) implies, for each $i$, that $\Omega \vdash \exists! v_i \theta_i(\overline{n}_1, \overline{n}_2, \ldots, \overline{n}_k, v_i)$, that is, there is exactly one such $v_i$, namely, $\overline{g_i(n_1, \ldots, n_k)}$. Now let $\varphi(x_1, \ldots, x_k, v)$ be the formula

$$\exists v_1 \cdots \exists v_j(\theta_1(x_1, \ldots, x_k, v_1) \wedge \cdots \wedge \theta_j(x_1, \ldots, x_k, v_j) \wedge \psi(v_1, \ldots, v_j, v)).$$

Since (6.17) holds for each $i$, where $1 \leq i \leq j$, and (6.18) holds, it follows that each of the following successive biconditionals are deducible from $\Omega$:

$$\begin{aligned}
&\varphi(\overline{n}_1, \ldots, \overline{n}_k, v) \\
&\quad \leftrightarrow \exists v_1 \cdots \exists v_j(\theta_1(\overline{n}_1, \ldots, \overline{n}_k, v_1) \wedge \cdots \wedge \theta_j(\overline{n}_1, \ldots, \overline{n}_k, v_j) \wedge \psi(v_1, \ldots, v_j, v)) \\
&\quad \leftrightarrow \exists v_1 \cdots \exists v_j(v_1 \doteq \overline{g_1(n_1, \ldots, n_k)} \wedge \cdots \wedge v_j \doteq \overline{g_j(n_1, \ldots, n_k)} \wedge \psi(v_1, \ldots, v_j, v)) \\
&\quad \leftrightarrow \psi(\overline{g_1(n_1, \ldots, n_k)}, \ldots, \overline{g_j(n_1, \ldots, n_k)}, v) \\
&\quad \leftrightarrow v \doteq \overline{f(g_1(n_1, \ldots, n_k), \ldots, g_j(n_1, \ldots, n_k))}.
\end{aligned}$$

Therefore,

$$\Omega \vdash \forall v(\varphi(\overline{n}_1, \ldots, \overline{n}_k, v) \leftrightarrow v \doteq \overline{f(g_1(n_1, \ldots, n_k), \ldots, g_j(n_1, \ldots, n_k))}).$$

Hence, the composition is functionally representable. $\square$

Theorems 6.3.15 and 6.3.14 imply that representable relations are also closed under composition with representable functions.

**Theorem 6.3.16.** *Let $g_1, \ldots, g_j$ be representable functions from $\mathbb{N}^k$ to $\mathbb{N}$ and let $R \subseteq \mathbb{N}^j$ be representable. Then the relation $S \subseteq \mathbb{N}^k$ defined by*

$$S(x_1, x_2, \ldots, x_k) \quad \textit{iff} \quad R(g_1(x_1, \ldots, x_k), \ldots, g_j(x_1, \ldots, x_k))$$

*is representable.*

Corollary 6.3.9 confirms that the projection functions are representable. Therefore, by Theorem 6.3.15, one can compose a representable function with a projection function and thereby obtain a representable function. Theorems 6.3.15 and 6.3.16 assume that the functions $g_1, \ldots, g_j$ have the same arity. However, the projection functions can be used to circumvent this apparent rigidity. By applying compositions with various projection functions, one can apply Theorems 6.3.15 and 6.3.16 when the functions $g_1, \ldots, g_j$ have different arity. For example, let $f\colon \mathbb{N}^2 \to \mathbb{N}$ and let $g_1\colon \mathbb{N}^2 \to \mathbb{N}$ and $g_2\colon \mathbb{N} \to \mathbb{N}$. Then the function

$$h(a, b) = f(g_1(a, b), g_2(b))$$

can be obtained by a composition with a projection function, namely,

$$h(a, b) = f(g_1(a, b), g_2(I_2^2(a, b))).$$

We have now shown that the set of representable functions contains all of the initial functions and is also closed under composition. The next theorem shows that the set of representable functions is closed under the total search operation (see Definition 5.3.5). Upon completing its proof, we will be closer to showing that all recursive functions are representable. To complete this task, we will need to show that the set of representable functions is closed under primitive recursion.

**Theorem 6.3.17.** *Let $g\colon \mathbb{N}^{k+1} \to \mathbb{N}$ be representable. Suppose that for all $\vec{x} \in \mathbb{N}^k$, there is a $y \in \mathbb{N}$ such that $g(\vec{x}, y) = 0$. Then the function $h\colon \mathbb{N}^k \to \mathbb{N}$ defined by*

$$h(\vec{x}) = \mu y(g(\vec{x}, y) = 0)$$

*is representable.*

*Proof.* We will show that $h$ is graph representable. Let $\psi(\vec{x}, y, z)$ affirm that $g$ is graph representable. Since

$$h(\vec{x}) = \text{the least } y \text{ such that } g(\vec{x}, y) = 0,$$

we see that

$$h(\vec{x}) = y \quad \text{iff} \quad g(\vec{x}, y) = 0 \wedge (\forall v < y) g(\vec{x}, v) \neq 0. \tag{6.19}$$

Now let $\theta(\vec{x}, y)$ be the $\mathcal{L}$-wff

$$\psi(\vec{x}, y, \dot{0}) \wedge (\forall v \mathrel{\dot{<}} y) \neg \psi(\vec{x}, v, \dot{0}).$$

This wff $\theta(\vec{x}, y)$ formalizes the right hand side of (6.19) and thus defines the graph of $h$ over $\mathcal{N}$. Since $\psi$ is representable, Theorem 6.3.4 implies that $\theta(\vec{x}, y)$ is numeralwise determined. Therefore, the graph of $h$ is representable by Lemma 6.3.3. □

**Corollary 6.3.18.** *Let $R \subseteq \mathbb{N}^{k+1}$ be a representable relation. Suppose that for all $\vec{x} \in \mathbb{N}^k$ there is a $y \in \mathbb{N}$ such that $R(\vec{x}, y)$. Then the function $f: \mathbb{N}^k \to \mathbb{N}$ defined by*

$$f(\vec{x}) = \textit{the least } y \in \mathbb{N} \textit{ such that } R(\vec{x}, y) \tag{6.20}$$

*is representable. Moreover, $R(\vec{x}, f(\vec{x}))$ for all $\vec{x} \in \mathbb{N}^k$.*

*Proof.* Let $R \subseteq \mathbb{N}^{k+1}$ be a representable relation. Let $\overline{R} = \mathbb{N}^{k+1} \setminus R$. By Exercise 2 and Theorem 6.3.14, the characteristic function $C_{\overline{R}}$ is representable. Since for all $\vec{x} \in \mathbb{N}^k$ there is a $y \in \mathbb{N}$ such that $R(\vec{x}, y)$, the function $f: \mathbb{N}^k \to \mathbb{N}$ defined by (6.20) satisfies $f(\vec{x}) = \mu y(C_{\overline{R}}(\vec{x}, y) = 0)$. Thus, $f$ is representable by Theorem 6.3.17. In addition, $R(\vec{x}, f(\vec{x}))$ for all $\vec{x} \in \mathbb{N}^k$. □

Recall that prime numbers and exponents are very useful for coding finite sequences of natural numbers. In Section 5.3, we showed how to encode any finite sequence $\langle x_0, x_1, \ldots, x_k \rangle$ of natural numbers by a single natural number which is denoted by $[x_0, x_1, \ldots, x_k]$ (see (5.9) on page 190). Any such natural number is called a *sequence number*. To show that representable functions are closed under primitive recursion, we will show that there exists a representable function that can decode any sequence number. We first formally repeat an observation that was made earlier.

**Proposition 6.3.19.** *The set of prime numbers is a representable set.*

*Proof.* Let $\pi(v)$ be the formula

$$\dot{S}\dot{0} \mathrel{\dot{<}} v \wedge (\forall x \mathrel{\dot{<}} v)(\forall y \mathrel{\dot{<}} v)(x \mathbin{\dot{\times}} y \mathrel{\dot{\neq}} v). \tag{6.21}$$

Theorem 6.3.4 implies that $\pi(v)$ is numeralwise determined. Since $\pi(v)$ defines the set of primes in $\mathcal{N}$, Lemma 6.3.3 shows that the set of primes is representable. □

Two prime numbers $x$ and $y$ are said to be *adjacent primes* when $x < y$ and no other prime number is strictly between $x$ and $y$.

**Proposition 6.3.20.** *Let $R = \{\langle x, y \rangle \in \mathbb{N}^2 : x \textit{ and } y \textit{ are adjacent primes}\}$. Then $R$ is representable.*

*Proof.* Let $\pi(v)$ be the formula (6.21) in the above proof that represents the set of prime numbers. Let $\alpha(x,y)$ be the formula

$$\pi(x) \land \pi(y) \land (\forall z \dot{<} y)((x \dot{<} z \land z \dot{<} y) \to \neg\pi(z)).$$

Clearly, $\alpha(x,y)$ defines the relation $R$ over $\mathcal{N}$. Proposition 6.3.19, Theorem 6.3.4, and Lemma 6.3.3 imply that $\alpha(x,y)$ represents $R$. □

For each $x \in \mathbb{N}$, let $p_x$ be the $(x+1)$-st prime number and let $g\colon \mathbb{N} \to \mathbb{N}$ be defined by $g(x) = p_x$. In the proof of Proposition 5.3.36, the function $g$ is defined by primitive recursion. Since we have not yet shown that such functions are representable, we will show that $g$ is representable by proving that its graph is representable.

Let $k \in \mathbb{N}$ and consider the natural number $z$ which is a product of adjacent primes,

$$z = p_0^0 \cdot p_1^1 \cdot p_2^2 \cdot p_3^3 \cdots p_k^k = 2^0 \cdot 3^1 \cdot 5^2 \cdot 7^3 \cdot 11^4 \cdots p_k^k,$$

where each succeeding exponent is the successor of the previous exponent. So, $2 \nmid z$. Let $p$ be a prime number such that $p^3 \mid z$ and $p^4 \nmid z$. What is $p$? Yes, $p = p_3 = 7$. In fact, by the definition of $z$, we see that $p_3 = 7$ if and only if $7^3 \mid z$ and $7^4 \nmid z$. Note that $2^0 \cdot 3^1 \cdot 5^2 \cdot 7^3 \le 7^1 \cdot 7^2 \cdot 7^3 \le 7^{3^2}$. One can show in general that

$$z = 2^0 \cdot 3^1 \cdot 5^2 \cdot 7^3 \cdot 11^4 \cdots p_k^k \le p_k^{k^2}. \tag{6.22}$$

Also, each of the above exponents on the left hand side of $\le$ are strictly less than $z$. These observations will be employed in the obscure proof of our next proposition.

**Proposition 6.3.21.** *Let function $g\colon \mathbb{N} \to \mathbb{N}$ defined by $g(x) = p_x$ be representable.*

*Proof.* We show that $g$ is graph representable. Consider the representable relations:
1. $\Pi(p)$: "$p$ is a prime,"
2. $A(q,r)$: "$q$ and $r$ are adjacent primes,"
3. $E(q^i, z, r^{i+1})$: "$q^i \mid z$ if and only if $r^{i+1} \mid z$,"
4. $B(p^x, z, p^{x+1})$: "$p^x \mid z$ and $p^{x+1} \nmid z$."

The divisibility relation $m \mid n$ is representable by Exercise 1. Also, items 3 and 4 are representable by Theorems 6.3.13 and 6.3.16. It now follows that $p_x = p$ if and only if

$$\begin{aligned}&\Pi(p) \land (\exists z \le p^{x^2})(2 \nmid z \land (\forall q < p)(\forall r \le p)(A(q,r) \to (\forall i < z)E(q^i, z, r^{i+1}))\\ &\land B(p^x, z, p^{x+1})).\end{aligned}$$

The conjunction above $B(p^x, z, p^{x+1})$ ensures that $z$ has the form in (6.22), at least up to $p_x^x$. Thus, $g$ is graph representable. So $g$ is representable by Theorem 6.3.12. □

Now, let $s$ be a natural number that codes a sequence, that is, $s = [a_0, a_1, a_2, \ldots, a_k]$. For $i \leq k$, we let $(s)_i = a_i$. Therefore, the function $\langle s, i\rangle \mapsto (s)_i$ acts as a "decoding" function. Proposition 5.3.38 shows this decoding function is primitive recursive. Our next proposition shows that this function is representable.

**Proposition 6.3.22.** *The function $\langle s, i\rangle \mapsto (s)_i$ is functionally representable.*

*Proof.* Let $R$ be the representable relation defined by

$$R(s,i,y) \quad \text{iff} \quad s = 0 \vee p_i^{y+2} \nmid s.$$

Clearly, for all natural numbers $s$ and $i$, there exists a $y$ such that $R(s,i,y)$. Therefore, $(s)_i = \mu y R(s,i,y)$ and is representable by Corollary 6.3.18. □

Note that $(s)_i$ is defined for all $s, i \in \mathbb{N}$, even when $s$ does not code a sequence.

Using Proposition 6.3.22, we can now prove that a function defined by primitive recursion, using representable functions, is itself representable. The proof is similar to item (2) in the proof of Theorem 5.4.10.

**Theorem 6.3.23.** *Let $f\colon \mathbb{N}^k \to \mathbb{N}$ and $g\colon \mathbb{N}^{k+2} \to \mathbb{N}$ be representable. Then the function $h\colon \mathbb{N}^{k+1} \to \mathbb{N}$ defined by the primitive recursion*
(a) $h(\vec{x}, 0) = f(\vec{x})$,
(b) $h(\vec{x}, n+1) = g(h(\vec{x}, n), \vec{x}, n)$, *for all $n \in \mathbb{N}$,*

*is also representable.*

*Proof.* Let $f\colon \mathbb{N}^k \to \mathbb{N}$ and $g\colon \mathbb{N}^{k+2} \to \mathbb{N}$ be graph representable. We will show that the same holds for $h$. Consider the relation $R \subseteq \mathbb{N}^{k+2}$ defined by

$$R(\vec{x}, n, s) \quad \text{iff} \quad (f(\vec{x}) = (s)_0 \quad \text{and} \quad (\forall i < n)(g((s)_i, \vec{x}, i) = (s)_{i+1})).$$

Proposition 6.3.22, Lemma 6.3.3, Theorem 6.3.4, and Theorem 6.3.16 imply that $R$ is representable. Clearly, for all $\vec{x} \in \mathbb{N}^k$ and $n \in \mathbb{N}$, there is an $s \in \mathbb{N}$ such that $R(\vec{x}, n, s)$. Corollary 6.3.18 thus implies that there is a representable function $\ell\colon \mathbb{N}^{k+1} \to \mathbb{N}$ such that $R(\vec{x}, n, \ell(\vec{x}, n))$, for every $\vec{x} \in \mathbb{N}^k$ and $n \in \mathbb{N}$. Because $h(\vec{x}, n) = (\ell(\vec{x}, n))_n$ for each $\vec{x} \in \mathbb{N}^k$ and $n \in \mathbb{N}$, Theorem 6.3.15 implies that $h$ is representable. □

**Representability Theorem 6.3.24.** *Every recursive function and recursive relation is representable.*

*Proof.* The initial functions are representable by Corollary 6.3.9. Furthermore, the set of representable functions is closed under composition, total search, and primitive recursion by Theorems 6.3.16, 6.3.17, and 6.3.23. Thus, as discussed on page 175, the set of representable functions contains all of the recursive functions. Since a recursive relation is one whose characteristic function is recursive, we also see that every recursive relation is representable. □

Lemma 6.3.3 implies that every representable relation is definable over $\mathcal{N}$. Thus, the following corollary follows from Theorem 6.3.24.

**Corollary 6.3.25.** *Every recursive relation is definable over* $\mathcal{N}$.

The converse of Theorem 6.3.24 also holds (see Theorem 6.4.26).

**Exercises 6.2.**

*1. Let $R = \{\langle m, n\rangle : m \mid n\}$. Show that $R$, the divisibility relation, is representable.

*2. Let $R \subseteq \mathbb{N}^k$. Show that $R$ is representable if and only if $\mathbb{N}^k \setminus R$ is representable.

3. Let $\sim$ be an equivalence relation on $\mathbb{N}$. Suppose that $\sim$ is representable. Show that each equivalence class is representable.

*4. Let $\varphi$ and $\psi$ be numeralwise determined. Show that each of the following formulas is also numeralwise determined:
   (a) $\neg\varphi$,
   (b) $\varphi \to \psi$,
   (c) $\varphi \wedge \psi$,
   (d) $\varphi \vee \psi$.

*5. Let $k \in \mathbb{N}$. Show that each formula listed below is numeralwise determined:
   (a) $v \doteq \overline{k}$,
   (b) $v \dot{<} \overline{k}$,
   (c) $\overline{k} \dot{<} v$.

*6. Let $R \subseteq \mathbb{N}^k$ be representable. Show that there is an $\mathcal{L}$-formula $\varphi(x_1, x_2, \ldots, x_k)$ such that for all $n_1, n_2, \ldots, n_k \in \mathbb{N}$,
   (1) $\langle n_1, n_2, \ldots, n_k\rangle \in R$ if and only if $\Omega \vdash \varphi(\overline{n}_1, \overline{n}_2, \ldots, \overline{n}_k)$,
   (2) $\langle n_1, n_2, \ldots, n_k\rangle \notin R$ if and only if $\Omega \vdash \neg\varphi(\overline{n}_1, \overline{n}_2, \ldots, \overline{n}_k)$.

*7. Let $\varphi(v_1, v_2, \ldots, v_k)$ represent a relation $R \subseteq \mathbb{N}^k$. Show that $\langle n_1, n_2, \ldots, n_k\rangle \in R$ if and only if $\mathcal{N} \models \varphi[\![n_1, n_2, \ldots, n_k]\!]$, for all $\langle n_1, n_2, \ldots, n_k\rangle \in \mathbb{N}^k$.

8. Let $R \subseteq \mathbb{N}^k$ and $S \subseteq \mathbb{N}^k$ be representable relations. Show that $R \cap S$ and $R \cup S$ are representable.

9. Let $\tau(x)$ be a term and let $g\colon \mathbb{N} \to \mathbb{N}$ be representable. Let $h\colon \mathbb{N} \to \mathbb{N}$ be defined by $h(x) = g(\tau^{\mathcal{N}}(x))$. Show that $h$ is representable.

10. Let $R \subseteq \mathbb{N}^k$ be a representable relation and let $f\colon \mathbb{N}^k \to \mathbb{N}$ and $g\colon \mathbb{N}^k \to \mathbb{N}$ be representable functions. Show that the function $h\colon \mathbb{N}^k \to \mathbb{N}$ is representable, where

$$h(\vec{x}) = \begin{cases} f(\vec{x}), & \text{if } \vec{x} \in R, \\ g(\vec{x}), & \text{if } \vec{x} \notin R. \end{cases}$$

11. Prove Corollary 6.3.25.

12. Let $R \subseteq \mathbb{N}^k$ be decidable. Show that $R$ is representable.

13. Show that if $\varphi$ is numeralwise determined, then so are $(\forall x \dot{\leq} y)\varphi$ and $(\exists x \dot{\leq} y)\varphi$.

## 6.4 Arithmetization of the formal language

The next important step which will allow us to address question (Q2) on page 205 is to take our formal language for number theory and define an effective correspondence between the formulas of this language and a recursive set of natural numbers. Such a correspondence is called an *arithmetization*, or *Gödel numbering*, of the language. Using our primitive recursive coding of sequences (via powers of primes) developed in Section 5.3, we can assign a unique natural number to each formula. We can then ask if the set of natural numbers that code all the valid formulas is decidable.

Again, let $\mathcal{L} = \{\dot{<}, \dot{0}, \dot{S}, \dot{+}, \dot{\times}, \dot{E}, \dot{=}\}$. We will assign natural numbers, called Gödel numbers, to all the terms and wffs of $\mathcal{L}$. We denote this assignment function by #. Thus, for every wff $\varphi$, $\#(\varphi)$ will be the Gödel number of $\varphi$. Recall that the operation $[x_0, x_1, \dots, x_k]$ denotes a sequence number (see Definition 5.3.39) which encodes the finite sequence of natural numbers $\langle x_0, x_1, x_2, \dots, x_k\rangle$ by a single natural number. Let Sq denote the set of sequence numbers.

**Definition 6.4.1.** Let $V = \{v_1, v_2, v_3, \dots\}$ be the set of all the variables of $\mathcal{L}$. Let $\mathcal{T}$ be the set of all the variables and the constant symbol in $\mathcal{L}$. Define $\#\colon \mathcal{T} \to \mathbb{N}$ by

$$\#(v) = \begin{cases} [2i], & \text{if } v = v_i, \\ [1], & \text{if } v = \dot{0}. \end{cases} \tag{6.23}$$

Let $V^{\#} = \{\#(v_i) : v_i \in V\}$. Since $m \in V^{\#}$ if and only if $m = [2i]$ for an $i$ such that $0 < i < m$, it follows that $V^{\#}$ is primitive recursive. Theorem 1.1.27 implies that we can extend the function # to all of the terms of $\mathcal{L}$ as in our next definition.

**Definition 6.4.2.** Let $T$ be the set of terms of $\mathcal{L}$. Define the function $\#\colon T \to \mathbb{N}$ by

$$\begin{aligned} \#(\dot{S}t) &= [3, \#(t)], \\ \#(t \mathbin{\dot{+}} \tau) &= [5, \#(t), \#(\tau)], \\ \#(t \mathbin{\dot{\times}} \tau) &= [7, \#(t), \#(\tau)], \\ \#(\dot{E}t\tau) &= [9, \#(t), \#(\tau)]. \end{aligned}$$

Let $T^{\#} = \{\#(t) : t \text{ is a term}\}$ be the set consisting of all the Gödel numbers of $\mathcal{L}$-terms. Now let $C$ be the characteristic function of $T^{\#}$ and let $m \in \mathbb{N}$. How can we evaluate $C(m)$? If $m = [1]$ or $m = \#(v_i)$, then $C(m) = 1$. If $m$ is a sequence number of length 2 and $(m)_0 = 3$, then $C(m) = C((m)_1)$. If $m$ is a sequence number of length 3 and either $(m)_0 = 5, 7$, or $9$, then $C(m) = C((m)_1) \cdot C((m)_2)$. If none of the previous conditions hold, then $C(m) = 0$. Note that $(m)_0, (m)_1, (m)_2$ are all strictly less than $m$. Thus, we can evaluate $C(m)$ using the values $C(0), C(1), \dots, C(m-1)$ and $m$.

We now expand the function # in Definition 6.4.2 to include the atomic formulas.

**Definition 6.4.3.** Let $A$ be the set of all the atomic formulas of $\mathcal{L}$ and define the function $\#: A \to \mathbb{N}$ by

$$\#(t \dot{<} \tau) = [11, \#(t), \#(\tau)],$$
$$\#(t \doteq \tau) = [13, \#(t), \#(\tau)].$$

Let $A^\# = \{\#(\varphi) : \varphi \text{ is an atomic formula}\}$ and $m \in \mathbb{N}$. How can we evaluate $C(m)$, when $C$ is the characteristic function of $A^\#$? If $m$ is a sequence number of length 3 and $(m)_0 = 11$ or $(m)_0 = 13$, and $(m)_1, (m)_2 \in T^\#$, then $C(m) = 1$. Otherwise, $C(m) = 0$.

Theorem 1.1.27 again implies that the function # given in Definition 6.4.3 can be extended, as in the next definition, to all the wffs of the language $\mathcal{L}$.

**Definition 6.4.4.** Let $W$ be the set of wffs of $\mathcal{L}$. Define the function $\#: W \to \mathbb{N}$ by

$$\#(\neg\psi) = [15, \#(\psi)],$$
$$\#(\psi \to \varphi) = [17, \#(\psi), \#(\varphi)],$$
$$\#(\forall v_i \psi) = [19, \#(v_i), \#(\psi)].$$

Let $W^\# = \{\#(\varphi) : \varphi \text{ is a wff}\}$ and $m \in \mathbb{N}$. How can we evaluate $C(m)$, when $C$ is the characteristic function of $W^\#$? If $m \in A^\#$, then $C(m) = 1$. If $m$ is a sequence number of length 2 and $(m)_0 = 15$, then $C(m) = C((m)_1)$. When $m$ is a sequence number of length 3 and $(m)_0 = 17$, then $C(m) = C((m)_1) \cdot C((m)_2)$. If $m$ is a sequence number of length 3, $(m)_0 = 19$, and $(m)_1 = [2i]$, where $i \geq 1$, then $C(m) = C((m)_2)$. If all of these conditions fail to hold, then $C(m) = 0$.

The observations we made on the characteristic functions of $T^\#$, $A^\#$, and $W^\#$ are formalized in the proof of our next result.

**Proposition 6.4.5.** *The following sets of natural numbers are primitive recursive:*
1. $T^\# = \{\#(t) : t \text{ is an } \mathcal{L}\text{-term}\}$;
2. $A^\# = \{\#(\varphi) : \varphi \text{ is an } \mathcal{L}\text{-atomic formula}\}$;
3. $W^\# = \{\#(\varphi) : \varphi \text{ is an } \mathcal{L}\text{-wff}\}$.

*Proof.* We will be applying Proposition 5.3.46, by defining a course-of-values recursion. Recall that if $C: \mathbb{N} \to \mathbb{N}$ and $s = \overline{C}(m) = [C(0), C(1), \ldots, C(m-1)]$, then $(s)_i = C(i)$ for $i < m$. Let Sq be the set of sequence numbers and let lh be the length function.

1. Let $C$ be the characteristic function of $T^\#$. We shall define a primitive recursive function $g$ such that $C(m) = g(\overline{C}(m), m)$ for all $m$. Define $g: \mathbb{N}^2 \to \mathbb{N}$ by

$$g(s,m) = \begin{cases} 1, & \text{if } m = [1] \text{ or } m \in V^\#, \\ (s)_{(m)_1}, & \text{if } m \in \mathrm{Sq}, \mathrm{lh}(m) = 2, \text{and } (m)_0 = 3, \\ (s)_{(m)_1} \times (s)_{(m)_2}, & \text{if } m \in \mathrm{Sq}, \mathrm{lh}(m) = 3, \text{and } (m)_0 = 5, 7, \text{or } 9, \\ 0, & \text{otherwise.} \end{cases}$$

By letting $s = \overline{C}(m) = [C(0), C(1), \ldots, C(m-1)]$ for each $m \in \mathbb{N}$, we infer (see the discussion after Definition 6.4.2) that

$$C(m) = g(\overline{C}(m), m). \tag{6.24}$$

One can formally prove (6.24) by induction on $m$. Since Sq and lh are primitive recursive by Propositions 5.3.40 and 5.3.41, one can now easily confirm that the conditions in the definition of $g$ are all primitive recursive. Proposition 5.3.46 thus implies that $C$ is primitive recursive.

2. Let $C$ be the characteristic function of $A^\#$. Then

$$C(m) = \begin{cases} 1, & \text{if } m \in \text{Sq}, \text{lh}(m) = 3, (m)_0 = 11 \text{ or } 13, (m)_1 \in T^\#, \text{and } (m)_2 \in T^\#, \\ 0, & \text{otherwise.} \end{cases}$$

Since $T^\#$ is primitive recursive and the conditions in the description of $C$ are all primitive recursive, we conclude that $A^\#$ is primitive recursive.

3. As in item 1, one can evaluate $C(m)$, the characteristic function of $W^\#$, using the values $C(0), C(1), \ldots, C(m-1)$ and $m$. See Exercise 4. □

When applying the function # to a term $\tau$ or formula $\varphi$, we may write $\#\tau$ and $\#\varphi$ to denote, respectively, $\#(\tau)$ and $\#(\varphi)$. In particular, $\#\tau_t^x = \#(\tau_t^x)$ and $\#\varphi_t^x = \#(\varphi_t^x)$. We note that Exercise 10 on page 16 implies that the function # is one-to-one.

Given a wff $\alpha$, a term $t$, and a variable $x$, Definition 3.3.13 presents a recursive definition of the substitution operation $\alpha_t^x$. Given the Gödel numbers of $\alpha$, $t$, and $x$, can we effectively get the Gödel number of $\alpha_t^x$? The answer is yes.

First we address this question when $\alpha$ is a term $\tau$. Consider the Gödel number $\#\tau_t^{v_i}$, where $\tau, t$ are terms and $v_i$ is a variable. How can we evaluate the natural number $\#\tau_t^{v_i}$ using the natural numbers $\#t, \#v_i, \#\tau$? Consider the following recursive attempt:

$$\#\tau_t^{v_i} = \begin{cases} \#t, & \text{if } \#\tau = \#v_i, \\ [3, \#q_t^{v_i}], & \text{if } \#\tau = [3, \#q], \text{ where } q \text{ is a term}, \\ [5, \#q_t^{v_i}, \#z_t^{v_i}], & \text{if } \#\tau = [5, \#q, \#z], \text{ where } q, z \text{ are terms}, \\ [7, \#q_t^{v_i}, \#z_t^{v_i}], & \text{if } \#\tau = [7, \#q, \#z], \text{ where } q, z \text{ are terms}, \\ [9, \#q_t^{v_i}, \#z_t^{v_i}], & \text{if } \#\tau = [9, \#q, \#z], \text{ where } q, z \text{ are terms}, \\ \#\tau, & \text{otherwise.} \end{cases}$$

This attempt motivates our subsequent proof where, in the definition of $g$, the inputs $u, i, m$ correspond, respectively, to the above $\#t, \#v_i, \#\tau$.

**Proposition 6.4.6.** *There is a primitive recursive function* $\text{Sb}': \mathbb{N}^3 \to \mathbb{N}$ *such that for any* $\mathcal{L}$*-term* $t$ *and variable* $v_i$*, we have* $\text{Sb}'(\#t, \#v_i, \#\tau) = \#\tau_t^{v_i}$ *for all terms* $\tau$*.*

*Proof.* Define $g\colon \mathbb{N}^4 \to \mathbb{N}$ by

$$g(s,u,i,m) = \begin{cases} u, & \text{if } m = \#v_i, & (1)\\ [(m)_0, (s)_{(m)_1}], & \text{if } m \in \mathrm{Sq},\ (m)_0 = 3, \text{ and } \mathrm{lh}(m) = 2, & (2)\\ [(m)_0, (s)_{(m)_1}, (s)_{(m)_2}], & \text{if } m \in \mathrm{Sq},\ (m)_0 = 5, \text{ and } \mathrm{lh}(m) = 3, & (3)\\ [(m)_0, (s)_{(m)_1}, (s)_{(m)_2}], & \text{if } m \in \mathrm{Sq},\ (m)_0 = 7, \text{ and } \mathrm{lh}(m) = 3, & (4)\\ [(m)_0, (s)_{(m)_1}, (s)_{(m)_2}], & \text{if } m \in \mathrm{Sq},\ (m)_0 = 9, \text{ and } \mathrm{lh}(m) = 3, & (5)\\ m, & \text{otherwise.} & (6) \end{cases}$$

Define $h\colon \mathbb{N}^3 \to \mathbb{N}$ by $h(u,i,m) = g(\overline{h}(u,i,m), u, i, m)$. Since $g$ is primitive recursive, $h$ is primitive recursive by Proposition 5.3.46. Let $t$ be a term and let $v_i$ be a variable.

**Claim.** *For all terms $\tau$, $h(\#t, \#v_i, \#\tau) = \#\tau_t^{v_i}$.*

*Proof of claim.* We prove that $h(\#t, \#v_i, \#\tau) = \#\tau_t^{v_i}$, by induction on terms $\tau$.

*Base step:* If $\tau = v_i$, then $\#\tau = \#v_i$. Thus, by (1) of the definition of $g$, we see that

$$h(\#t, \#v_i, \#\tau) = g(\overline{h}(\#t, \#v_i, \#v_i), \#t, \#v_i, \#v_i) = \#t.$$

If $\tau = v_j$, where $j \neq i$, then $\#\tau = \#v_j$. Thus, by (6) of the definition of $g$,

$$h(\#t, \#v_i, \#v_j) = g(\overline{h}(\#t, \#v_i, \#v_j), \#t, \#v_i, \#v_j) = \#v_j.$$

If $\tau = \dot{0}$, then $\#\tau = \#\dot{0}$, and by (6) of the definition of $g$,

$$h(\#t, \#v_i, \#\dot{0}) = g(\overline{h}(\#t, \#v_i, \#\dot{0}), \#t, \#v_i, \#\dot{0}) = \#\dot{0}.$$

*Inductive step:* Let $\tau$ and $\widehat{\tau}$ be arbitrary terms. Assume the induction hypothesis

$$h(\#t, \#v_i, \#\tau) = \#\tau_t^{v_i} \quad \text{and} \quad h(\#t, \#v_i, \#\widehat{\tau}) = \#\widehat{\tau}_t^{v_i}. \tag{IH}$$

We must prove that the same holds for each of the terms $\dot{S}\tau$, $\tau \dotplus \widehat{\tau}$, $\tau \dot{\times} \widehat{\tau}$, and $\dot{E}\tau\widehat{\tau}$. It is important to note that, by the definition of $\overline{h}$, we have

$$(\overline{h}(\#t, \#v_i, m))_k = h(\#t, \#v_i, k), \quad \text{whenever } k < m. \tag{6.25}$$

We now show that $h(\#t, \#v_i, \#\dot{S}\tau) = \#(\dot{S}\tau)_t^{v_i}$. By Definition 6.4.2, $\#\dot{S}\tau = [3, \#\tau]$. Also note that $(\star)$ $([3, \#\tau])_1 = \#\tau$. Thus,

$$\begin{aligned} h(\#t, \#v_i, \#\dot{S}\tau) &= g(\overline{h}(\#t, \#v_i, [3,\#\tau]), \#t, \#v_i, [3,\#\tau]) && \text{by definition of } h,\\ &= [3, (\overline{h}(\#t, \#v_i, [3,\#\tau]))_{\#\tau}] && \text{by } (\star) \text{ and (2) of the definition of } g,\\ &= [3, h(\#t, \#v_i, \#\tau)] && \text{by (6.25) as } \#\tau < [3,\#\tau],\\ &= [3, \#\tau_t^{v_i}] = \#(\dot{S}\tau)_t^{v_i} && \text{by (IH) and Definition 6.4.2.} \end{aligned}$$

Therefore, $h(\#t, \#v_i, \#\dot{S}\tau) = \#(\dot{S}\tau)_t^{v_i}$. The proofs for the terms $\tau \dotplus \hat{\tau}$, $\tau \dot{\times} \hat{\tau}$, and $\dot{E}\tau\hat{\tau}$ are similar (see Exercise 5). (Claim) □

Let $\mathrm{Sb}' = h$. Thus, $\mathrm{Sb}'$ is as required. (Proposition) □

Therefore, $\mathrm{Sb}'(\#t, \#v, \#\tau)$ is the Gödel number obtained when "$t$ replaces $v$ in $\tau$."

**Corollary 6.4.7.** *There is a primitive recursive relation $O \subseteq \mathbb{N}^2$ such that for every term $\tau$ and variable $v_i$,*

$$\langle \#v_i, \#\tau\rangle \in O \quad \textit{if and only if} \quad v_i \textit{ occurs in the term } \tau.$$

*Proof.* Clearly, $\langle \#v_i, \#\tau\rangle \in O$ if and only if $\mathrm{Sb}'(\#\dot{0}, \#v_i, \#\tau) \neq \#\tau$. □

Now let $\mathrm{Sb}(\#t, \#v_i, \#\alpha)$ denote the Gödel number $\#\alpha_t^{v_i}$, where $\alpha$ is a wff, $v_i$ is a variable, and $t$ is a term. Can we evaluate $\#\alpha_t^{v_i}$ using only the natural numbers $\#t$, $\#v_i$, and $\#\alpha$? Consider the recursive attempt, which is modeled after Definition 3.3.13:

$$\#\alpha_t^{v_i} = \begin{cases} [11, \#q_t^{v_i}, \#z_t^{v_i}], & \text{if } \#\alpha = [11, \#q, \#z], \text{ where } q, z \text{ are terms,} \\ [13, \#q_t^{v_i}, \#z_t^{v_i}], & \text{if } \#\alpha = [13, \#q, \#z], \text{ where } q, z \text{ are terms,} \\ [15, \#\psi_t^{v_i}], & \text{if } \#\alpha = [15, \#\psi], \text{ where } \psi \text{ is a formula,} \\ [17, \#\psi_t^{v_i}, \#\varphi_t^{v_i}], & \text{if } \#\alpha = [17, \#\psi, \#\varphi], \text{ where } \psi, \varphi \text{ are formulas,} \\ [19, \#v_j, \#\psi_t^{v_i}], & \text{if } \#\alpha = [19, \#v_j, \#\psi], \text{ where } \#v_i \neq \#v_j \text{ and } \psi \text{ is a wff,} \\ \#\alpha, & \text{otherwise.} \end{cases}$$

The above "recursive attempt" is presented only to assist the reader in extending the proof of Proposition 6.4.6 and thereby prove the following result (see Exercise 6).

**Proposition 6.4.8.** *There is a primitive recursive function* $\mathrm{Sb}\colon \mathbb{N}^3 \to \mathbb{N}$ *such that for any $\mathcal{L}$-term $t$ and variable $v_i$, we have* $\mathrm{Sb}(\#t, \#v_i, \#\alpha) = \#\alpha_t^{v_i}$ *for all wffs $\alpha$.*

$\mathrm{Sb}(\#t, \#v, \#\alpha)$ is the Gödel number obtained when "$t$ replaces $v$ in $\alpha$," if $v$ is free in $\alpha$.

**Corollary 6.4.9.** *There is a primitive recursive relation $F \subseteq \mathbb{N}^2$ such that for every term or formula $\alpha$ and variable $v_i$,*

$$\langle \#v_i, \#\alpha\rangle \in F \quad \textit{if and only if} \quad v_i \textit{ occurs free in } \alpha.$$

*Proof.* Define $F$ by

$$\langle k, a\rangle \in F \quad \text{iff} \quad (a \in T^{\#} \text{ and } \mathrm{Sb}'(\#\dot{0}, k, a) \neq a) \quad \text{or} \quad (a \in W^{\#} \text{ and } \mathrm{Sb}(\#\dot{0}, k, a) \neq a).$$

Thus, $F$ is primitive recursive and satisfies the required condition. □

Let $S^{\#} = \{\#\varphi : \varphi \text{ is an } \mathcal{L}\text{-sentence}\}$. We can now prove that $S^{\#}$ is primitive recursive.

**Proposition 6.4.10.** *The set $S^{\#}$ of Gödel numbers of sentences is primitive recursive.*

*Proof.* Let $S^\#$ be the set of Gödel numbers of sentences. Then

$$a \in S^\# \quad \text{iff} \quad a \in W^\# \quad \text{and} \quad (\forall b < a)(\text{if } b \in V^\#, \text{then } \langle b, a\rangle \notin F).$$

Since $W^\#$, $V^\#$, and $F$ are primitive recursive, $S^\#$ is primitive recursive. □

Recall that a term $t$ is substitutable for a variable $x$ in a formula $\alpha$ if the resulting substitution $\alpha^x_t$ does not produce a captured variable. We will next show that, with respect to Gödel numbers, the substitutability relation is primitive recursive. This is done by a course-of-values recursion (one should review Proposition 5.3.46).

As you may remember, Definition 3.3.15 recursively defines the notion of a term being substitutable for a variable in a formula (one should also review Definition 3.3.15). This definition has multiple technical cases. The proof of our next proposition will therefore also have multiple technical cases.

To prepare for the proof, let us revisit the definition of substitutability via Gödel numbers. Let $t$ be a term and let $v_i$ be a variable. Let $K\colon \mathbb{N} \to \{0, 1\}$ be such that for any wff $\alpha$, $K(\#\alpha) = 1$ if and only if "$t$ is *substitutable* for $v_i$ in $\alpha$." For each wff $\alpha$, $K(\#\alpha)$ can be defined recursively, as in Definition 3.3.15, by

$$K(\#\alpha) = \begin{cases} 1, & \text{if } \#\alpha \in A^\#, \\ K(\#\psi), & \text{if } \#\alpha = [15, \#\psi], \\ K(\#\psi) \times K(\#\varphi), & \text{if } \#\alpha = [17, \#\psi, \#\varphi], \\ 1, & \text{if } \#\alpha = [19, \#v_j, \#\psi] \text{ and } \langle \#v_i, \#\alpha\rangle \notin F, \\ K(\#\psi), & \text{if } \#\alpha = [19, \#v_j, \#\psi],\ \langle \#v_i, \#\alpha\rangle \in F, \\ & \text{and } \langle \#v_j, \#t\rangle \notin O, \\ 0, & \text{otherwise.} \end{cases} \tag{6.26}$$

Since $K$ is defined recursively, one might reasonably infer that $K$ is recursive, as defined by Definition 5.3.6. In the proof of our next proposition, we will prove that $K$ is, in fact, primitive recursive. However, the proof is rather technical because it applies a "course-of-values" argument.

Here is a summary of the proof: The above recursive definition of $K$ will guide us in defining a primitive recursive function $g(s, x, y, m)$ where the variables $x, y, m$ correspond to the above $\#t, \#v_i, \#\alpha$. Using $g$, we will apply a course-of-values recursion to show that the notion of being substitutable can be interpreted, via the Gödel numbering, as being primitive recursive.

**Proposition 6.4.11.** *Let* Sbl *be the 3-place relation on Gödel numbers defined by*

$$\langle \#t, \#v_i, \#\alpha\rangle \in \text{Sbl} \quad \textit{if and only if} \quad t \textit{ is substitutable for } v_i \textit{ in } \alpha, \tag{6.27}$$

*whenever $t$ is a term, $v_i$ is a variable, and $\alpha$ is a formula. Then the relation* Sbl *is primitive recursive.*

*Proof.* Define the primitive recursive function $g: \mathbb{N}^4 \to \mathbb{N}$ by

$$g(s,x,y,m) = \begin{cases} 1, & \text{if } m \in A^{\#}, & (1) \\ (s)_{(m)_1}, & \text{if } m \in \mathrm{Sq}, \mathrm{lh}(m) = 2, \text{ and } (m)_0 = 15, & (2) \\ (s)_{(m)_1} \times (s)_{(m)_2}, & \text{if } m \in \mathrm{Sq}, \mathrm{lh}(m) = 3, \text{ and } (m)_0 = 17, & (3) \\ 1, & \text{if } m \in \mathrm{Sq}, \mathrm{lh}(m) = 3, (m)_0 = 19, & (4) \\ & \text{and } \langle y, m\rangle \notin F, & \\ (s)_{(m)_2}, & \text{if } m \in \mathrm{Sq}, \mathrm{lh}(m) = 3, (m)_0 = 19, & (5) \\ & \langle y, m\rangle \in F, \text{ and } \langle (m)_1, x\rangle \notin O, & \\ 0, & \text{otherwise.} & (6) \end{cases}$$

The cases in the definition of $g$ are exclusive. Define $h: \mathbb{N}^3 \to \mathbb{N}$

$$h(x,y,m) = g(\overline{h}(x,y,m), x, y, m). \tag{6.28}$$

Since $g$ is primitive recursive, $h$ is also primitive recursive by Proposition 5.3.46. Let $K: \mathbb{N}^3 \to \{0,1\}$ be the characteristic function of Sbl. Let $t$ be a term and let $v_i$ be a variable. By induction on $\alpha$, we prove that for every formula $\alpha$,

$$h(\#t, \#v_i, \#\alpha) = K(\#t, \#v_i, \#\alpha). \tag{6.29}$$

*Base step:* Let $\alpha$ be an atomic formula. By Definition 3.3.15 and (1) of the definition of $g$, we see that $h(\#t, \#v_i, \#\alpha) = K(\#t, \#v_i, \#\alpha)$.

*Inductive step:* Let $\psi$ and $\varphi$ be arbitrary wffs. Assume the induction hypothesis

$$\begin{aligned} h(\#t, \#v_i, \#\psi) &= K(\#t, \#v_i, \#\psi), \\ h(\#t, \#v_i, \#\varphi) &= K(\#t, \#v_i, \#\varphi). \end{aligned} \tag{IH}$$

We must prove that the same holds for the wffs $\neg\psi$, $\psi \to \varphi$, and $\forall v_j \psi$. Let us first consider the formula $\forall v_j \psi$. Thus, (▼) $\#(\forall v_j \psi) = [19, \#v_j, \#\psi]$. Thus, either (4), (5), or (6) in the definition of $g$ holds. Note that $([19, \#v_j, \#\psi])_2 = \#\psi$. Hence,

$$\begin{aligned} h(\#t, \#v_i, \#(\forall v_j \psi)) &= g(\overline{h}(\#t, \#v_i, \#(\forall v_j \psi)), \#t, \#v_i, \#(\forall v_j \psi)) && \text{by (6.28)} \\ &= g(\overline{h}(\#t, \#v_i, \#(\forall v_j \psi)), \#t, \#v_i, [19, \#v_j, \#\psi]) && \text{by (▼)} \\ &= \begin{cases} 1, & \text{if (4) of the definition of } g \text{ holds,} \\ \overline{h}(\#t, \#v_i, \#(\forall v_j \psi))_{\#\psi}, & \text{if (5) of the definition of } g \text{ holds,} \\ 0, & \text{if (6) of the definition of } g \text{ holds} \end{cases} \\ &= \begin{cases} 1, & \text{if (4) holds,} \\ h(\#t, \#v_i, \#\psi), & \text{if (5) holds, as } \#\psi < \#(\forall v_j \psi), \\ 0, & \text{if (6) holds} \end{cases} \end{aligned}$$

$$
\begin{aligned}
&= \begin{cases} 1, & \text{if (4) holds,} \\ K(\#t, \#v_i, \#\psi), & \text{if (5) holds, by (IH),} \\ 0, & \text{if (6) holds} \end{cases} \\
&= K(\#t, \#v_i, \#(\forall v_j \psi)),
\end{aligned}
$$

where the last equality follows from (6.26). Hence,

$$h(\#t, \#v_i, \#(\forall v_j \psi)) = K(\#t, \#v_i, \#(\forall v_j \psi)).$$

The completion of the induction proof for the two wffs $\neg\psi$ and $(\psi \to \varphi)$ is an exercise. Thus, (6.29) holds for all terms $t$, variables $v_i$, and wffs $\alpha$. Hence, for all $x, y, m \in \mathbb{N}$,

$$\langle x, y, m\rangle \in \text{Sbl} \quad \text{if and only if} \quad x \in T^{\#}, y \in V^{\#}, m \in W^{\#}, \text{ and } h(x, y, m) = 1.$$

Therefore, Sbl is primitive recursive. □

**Proposition 6.4.12.** *Define the 2-place relation* Gen *on the Gödel numbers by* Gen$(\#\psi, \#\varphi)$ *if and only if* $\varphi$ *is a generalization of* $\psi$*. Then* Gen *is primitive recursive.*

*Proof.* Let $K\colon \mathbb{N}^2 \to \{0, 1\}$ be such that $K(\#\psi, \#\varphi) = 1$ if and only if Gen$(\#\psi, \#\varphi)$, whenever $\psi$ and $\varphi$ are wffs. By applying Definition 3.3.12, the function $K$ satisfies the following recursive definition:

$$K(\#\psi, \#\varphi) = \begin{cases} 1, & \text{if } \#\varphi = \#\psi, \\ K(\#\psi, \#\beta), & \text{if } \#\varphi = [19, \#v_i, \#\beta] \text{ and } \#\psi < \#\varphi, \\ 0, & \text{otherwise,} \end{cases}$$

for all wffs $\psi$ and $\varphi$. Using the above recursive definition as a guide, one can now prove that Gen is primitive recursive using a course-of-values recursion (see Exercise 8). □

### 6.4.1 The logical axioms revisited

Recall that the logical axioms are generalizations of the following, where $x$ and $y$ denote any of the variables in $V$:

1. first-order tautologies;
2. $\forall x \alpha \to \alpha_t^x$, where $t$ is substitutable for $x$ in $\alpha$;
3. $\forall x(\alpha \to \beta) \to (\forall x \alpha \to \forall x \beta)$;
4. $\alpha \to \forall x \alpha$, where $x$ does not occur free in $\alpha$;
5. $x \doteq x$;
6. $x \doteq y \to (\alpha \to \alpha_y^x)$, where $\alpha$ is an atomic formula.

In this section, we will show that the set of Gödel numbers of each of the six logical axiom groups is primitive recursive. First we address tautologies.

The *subformulas* of a wff $\alpha$ are defined by the following recursion:
1. If $\alpha$ is atomic, then $\alpha$ is its only subformula.
2. If $\alpha$ is $\neg\psi$, then the subformulas of $\alpha$ are $\alpha$ and the subformulas of $\psi$.
3. If $\alpha$ is $\psi \to \varphi$, then the subformulas of $\alpha$ are $\alpha$ and the subformulas of $\psi$ and $\varphi$.
4. If $\alpha$ is $\forall x\psi$, then the subformulas of $\alpha$ are $\alpha$ and the subformulas of $\psi$.

Let $\alpha$ be a wff. It follows that if $\beta$ is a subformula of $\alpha$, then $\#\beta \le \#\alpha$. Thus, as # is one-to-one, the number of subformulas of $\alpha$ is less than or equal to $\#\alpha$.

It also follows from Exercise 1 on page 116 that every wff $\alpha$ is generated by the prime subformulas of $\alpha$ (see Definition 3.3.2). Hence, to determine whether or not a wff $\alpha$ is a tautology, it is sufficient to assign truth values to every subformula of $\alpha$ such that the assignment agrees with the truth table assignments for negations and conditionals. More specifically, let $S$ be the set of all the subformulas of $\alpha$ and let $v\colon S \to \{T, F\}$ be a truth assignment. Suppose that $\neg\psi$ and $\beta \to \varphi$ are subformulas of $\alpha$. In order for $v$ to agree with the truth table assignments for negation and the conditional, we must have $v(\neg\psi) = T$ if and only if $v(\psi) = F$ and $v(\beta \to \varphi) = T$ if and only if $v(\beta) = T$ implies $v(\varphi) = T$. Given any such truth assignment $v$, $v(\alpha)$ determines the truth value of $\alpha$. So if $v(\alpha) = T$ for every such truth assignment, then $\alpha$ is a tautology. This conclusion depends only on the values that these truth assignments give to the prime subformulas of $\alpha$. Thus, having $S$ contain all the subformulas of $\alpha$ (or more) is a harmless overkill (see Theorem 2.2.11).

Since we are involved with number theory, we will now view a truth assignment as a sequence number $v$ of a finite sequence of 0's and 1's, where 0, 1 represent $F$, $T$, respectively. For example, $v = [1, 0, 1, 0]$ will be understood to be a truth assignment, where $(v)_2 = 1$. Now let $\alpha$ be a wff. Since $\#\beta \le \#\alpha$ for every subformula $\beta$ of $\alpha$, we need to use truth assignments $v$ of length $\#\alpha + 1$ in order to evaluate the value of $(v)_{\#\alpha}$. Such a truth assignment that always assigns the value of 1 satisfies an inequality:

$$\overbrace{[1, 1, 1, \ldots, 1]}^{\#\alpha + 1 \text{ times}} = p_0^2 \cdot p_1^2 \cdots p_{\#\alpha}^2 < p_{\#\alpha}^2 \cdot p_{\#\alpha}^2 \cdots p_{\#\alpha}^2 = p_{\#\alpha}^{2(\#\alpha+1)}.$$

Moreover, this truth assignment is larger than all other truth assignments $v$ of length $\#\alpha + 1$. These inequalities justify an upper bound used in the proof of our next result. Let $\mathrm{Taut} = \{\#\alpha \in W^{\#} : \alpha \text{ is a tautology}\}$, the set of Gödel numbers of tautologies.

**Proposition 6.4.13.** *The set* Taut *is primitive recursive.*

*Proof.* Using logical notation, let $\mathrm{Ta}(v)$ be the primitive recursive relation

$$\begin{aligned} &v \in \mathrm{Sq} \wedge (\forall i < \mathrm{lh}(v))((v)_i \le 1) \\ &\quad \wedge (\forall i < \mathrm{lh}(v))([15, i] < \mathrm{lh}(v) \to ((v)_{[15,i]} = 1 \leftrightarrow (v)_i = 0)) \\ &\quad \wedge (\forall i, j < \mathrm{lh}(v))([17, i, j] < \mathrm{lh}(v) \to ((v)_{[17,i,j]} = 1 \leftrightarrow ((v)_i = 1 \to (v)_j = 1))). \end{aligned}$$

$\mathrm{Ta}(v)$ asserts that $v$ is a truth assignment that agrees with the truth table assignments for the Gödel numbers of the negations and conditionals in its domain. Thus, again using logical notation, we have

$$\#\alpha \in \mathrm{Taut} \quad \text{iff} \quad \#\alpha \in W^{\#} \wedge (\forall v < p_{\#\alpha}^{2(\#\alpha+1)})((\mathrm{Ta}(v) \wedge (\mathrm{lh}(v) = \#\alpha + 1)) \rightarrow (v)_{\#\alpha} = 1).$$

Therefore, Taut is primitive recursive. □

**Proposition 6.4.14.** *The set of Gödel numbers of formulas of the form* $\forall v_i\alpha \rightarrow \alpha_t^{v_i}$, *where* $t$ *is substitutable for* $v_i$ *in* $\alpha$, *is primitive recursive.*

*Proof.* If a wff $\varphi$ has the form $\forall v_i\alpha \rightarrow \alpha_t^{v_i}$, then $\#\varphi = \#(\forall v_i\alpha \rightarrow \alpha_t^{v_i})$. Hence, by Definition 6.4.4 and Proposition 6.4.8, we have

$$\#\varphi = [17, \#\forall v_i\alpha, \#\alpha_t^{v_i}] = [17, [19, \#v_i, \#\alpha], \mathrm{Sb}(\#t, \#v_i, \#\alpha)].$$

It thus follows that $\#v_i, \#t, \#\alpha < \#\varphi$. By using logical notation and Proposition 6.4.11, we therefore see that $\#\varphi$ is the Gödel number of a wff of the desired form if and only if

$$(\exists v < \#\varphi)(\exists x < \#\varphi)(\exists a < \#\varphi)(v \in V^{\#} \wedge x \in T^{\#} \wedge a \in W^{\#}$$
$$\wedge \#\varphi = [17, [19, v, a], \mathrm{Sb}(x, v, a)] \wedge \langle x, v, a\rangle \in \mathrm{Sbl}).$$

Therefore, the set of such Gödel numbers is primitive recursive. □

**Proposition 6.4.15.** *The set of Gödel numbers of formulas of the form*

$$\forall v_i(\alpha \rightarrow \beta) \rightarrow (\forall v_i\alpha \rightarrow \forall v_i\beta)$$

*is primitive recursive.*

*Proof.* If a wff $\varphi$ has the form $\forall v_i(\alpha \rightarrow \beta) \rightarrow (\forall v_i\alpha \rightarrow \forall v_i\beta)$, then

$$\#\varphi = \#(\forall v_i(\alpha \rightarrow \beta) \rightarrow (\forall v_i\alpha \rightarrow \forall v_i\beta)).$$

Hence, by Definition 6.4.4, we have

$$\#\varphi = [17, \#\forall v_i(\alpha \rightarrow \beta), \#(\forall v_i\alpha \rightarrow \forall v_i\beta)], \tag{6.30}$$

where
(a) $\#\forall v_i(\alpha \rightarrow \beta) = [19, \#v_i, [17, \#\alpha, \#\beta]]$,
(b) $\#(\forall v_i\alpha \rightarrow \forall v_i\beta) = [17, \#\forall v_i\alpha, \#\forall v_i\beta] = [17, [19, \#v_i, \#\alpha], [19, \#v_i, \#\beta]]$.

After substituting (a) and (b) into (6.30), it follows that the set of Gödel numbers $\#\varphi$ satisfying (6.30) is primitive recursive, as in the proof of Proposition 6.4.14. □

**Proposition 6.4.16.** *The set of Gödel numbers of formulas of the form $\alpha \to \forall v_i \alpha$, where $v_i$ does not occur free in $\alpha$, is primitive recursive.*

*Proof.* The proof is similar to that of Proposition 6.4.14, using the primitive recursive relation $F$ in Corollary 6.4.9. □

**Proposition 6.4.17.** *The set of Gödel numbers of formulas of the form $v_i \doteq v_j$ is primitive recursive.*

*Proof.* By Definition 6.4.3, $\#\varphi$ is the Gödel number of a wff of the form $v_i \doteq v_j$ if and only if $(\exists v < \#\varphi)(\exists v' < \#\varphi)(v \in V^\# \wedge v' \in V^\# \wedge \#\varphi = [13, v, v'])$. Therefore, the set of Gödel numbers of formulas of this form is primitive recursive. □

**Proposition 6.4.18.** *The set of Gödel numbers of formulas of the form*

$$v_i \doteq v_j \to (\alpha \to \alpha_{v_j}^{v_i}),$$

*where $\alpha$ is an atomic formula, is primitive recursive.*

*Proof.* The proof is similar to that of Proposition 6.4.14, using the primitive recursive function Sb in Corollary 6.4.8. □

**Definition 6.4.19.** Let $\Omega^\#$ be the set of Gödel numbers of all the sentences in $\Omega$ and let MP be the relation defined by

$$\mathrm{MP} = \{\langle \#(\varphi \to \psi), \#\varphi, \#\psi \rangle : \varphi \to \psi, \varphi, \psi \text{ are wffs}\}.$$

Let $\mathrm{LA}_i'$ be the set of Gödel numbers of the wffs in logical axiom group $i$, where $i = 1, 2, \ldots, 6$. For each such $i$, let

$$\mathrm{LA}_i = \mathrm{LA}_i' \cup \{\#\varphi : (\exists \#\alpha < \#\varphi)(\mathrm{Gen}(\alpha, \varphi) \wedge \alpha \in \mathrm{LA}_i')\}.$$

Now let $\mathrm{LA} = \bigcup_{1 \le i \le 6} \mathrm{LA}_i$.

Therefore, the set LA consists of the Gödel numbers of all the logical axioms and their generalizations.

**Proposition 6.4.20.** *The sets* LA, $\Omega^\#$ *and the relation* MP *are primitive recursive.*

*Proof.* Propositions 6.4.13–6.4.18 imply, respectively, that $\mathrm{LA}_i'$ is primitive recursive for each $i = 1, 2, \ldots, 6$. Proposition 6.4.12 implies that the relation Gen is primitive recursive. Thus, for each such $i$, $\mathrm{LA}_i$ is primitive recursive. Since $\mathrm{LA} = \bigcup_{1 \le i \le 6} \mathrm{LA}_i$, Corollary 5.3.22(2) implies that LA is primitive recursive. Because $\Omega$ is a finite set, Proposition 5.3.24 implies that $\Omega^\#$ is primitive recursive. Finally, since

$$\langle \#(\varphi \to \psi), \#\varphi, \#\psi \rangle \in \mathrm{MP} \quad \text{iff} \quad \#\varphi \in W^\# \text{ and } \#\psi \in W^\# \text{ and } \#(\varphi \to \psi) = [17, \#\varphi, \#\psi],$$

it follows that MP is primitive recursive. □

Let $\Gamma$ be a set of $\mathcal{L}$-wffs and let $\Lambda$ be the set of logical axioms and their generalizations. Recalling Definition 3.3.20, a deduction is a sequence of formulas $\langle \alpha_1, \ldots, \alpha_n \rangle$ such that for all $1 \le k \le n$, either

(a) $\alpha_k$ is in $\Gamma \cup \Lambda$, or
(b) $\alpha_k$ is obtained by modus ponens from two earlier wffs in the sequence, that is, for some $i$ and $j$ less than $k$, the wffs $\alpha_i$ and $\alpha_j = (\alpha_i \to \alpha_k)$ are in the sequence.

We now rephrase this definition in terms of sequence numbers consisting of Gödel numbers. We then show that the set of such sequence numbers is primitive recursive.

**Definition 6.4.21.** Let $\Gamma$ be a set of $\mathcal{L}$-wffs. Then:

(1) $\mathbb{D}_\Gamma = \{[\#\alpha_1, \ldots, \#\alpha_n] : n \ge 1 \text{ and } \langle \alpha_1, \ldots, \alpha_n \rangle \text{ is a deduction from } \Gamma\}$;
(2) $\mathbb{P}_\Gamma = \{\langle s, m \rangle : \mathbb{D}_\Gamma(s) \text{ and } (s)_{\mathrm{lh}(s)-1} = m\}$.

In Definition 6.4.21, $\mathbb{D}_\Gamma(s)$ asserts that "$s$ codes a deduction from $\Gamma$," and $\mathbb{P}_\Gamma(s, m)$ asserts that "$s$ codes a deduction (proof) of the formula with Gödel number $m$." For any set $\Gamma$ of wffs, we will let $\Gamma^\# = \{\#\varphi : \varphi \in \Gamma\}$. Note that if $\mathbb{D}_\Gamma(s)$, then $(s)_i \in W^\#$ for all $i < \mathrm{lh}(s)$. Of course, $\mathbb{D}_\Gamma(s)$ means that $s \in \mathbb{D}_\Gamma$.

**Theorem 6.4.22.** *For any set $\Gamma$ of wffs, if $\Gamma^\#$ is (primitive) recursive, then the set $\mathbb{D}_\Gamma$ and the relation $\mathbb{P}_\Gamma$ are (primitive) recursive.*

*Proof.* Assume that $\Gamma^\#$ is (primitive) recursive. Let $U = \Gamma^\# \cup \mathrm{LA}$. Since

$$\mathbb{D}_\Gamma(s) \quad \text{iff} \quad \mathrm{Sq}(s) \wedge (\forall k < \mathrm{lh}(s))(((s)_k \in U) \vee (\exists i < k)(\exists j < k)\mathrm{MP}((s)_i, (s)_j, (s)_k)),$$

we see that $\mathbb{D}_\Gamma$ is (primitive) recursive. Hence, $\mathbb{P}_\Gamma$ is (primitive) recursive. □

**Corollary 6.4.23.** *For any set $\Gamma$ of wffs, if $\Gamma^\#$ is (primitive) recursive, then the set of Gödel numbers $\{\#\varphi : \Gamma \vdash \varphi\}$ is recursively enumerable.*

*Proof.* Assume that $\Gamma^\#$ is (primitive) recursive. Let $D = \{\#\varphi : \Gamma \vdash \varphi\}$. Then $m \in D$ if and only if $\exists s \mathbb{P}_\Gamma(s, m)$. Thus, $D$ is recursively enumerable. □

Theorem 6.3.24 shows that every recursive function and relation is representable. We will soon establish the converse. Recall that for any natural number $n$, we let $\overline{n}$ denote the $\mathcal{L}$-term $\dot{S}^n\dot{0}$, that is,

$$\overline{n} = \dot{S}^n\dot{0} = \overbrace{\dot{S}\dot{S}\cdots\dot{S}}^{n \text{ times}}\,\dot{0}.$$

So $\overline{n+1} = \dot{S}\overline{n}$, $\overline{1} = \dot{S}\dot{0}$, and $\overline{0} = \dot{0}$. Given any natural number $n$, can we recursively compute the Gödel number $\#\overline{n}$ of the term $\overline{n}$? Yes.

**Lemma 6.4.24.** *The function $f\colon \mathbb{N} \to \mathbb{N}$ defined by $f(n) = \#\overline{n}$ is primitive recursive.*

*Proof.* Since $f$ satisfies the primitive recursion

(1) $f(0) = \#\dot{0}$,

(2) $f(n+1) = [3, f(n)]$ for all $n \in \mathbb{N}$,

we conclude that $f$ is primitive recursive. □

**Lemma 6.4.25.** *Let $\theta(x_1, x_2, \ldots, x_k)$ be a wff whose free variables are among $x_1, x_2, \ldots, x_k$. The function $g\colon \mathbb{N}^k \to \mathbb{N}$ defined by*

$$g(n_1, n_2, \ldots, n_k) = \#\theta(\overline{n}_1, \overline{n}_2, \ldots, \overline{n}_k)$$

*is primitive recursive.*

*Proof.* Let $f\colon \mathbb{N} \to \mathbb{N}$ be as in Lemma 6.4.24. Thus, $f(n_i) = \#\overline{n}_i$ for $1 \le i \le k$. Using Proposition 6.4.8, we observe that

1. $\#\theta(\overline{n}_1, x_2, \ldots, x_k) = \#\theta^{x_1}_{\overline{n}_1} = \mathrm{Sb}(f(n_1), \#x_1, \#\theta)$,
2. $\#\theta(\overline{n}_1, \overline{n}_2, \ldots, x_k) = \#\theta^{x_1 x_2}_{\overline{n}_1 \overline{n}_2} = \mathrm{Sb}(f(n_2), \#x_2, \#\theta^{x_1}_{\overline{n}_1})$
$= \mathrm{Sb}(f(n_2), \#x_2, \mathrm{Sb}(f(n_1), \#x_1, \#\theta))$,

$\vdots$

$k$. $\#\theta(\overline{n}_1, \overline{n}_2, \ldots, \overline{n}_k) = \#\theta^{x_1}_{\overline{n}_1} \cdots {}^{x_k}_{\overline{n}_k} = \mathrm{Sb}(f(n_k), \#x_k, \#\theta^{x_1 x_2}_{\overline{n}_1 \overline{n}_2} \cdots {}^{x_{k-1}}_{\overline{n}_{k-1}})$
$= \mathrm{Sb}(f(n_k), \#x_k, \mathrm{Sb}(f(n_{k-1}), \#x_{k-1}, \mathrm{Sb}(\ldots\ldots)))$.

As Sb and $f$ are primitive recursive, it follows, for each $i = 1, 2, \ldots, k$, that the natural number $\#\theta^{x_1 x_2}_{\overline{n}_1 \overline{n}_2} \cdots {}^{x_i}_{\overline{n}_i}$ is the value of a composition of primitive recursive functions. In particular, $g(n_1, n_2, \ldots, n_k) = \#\theta(\overline{n}_1, \overline{n}_2, \ldots, \overline{n}_k) = \#\theta^{x_1 x_2}_{\overline{n}_1 \overline{n}_2} \cdots {}^{x_k}_{\overline{n}_k}$ is the value of a (finite) composition of primitive recursive functions. Thus, $g$ is primitive recursive. □

**Theorem 6.4.26.** *Every representable relation and function is recursive.*

*Proof.* Let $P \subseteq \mathbb{N}^k$ be a representable relation. Thus, by Exercise 6 on page 224, there is a formula $\varphi(x_1, x_2, \ldots, x_k)$ such that for all $n_1, n_2, \ldots, n_k \in \mathbb{N}$,

(1) $\langle n_1, n_2, \ldots, n_k\rangle \in P$ if and only if $\Omega \vdash \varphi(\overline{n}_1, \overline{n}_2, \ldots, \overline{n}_k)$,

(2) $\langle n_1, n_2, \ldots, n_k\rangle \notin P$ if and only if $\Omega \vdash \neg\varphi(\overline{n}_1, \overline{n}_2, \ldots, \overline{n}_k)$.

Define $g\colon \mathbb{N}^k \to \mathbb{N}$ by (▲) $g(n_1, n_2, \ldots, n_k) = \#\varphi(\overline{n}_1, \overline{n}_2, \ldots, \overline{n}_k)$. Hence,

$$\begin{aligned} \langle n_1, n_2, \ldots, n_k\rangle \in P \quad &\text{iff} \quad \Omega \vdash \varphi(\overline{n}_1, \overline{n}_2, \ldots, \overline{n}_k) && \text{by (1)},\\ &\text{iff} \quad \exists s \mathbb{P}_\Omega(s, \#\varphi(\overline{n}_1, \overline{n}_2, \ldots, \overline{n}_k)) && \text{by Definition 6.4.21},\\ &\text{iff} \quad \exists s \mathbb{P}_\Omega(s, g(n_1, \ldots, n_k)) && \text{by (▲)}. \end{aligned}$$

Since $\Omega^\#$ is primitive recursive (by Proposition 6.4.20), Theorem 6.4.22 implies that the relation $\mathbb{P}_\Omega$ is primitive recursive. Lemma 6.4.25 implies that the function $g$ is primitive recursive. Lemma 5.4.5 and the above equivalences now allow us to conclude that $P$ is recursively enumerable. A similar argument shows that $\mathbb{N}^k \setminus P$ is recursively enumerable. Therefore, by Theorem 5.4.16, $P$ is recursive.

If a function $f$ is representable, then Theorem 6.3.12 and the above argument shows that the graph of $f$ is recursive. Thus, by Exercise 14 on page 195, $f$ is recursive. □

Theorem 6.3.24 and Theorem 6.4.26 show that a function (relation) is representable if and only if the function (relation) is recursive.

**Exercises 6.3.**

1. Suppose that $\#\forall v_i\alpha = \#\forall v_i\beta$, where $\alpha$ and $\beta$ are wffs in $\mathcal{L} = \{\dot{<}, \dot{0}, \dot{S}, \dot{+}, \dot{\times}, \dot{E}, \dot{=}\}$. Show that $\#\alpha = \#\beta$.
2. Show that there is a primitive recursive function $f$ such that for all wffs $\alpha$, if $\#\alpha$ is the Gödel number of $\alpha$, then $f(\#\alpha)$ is the Gödel number of $\forall v_i\alpha$.
3. Complete the proof of item 1 of Proposition 6.4.5 by proving that $C$ satisfies (6.24) for all $m \in \mathbb{N}$.
*4. Prove item 3 of Proposition 6.4.5.
*5. Complete the proof of the claim in the proof of Proposition 6.4.6.
*6. Prove Proposition 6.4.8.
7. Complete the induction proof of Proposition 6.4.11 for the wffs $\neg\psi$ and $\psi \to \varphi$.
*8. Prove Proposition 6.4.12.
9. Let $R$ be the 2-place relation on Gödel numbers defined by $R(\#\psi, \#\alpha)$ if and only if $\psi$ is a subformula of $\alpha$. Show that $R$ is primitive recursive.
10. Let $P \subseteq \mathbb{N}^k$ be recursively enumerable. Show that there is a $(k+1)$-place primitive recursive relation $R$ such that $P(\vec{x})$ if and only if $\exists p R(\vec{x}, p)$, for all $\vec{x} \in \mathbb{N}^k$.
11. Show that every representable relation is decidable.

**Exercise Notes:** For Exercise 10, first show that it holds when $P$ is just recursive. Consider applying Theorem 6.3.24 and the proof of Theorem 6.4.26.

## 6.5 The incompleteness theorems

We are working with the language $\mathcal{L} = \{\dot{<}, \dot{0}, \dot{S}, \dot{+}, \dot{\times}, \dot{E}, \dot{=}\}$ and the standard model of arithmetic

$$\mathcal{N} = \langle \mathbb{N}; 0, S, <, +, \times, E \rangle.$$

We now repeat the question that opened this chapter:

$$\text{Is there a decidable set of } \mathcal{L}\text{-sentences } \Gamma \text{ such that } \mathcal{N} \models \Gamma \text{ and } \mathcal{N} \models \varphi \text{ if and only if } \Gamma \vdash \varphi, \text{ for each } \mathcal{L}\text{-sentence } \varphi? \tag{Q2}$$

Kurt Gödel was the first to answer this question. His resulting theorems in this area are among the most significant results produced in modern mathematical logic.

The next amazing lemma follows from arguments first presented by Kurt Gödel in [4]. The proof of this lemma is rather tricky, but the following proof is designed to ensure validity and readability. With this result, we will have sentences that indirectly refer to themselves. We let $x$ denote an arbitrary variable.

**Fixed-Point Lemma 6.5.1.** *For any formula $\theta(x)$ in which only $x$ occurs free, there is a sentence $\sigma$ such that*

$$\Omega \vdash \sigma \leftrightarrow \theta(\overline{\#\sigma}).$$

*Proof.* Let the variable $x$ be fixed. Define the function $f\colon \mathbb{N}^2 \to \mathbb{N}$ by

$$f(m,n) = \begin{cases} \mathrm{Sb}(\#\overline{n}, \#x, m), & \text{if } m \in W^{\#}, \\ 0, & \text{otherwise.} \end{cases}$$

It follows from Proposition 6.4.8 that the function $f$ is primitive recursive. Recall that $\mathrm{Sb}(\#\overline{n}, \#x, \#\alpha) = \#\alpha^x_{\overline{n}}$ and $\#\alpha^x_{\overline{n}} = \#\alpha(\overline{n})$ when $x$ is the only free variable in the wff $\alpha$. Hence, in such a case,

$$f(\#\alpha, \#\alpha) = \mathrm{Sb}(\#\overline{\#\alpha}, \#x, \#\alpha) = \#\alpha^x_{\overline{\#\alpha}} = \#\alpha(\overline{\#\alpha}), \tag{6.31}$$

where $\overline{\#\alpha}$ is the term $\dot{S}^{\#\alpha}\dot{0}$ in the language $\mathcal{L}$ (see page 236). Theorem 6.3.24 implies that $f$ is functionally representable. Therefore, there is an $\mathcal{L}$-formula $\varphi(v_1, v_2, v)$ whose free variables are among $v_1, v_2, v$, such that for all $n_1, n_2 \in \mathbb{N}$,

$$\Omega \vdash \forall v\Big(\varphi(\overline{n}_1, \overline{n}_2, v) \leftrightarrow v \doteq \overline{f(n_1, n_2)}\Big). \tag{6.32}$$

By means of an alphabetic variant, we can assume that the variable $x$ does not appear in $\varphi(v_1, v_2, v)$. Now let $\alpha(x)$ be the formula

$$\forall v(\varphi(x, x, v) \to \theta(v)), \tag{6.33}$$

where we assume that $v$ is substitutable for $x$ in $\theta$ (otherwise replace $v$ with a variable that is substitutable for $x$). Let $e = \#\alpha$ be the Gödel number of the formula $\alpha(x)$ in (6.33). Note that this formula has only $x$ as a free variable. Hence, $\alpha(\overline{e})$ is the sentence

$$\forall v(\varphi(\overline{e}, \overline{e}, v) \to \theta(v)). \tag{6.34}$$

Since $e = \#\alpha$, (6.31) implies that $\overline{f(e,e)} = \overline{\#\alpha(\overline{e})}$. So from (6.32), we conclude that

$$\Omega \vdash \forall v\Big(\varphi(\overline{e}, \overline{e}, v) \leftrightarrow v \doteq \overline{\#\alpha(\overline{e})}\Big). \tag{6.35}$$

Furthermore, by Logical axiom 3.3.17(2), (6.35) implies that

$$\Omega \vdash \varphi\left(\overline{e}, \overline{e}, \overline{\#\alpha(\overline{e})}\right) \leftrightarrow \overline{\#\alpha(\overline{e})} \doteq \overline{\#\alpha(\overline{e})}. \tag{6.36}$$

Therefore, from (6.36) and Proposition 3.3.48, it follows that (▲) $\Omega \vdash \varphi(\overline{e}, \overline{e}, \overline{\#\alpha(\overline{e})})$.

**Claim.** *We have* $\Omega \vdash \alpha(\overline{e}) \leftrightarrow \theta\left(\overline{\#\alpha(\overline{e})}\right)$.

*Proof.* We first show that $\Omega \vdash \alpha(\overline{e}) \to \theta\left(\overline{\#\alpha(\overline{e})}\right)$. Since $\alpha(\overline{e})$ is the formula in (6.34), Logical axiom 3.3.17(2) and (6.34) imply that

$$\Omega; \alpha(\overline{e}) \vdash \varphi\left(\overline{e}, \overline{e}, \overline{\#\alpha(\overline{e})}\right) \to \theta\left(\overline{\#\alpha(\overline{e})}\right).$$

Thus, by (▲) and modus ponens, $\Omega; \alpha(\overline{e}) \vdash \theta\left(\overline{\#\alpha(\overline{e})}\right)$. Hence, $\Omega \vdash \alpha(\overline{e}) \to \theta\left(\overline{\#\alpha(\overline{e})}\right)$, by the deduction theorem (Theorem 3.3.33).

Now we show that $\Omega \vdash \theta\left(\overline{\#\alpha(\overline{e})}\right) \to \alpha(\overline{e})$. Since $\alpha(\overline{e})$ is the formula in (6.34), we must show, by the deduction theorem, that

$$\Omega; \theta\left(\overline{\#\alpha(\overline{e})}\right) \vdash \forall v(\varphi(\overline{e}, \overline{e}, v) \to \theta(v)). \tag{6.37}$$

By the generalization theorem (Theorem 3.3.29) and the deduction theorem (Theorem 3.3.33), to prove (6.37), it is sufficient to show that

$$\Omega \cup \left\{\theta\left(\overline{\#\alpha(\overline{e})}\right), \varphi(\overline{e}, \overline{e}, v)\right\} \vdash \theta(v). \tag{6.38}$$

Note that (6.35) implies that

$$\Omega \cup \{\varphi(\overline{e}, \overline{e}, v)\} \vdash v \doteq \overline{\#\alpha(\overline{e})}. \tag{6.39}$$

Moreover, Exercise 7 on page 135 implies that

$$\left\{\theta\left(\overline{\#\alpha(\overline{e})}\right), v \doteq \overline{\#\alpha(\overline{e})}\right\} \vdash \theta(v). \tag{6.40}$$

Clearly, (6.39) and (6.40) imply (6.38). Therefore, (6.37) holds. (Claim) □

Now let $\sigma$ be the sentence $\alpha(\overline{e})$. The claim implies that $\Omega \vdash \sigma \leftrightarrow \theta(\overline{\#\sigma})$. (Lemma) □

Lemma 6.5.1 and Theorem 4.1.5 imply the following corollary.

**Corollary 6.5.2.** *For any formula $\theta(x)$ in which only $x$ occurs free, there is a sentence $\sigma$ such that $\mathcal{N} \vDash \sigma \leftrightarrow \theta(\overline{\#\sigma})$.*

The above corollary, together with the fixed-point lemma, will allow us to prove a significant theorem due to Alfred Tarski. As you may recall, Tarski provided, in English, a definition of truth in a structure for a first-order language (see Definition 3.2.6). Using this definition, one can determine which $\mathcal{L}$-sentences are true in the standard model $\mathcal{N}$ and which are false. The next theorem shows that Tarski's definition of truth, when applied to the structure $\mathcal{N}$, cannot be translated into the language $\mathcal{L}$.

Recall that the theory of $\mathcal{N}$, denoted by $\text{Th}(\mathcal{N})$, is the set of all $\mathcal{L}$-sentences that are true in $\mathcal{N}$. Consequently, $\text{Th}(\mathcal{N})^{\#} = \{\#\varphi : \varphi \in \text{Th}(\mathcal{N})\}$ is the set of Gödel numbers of $\mathcal{L}$-sentences that are true in $\mathcal{N}$.

**Tarski Undefinability Theorem 6.5.3.** *The set $Th(\mathcal{N})^{\#}$ is not definable over $\mathcal{N}$.*

*Proof.* Suppose, for a contradiction, that the set $\text{Th}(\mathcal{N})^{\#}$ is definable over $\mathcal{N}$. Thus, let $\theta(v_1)$ be a formula with one free variable such that for all $\mathcal{L}$-sentences $\varphi$,

$$\mathcal{N} \models \varphi \quad \text{iff} \quad \mathcal{N} \models \theta[\![\#\varphi]\!]. \tag{6.41}$$

Since $\neg\theta(v_1)$ has only one free variable, Corollary 6.5.2 asserts that there exists an $\mathcal{L}$-sentence $\sigma$ such that $\mathcal{N} \models \sigma \leftrightarrow \neg\theta(\overline{\#\sigma})$. Since $\overline{\#\sigma}^{\mathcal{N}} = \#\sigma$, we conclude that

$$\mathcal{N} \models \sigma \quad \text{iff} \quad \mathcal{N} \models \neg\theta[\![\#\sigma]\!],$$

which contradicts (6.41) by letting $\varphi = \sigma$. □

Theorem 5.4.14 justifies our next definition.

**Definition 6.5.4.** Let $\Gamma$ be a set of wffs. Then $\Gamma$ is decidable if and only if $\Gamma^{\#}$ is recursive.

Definition 6.5.4 and Theorem 6.5.3 imply that the theory of $\mathcal{N}$ is undecidable.

**Lemma 6.5.5.** *The set $Th(\mathcal{N})^{\#}$ is not recursive.*

*Proof.* If $\text{Th}(\mathcal{N})^{\#}$ were recursive, then, by Corollary 6.3.25, it would be definable over $\mathcal{N}$, contradicting Theorem 6.5.3. □

### 6.5.1 Gödel's first incompleteness theorem

We are now in a position to answer the question that was posed at the beginning of this chapter, namely:

> Is there a decidable set of $\mathcal{L}$-sentences $\Gamma$ such that $\mathcal{N} \models \Gamma$ and $\mathcal{N} \models \varphi$ if and only if $\Gamma \vdash \varphi$, for each $\mathcal{L}$-sentence $\varphi$? (Q2)

We begin by reviewing some relevant definitions. A set $T$ of $\mathcal{L}$-sentences is said to be a *theory* if and only if $T$ is closed under logical implication, that is, by the completeness theorem (Theorem 4.2.8), for any sentence $\sigma$,

$$\text{if } T \vdash \sigma, \quad \text{then } \sigma \in T.$$

A theory $T$ is *complete* if for every sentence $\varphi$, either $T \vdash \varphi$ or $T \vdash \neg\varphi$.

In Definition 4.3.14, we defined $\mathrm{Cn}(\Gamma)$ to be the set of *consequences* of $\Gamma$, where $\Gamma$ is a set of $\mathcal{L}$-sentences. By the completeness theorem (Theorem 4.2.8), we can also define $\mathrm{Cn}(\Gamma)$ to be the set of sentences $\varphi$ such that $\Gamma \vdash \varphi$. We let $\mathrm{Cn}(\Gamma)^{\#} = \{\#\varphi : \Gamma \vdash \varphi\}$. Thus, if $\Gamma$ is a theory, then $\mathrm{Cn}(\Gamma) = \Gamma$. Finally, recall that $\Gamma$ is *consistent* if there is no formula $\beta$ such that $\Gamma \vdash \beta$ and $\Gamma \vdash \neg\beta$.

The next definition is a relevant extension of Definition 4.3.15.

**Definition 6.5.6.** A theory $T$ is said to be *axiomatizable* if $T = \mathrm{Cn}(\Gamma)$ for some set of sentences $\Gamma$ such that $\Gamma^{\#}$ is recursive.

Definitions 6.5.4 and 6.5.6 allows us to rephrase question (Q2) as follows:

Is the theory $\mathrm{Th}(\mathcal{N})$ axiomatizable?

**Lemma 6.5.7.** *Let $\Gamma$ be a consistent set of $\mathcal{L}$-sentences. If $\Gamma^{\#}$ is recursive and $Cn(\Gamma)$ is a complete theory, then $Cn(\Gamma)^{\#}$ is recursive.*

*Proof.* Assume that $\Gamma$ is consistent, $\Gamma^{\#}$ is recursive, and $\mathrm{Cn}(\Gamma)$ is also a complete theory. Let $S^{\#}$ be the set of Gödel numbers of sentences. By Proposition 6.4.10, $S^{\#}$ is recursive. Let $A = \mathrm{Cn}(\Gamma)^{\#}$. As in the proof of Corollary 6.4.23, we have $\#\varphi \in A$ if and only if $\#\varphi \in S^{\#}$ and $\exists s \mathbb{P}_{\Gamma}(s, \#\varphi)$. Therefore, by Theorem 6.4.22 and Proposition 6.4.10, $A$ is recursively enumerable. Now let $B = \{\#\varphi \in S^{\#} : \Gamma \vdash \neg\varphi\}$. By similar reasoning, we see that $B$ is recursively enumerable. Since $\Gamma$ is consistent, we see that $A \cap B = \varnothing$. Moreover, because $\mathrm{Cn}(\Gamma)$ is a complete theory, it follows that $A \cup B = S^{\#}$. Since $S^{\#}$ is recursive, Exercise 4 implies that $A$ is recursive, that is, $\mathrm{Cn}(\Gamma)^{\#}$ is recursive. □

Lemma 6.5.7 demonstrates that a complete axiomatizable theory is recursive.

**Gödel Incompleteness Theorem 6.5.8.** *Let $\Gamma$ be a set of $\mathcal{L}$-sentences such that $\mathcal{N} \models \Gamma$. If $\Gamma^{\#}$ is recursive, then $Cn(\Gamma)$ is not a complete theory.*

*Proof.* Let $\Gamma$ be a set of $\mathcal{L}$-sentences such that $\mathcal{N} \models \Gamma$ and $\Gamma^{\#}$ is recursive. Suppose, for a contradiction, that $\mathrm{Cn}(\Gamma)$ is a complete theory. Lemma 6.5.7 thus implies that $\mathrm{Cn}(\Gamma)^{\#}$ is recursive. Moreover, Exercise 5 on page 146 implies that $\mathrm{Cn}(\Gamma) = \mathrm{Th}(\mathcal{N})$. Hence, $\mathrm{Cn}(\Gamma)^{\#} = \mathrm{Th}(\mathcal{N})^{\#}$. Thus, $\mathrm{Th}(\mathcal{N})^{\#}$ is recursive, which contradicts Lemma 6.5.5. □

Theorem 6.5.8 provides a clear negative answer to question (Q2) (see page 241). The theorem implies there is no decidable set $\Gamma$ of $\mathcal{L}$-sentences such that (1) $\mathcal{N} \models \Gamma$ and (2) every sentence that is true in $\mathcal{N}$ is also deducible from $\Gamma$. Thus, for any decidable set $\Gamma$ such that $\mathcal{N} \models \Gamma$, there is a sentence $\beta$ such that $\Gamma \nvdash \beta$ and $\Gamma \nvdash \neg\beta$. Since either $\mathcal{N} \models \beta$ or $\mathcal{N} \models \neg\beta$, it follows that there is a sentence that is true in $\mathcal{N}$ and yet, it is not deducible from $\Gamma$.

**Corollary 6.5.9.** *The theory $Cn(\Omega)$ is not complete.*

**Lemma 6.5.10.** *Let $\Gamma$ be a set of $\mathcal{L}$-sentences and let $\sigma$ be an $\mathcal{L}$-sentence. If $Cn(\Gamma)^{\#}$ is recursive, then $Cn(\Gamma; \sigma)^{\#}$ is recursive.*

*Proof.* Let $\Gamma$ and $\sigma$ be as stated and assume that $\mathrm{Cn}(\Gamma)^{\#}$ is recursive. By the deduction theorem (Theorem 3.3.33), for any wff $\varphi$, we have

$$\Gamma; \sigma \vdash \varphi \quad \text{iff} \quad \Gamma \vdash (\sigma \to \varphi).$$

Hence,

$$\#\varphi \in \mathrm{Cn}(\Gamma; \sigma)^{\#} \quad \text{iff} \quad \#(\sigma \to \varphi) \in \mathrm{Cn}(\Gamma). \tag{6.42}$$

Let $g \colon \mathbb{N} \to \mathbb{N}$ be defined by $g(m) = [17, \#\sigma, m]$ (see Definition 6.4.4). Clearly, $g$ is primitive recursive and (6.42) implies that

$$m \in \mathrm{Cn}(\Gamma; \sigma)^{\#} \quad \text{iff} \quad g(m) \in \mathrm{Cn}(\Gamma)^{\#}.$$

Thus, by the substitution rule (5.5), we conclude that $\mathrm{Cn}(\Gamma; \sigma)^{\#}$ is recursive. □

By induction, we have the following corollary.

**Corollary 6.5.11.** *Let $\Gamma$ be a set of $\mathcal{L}$-sentences and let $\Sigma$ be a finite set of $\mathcal{L}$-sentences. If $Cn(\Gamma)^{\#}$ is recursive, then $Cn(\Gamma \cup \Sigma)^{\#}$ is recursive.*

**Theorem 6.5.12.** *Let $\Gamma$ be a set of $\mathcal{L}$-sentences such that $\Gamma \cup \Omega$ is consistent. Then $Cn(\Gamma)^{\#}$ is not recursive.*

*Proof.* Let $\Gamma$ be such that $\Gamma \cup \Omega$ is consistent. Suppose, to the contrary, that $\mathrm{Cn}(\Gamma)^{\#}$ is recursive. Hence, by Corollary 6.5.11, $\mathrm{Cn}(\Gamma \cup \Omega)^{\#}$ is recursive. Thus, $\mathrm{Cn}(\Gamma \cup \Omega)^{\#}$ is representable by Theorem 6.3.24. Therefore, by Definition 6.3.1, there exists a formula $\theta(x)$ such that for each $\mathcal{L}$-sentence $\sigma$,

$$\Gamma \cup \Omega \vdash \sigma \text{ implies } \Omega \vdash \theta(\overline{\#\sigma}), \tag{6.43}$$

$$\Gamma \cup \Omega \nvdash \sigma \text{ implies } \Omega \vdash \neg\theta(\overline{\#\sigma}). \tag{6.44}$$

By Lemma 6.5.1, there is a particular sentence $\sigma$ such that

$$\Omega \vdash \sigma \leftrightarrow \neg\theta(\overline{\#\sigma}). \tag{6.45}$$

Logical axiom 3.3.17(1) and (6.45) imply that

$$(\blacktriangle)\ \Omega \vdash \theta(\overline{\#\sigma}) \to \neg\sigma \quad \text{and} \quad (\blacktriangledown)\ \Omega \vdash \neg\theta(\overline{\#\sigma}) \to \sigma.$$

There are two cases to consider: Either $\Gamma \cup \Omega \vdash \sigma$ or $\Gamma \cup \Omega \nvdash \sigma$. If $\Gamma \cup \Omega \vdash \sigma$, then we infer that

$$
\begin{aligned}
\Gamma \cup \Omega \vdash \sigma &\Rightarrow \Omega \vdash \theta(\overline{\#\sigma}) && \text{by (6.43),}\\
&\Rightarrow \Omega \vdash \neg\sigma && \text{by } (\blacktriangle) \text{ and modus ponens.}
\end{aligned}
$$

So $\Gamma \cup \Omega \vdash \sigma$ and $\Omega \vdash \neg\sigma$. This contradicts our assumption that $\Gamma \cup \Omega$ is consistent.

If $\Gamma \cup \Omega \nvdash \sigma$, then we conclude that

$$
\begin{aligned}
\Gamma \cup \Omega \nvdash \sigma &\Rightarrow \Omega \vdash \neg\theta(\overline{\#\sigma}) && \text{by (6.44),}\\
&\Rightarrow \Omega \vdash \sigma && \text{by } (\blacktriangledown) \text{ and modus ponens.}
\end{aligned}
$$

So $\Gamma \cup \Omega \nvdash \sigma$ and $\Omega \vdash \sigma$, which are contradictory. Hence, $\mathrm{Cn}(\Gamma)^{\#}$ is not recursive. □

**Corollary 6.5.13.** *The set $Cn(\Omega)^{\#}$ is not recursive.*

The following corollary is an extension of Gödel's incompleteness theorem. It is the result of replacing "$\mathcal{N} \vDash \Gamma$" in Theorem 6.5.8 with "$\Gamma \cup \Omega$ is consistent."

**Corollary 6.5.14.** *Let $\Gamma$ be a set of $\mathcal{L}$-sentences. If the set $\Gamma^{\#}$ is recursive and $\Gamma \cup \Omega$ is consistent, then the theory $Cn(\Gamma)$ is not complete.*

The completeness theorem implies that an $\mathcal{L}$-sentence $\varphi$ is logically valid if and only if $\vdash \varphi$, that is, $\varphi$ is deducible from the logical axioms. Let $\mathbb{V}$ be the set of all $\mathcal{L}$-sentences that are logically valid. It follows that $\mathrm{Cn}(\mathbb{V}) = \mathbb{V}$.

**Corollary 6.5.15** (Alonzo Church). *The set $\mathbb{V}^{\#}$ is not recursive.*

Gödel's (1931) incompleteness theorem (Theorem 6.5.8) shows that no matter how one selects a consistent recursive set of axioms, which hold in $\mathcal{N}$, these axioms will be incomplete, that is, there will always be a true statement that cannot be deduced from the axioms. Gödel's incompleteness theorem shows that there exists a limitation to the axiomatic method. In the next section we will explore another limitation that Gödel discovered.

### 6.5.2 Gödel's second incompleteness theorem

After proving his first incompleteness theorem, Gödel pursued the following question: Can an axiomatic system prove its own consistency? Before we investigate Gödel's answer to this question, we revisit some pertinent topics. Recalling Definition 6.4.21, we let:
(1) $\mathbb{D}_\Gamma = \{[\#\alpha_1, \ldots, \#\alpha_n] : n \geq 1 \text{ and } \langle \alpha_1, \ldots, \alpha_n \rangle \text{ is a deduction from } \Gamma\}$,
(2) $\mathbb{P}_\Gamma = \{\langle s, m \rangle : \mathbb{D}_\Gamma(s) \text{ and } (s)_{\mathrm{lh}(s)-1} = m\}$,

for a set $\Gamma$ of $\mathcal{L}$-wffs. In Theorem 6.4.22 we showed that if $\Gamma^{\#}$ is (primitive) recursive, then the relations $\mathbb{D}_\Gamma(s)$ and $\mathbb{P}_\Gamma(s, m)$ are (primitive) recursive. Thus, if $\Gamma$ is a set of $\mathcal{L}$-sentences and $\Gamma^{\#}$ is recursive, then for any $\mathcal{L}$-sentence $\sigma$,

$$\Gamma \vdash \sigma \quad \text{iff} \quad \exists s \mathbb{P}_\Gamma(s, \#\sigma). \tag{6.46}$$

Therefore, $\mathrm{Cn}(\Gamma)^\#$ is recursively enumerable. Thus, in particular, $\mathrm{Cn}(\Omega)^\#$ is recursively enumerable and, by Corollary 6.5.13, $\mathrm{Cn}(\Omega)^\#$ is not recursive. This provides an example of a recursively enumerable set that is not recursive.

Now let $\Gamma^\#$ be recursive, where $\Gamma$ is a set of $\mathcal{L}$-sentences. Since the relation $\mathbb{P}_\Gamma$ is recursive, it is also representable. So, for any such set $\Gamma$, we will let $\psi_\Gamma(x, y)$ be the formula that represents $\mathbb{P}_\Gamma$. If a deduction exists, then can one deduce that it exists?

**Lemma 6.5.16.** *Let $\Gamma$ be a set of $\mathcal{L}$-sentences such that $\Omega \subseteq \Gamma$ and $\Gamma^\#$ is recursive. Then, for any $\mathcal{L}$-sentence $\sigma$, if $\Gamma \vdash \sigma$, then $\Gamma \vdash \exists s \psi_\Gamma(s, \overline{\#\sigma})$.*

*Proof.* Let $\Gamma$ be as stated and let $\sigma$ be an $\mathcal{L}$-sentence. Assume that $\Gamma \vdash \sigma$. Thus, there exists a deduction of $\sigma$ from $\Gamma$. Let $d = [\#\alpha_1, \ldots, \#\alpha_n]$ be the sequence number that consists of the Gödel numbers of the formulas as they appear in the deduction of $\sigma$ from $\Gamma$. Hence, $\mathbb{P}_\Gamma(d, \#\sigma)$. Since $\psi_\Gamma(x, y)$ represents $\mathbb{P}_\Gamma$, we conclude that

$$\Omega \vdash \psi_\Gamma(\overline{d}, \overline{\#\sigma}).$$

Since $\overline{d}$ is a term with no variables, Logical axioms 3.3.17(1)(2) imply that

$$\Omega \vdash \psi_\Gamma(\overline{d}, \overline{\#\sigma}) \to \exists s \psi_\Gamma(s, \overline{\#\sigma}).$$

Therefore, by modus ponens, we have $\Gamma \vdash \exists s \psi_\Gamma(s, \overline{\#\sigma})$. □

By Lemma 6.5.1, there is an $\mathcal{L}$-sentence $\sigma$ such that

$$\Omega \vdash \sigma \leftrightarrow \neg \exists s \psi_\Gamma(s, \overline{\#\sigma}). \tag{6.47}$$

The sentence $\sigma$ indirectly asserts that "I am not deducible from $\Gamma$." If $\Gamma$ is consistent, then our next lemma shows that "$\sigma$ is not lying."

**Lemma 6.5.17.** *Let $\Gamma$ be a set of $\mathcal{L}$-sentences such that $\Omega \subseteq \Gamma$ and $\Gamma^\#$ is recursive. Let $\sigma$ be as in* (6.47). *If $\Gamma$ is consistent, then $\Gamma \nvdash \sigma$.*

*Proof.* Let $\Gamma$ and $\sigma$ be as stated above and assume that $\Gamma$ is consistent. Suppose, for a contradiction, that $\Gamma \vdash \sigma$. Then we conclude that

$$\begin{aligned} \Gamma \vdash \sigma &\Rightarrow \Gamma \vdash \exists s \psi_\Gamma(s, \overline{\#\sigma}) && \text{by Lemma 6.5.16,} \\ &\Rightarrow \Gamma \vdash \neg\sigma && \text{by (6.47) as } \Omega \subseteq \Gamma. \end{aligned}$$

Hence, $\Gamma$ is inconsistent, contradicting our assumption. Therefore, $\Gamma \nvdash \sigma$. □

Let $\Gamma$ and $\sigma$ be as in Lemma 6.5.17. We now show that the conditional in Lemma 6.5.17

$$\text{"if } \Gamma \text{ is consistent, then } \Gamma \nvdash \sigma\text{"} \tag{6.48}$$

can be translated into the first-order language $\mathcal{L}$. Note that

$$\Gamma \text{ is consistent if and only if } \Gamma \nvdash \dot{0} \neq \dot{0}.$$

Let Con(Γ) be the $\mathcal{L}$-sentence $\neg\exists s\psi_\Gamma\big(s, \overline{\#(\dot{0} \neq \dot{0})}\big)$. Thus, our translation of (6.48) is

$$\mathrm{Con}(\Gamma) \to \neg\exists s\psi_\Gamma(s, \overline{\#\sigma}). \tag{6.49}$$

The proof of Lemma 6.5.17 was quite straightforward. Thus, one may suspect that

$$\Gamma \vdash (\mathrm{Con}(\Gamma) \to \neg\exists s\psi_\Gamma(s, \overline{\#\sigma})).$$

How strong must Γ be so that the above holds?

**Definition 6.5.18.** Let Γ be a set of $\mathcal{L}$-sentences such that $\Omega \subseteq \Gamma$ and Γ contains all the universal closures of formulas having the form

$$(\varphi(\dot{0}) \land \forall x(\varphi(x) \to \varphi(\dot{S}x))) \to \forall x\varphi(x). \tag{6.50}$$

Then Γ is called an *extension of Peano arithmetic*.

An extension of Peano arithmetic contains all of the “induction axioms” that have the form (6.50). These axioms all hold in the standard model $\mathcal{N}$ and allow one to deduce many additional valid properties about the natural numbers.

The proof of our next lemma is quite technical and, as a result, we will not present the proof here. For a complete proof, see Section 5.3 of [5].

**Lemma 6.5.19.** *Let Γ be a set of $\mathcal{L}$-sentences such that $\Gamma^\#$ is recursive and Γ is a consistent extension of Peano arithmetic. Also, let σ be as in* (6.47). *Then*

$$\Gamma \vdash (\mathit{Con}(\Gamma) \to \neg\exists s\psi_\Gamma(s, \overline{\#\sigma})).$$

We can now state and prove Gödel’s second incompleteness theorem, which shows that the $\mathcal{L}$-sentence Con(Γ) is not deducible from Γ whenever Γ is a consistent extension of Peano arithmetic and $\Gamma^\#$ is recursive.

**Gödel’s Second Incompleteness Theorem 6.5.20.** *Let Γ be a set of $\mathcal{L}$-sentences. If Γ is a consistent extension of Peano arithmetic and $\Gamma^\#$ is recursive, then $\Gamma \nvdash \mathit{Con}(\Gamma)$.*

*Proof.* Let Γ be a consistent extension of Peano arithmetic such that $\Gamma^\#$ is recursive. Suppose, for a contradiction, that $\Gamma \vdash \mathrm{Con}(\Gamma)$ and let σ be as in (6.47). Therefore, by Lemma 6.5.19 and modus ponens, we have $\Gamma \vdash \neg\exists s\psi_\Gamma(s, \overline{\#\sigma})$. Since $\Omega \subseteq \Gamma$, (6.47) implies that $\Gamma \vdash \sigma$, and this contradicts Lemma 6.5.17. □

### 6.5.3 Epilogue

In the early 1920s, the German mathematician David Hilbert (1862–1943) put forward a new proposal for the foundation of classical mathematics. This program is referred to as Hilbert's Program. With respect to number theory, David Hilbert proposed that mathematicians find a single formal system from which one can deduce all of the true statements in number theory and also prove that the system is consistent.

Kurt Gödel's theorems show that Hilbert's program is unattainable. In his first theorem, Gödel proves that any consistent recursive set of axioms for Peano arithmetic can never be complete as there will always exist true statements about the natural numbers that cannot be deduced from the given axioms. In his second incompleteness theorem, Gödel shows that such a system cannot prove its own consistency.

Moreover, Gödel's theorems can be extended to other recursive axiomatic systems in which an extension of Peano arithmetic can be interpreted. In particular, one can show that the standard axiomatic system for set theory (ZFC) is incomplete and cannot prove its own consistency, unless it is inconsistent.

Gödel's two incompleteness theorems transformed research in the foundations of mathematics, both in mathematics and in philosophy. Moreover, these theorems have also become important in theoretical computer science, as they show that there are limitations on the kinds of problems that can be solved computationally.

**Exercises 6.4.**

1. Let $\theta(x)$ and $\psi(x)$ be wffs in which only $x$ occurs free. Show that there is a sentence $\sigma$ such that $\Omega \vdash \sigma \to \theta(\overline{\#\sigma})$ and $\Omega \vdash \sigma \to \psi(\overline{\#\sigma})$.
2. Let $\theta(x)$ be a wff in which only $x$ occurs free. Suppose that $\Omega \vdash \forall x\theta(x)$.
   (a) By Lemma 6.5.1 there is a $\sigma'$ such that $\Omega \vdash \sigma' \leftrightarrow \theta(\overline{\#\sigma'})$. Show that $\Omega \vdash \sigma'$.
   (b) By Lemma 6.5.1 there is a $\sigma$ such that $\Omega \vdash \sigma \leftrightarrow \neg\theta(\overline{\#\sigma})$. Show that $\Omega \vdash \neg\sigma$.
3. Prove Corollary 6.5.2.
*4. Let $A$ and $B$ be recursively enumerable sets of natural numbers. Suppose that $A \cup B = C$ is recursive and $A \cap B = \varnothing$. Prove that $A$ is recursive.
5. Prove Corollary 6.5.9.
6. Show that $\mathrm{Th}(\mathcal{N})$ is not finitely axiomatizable.
7. Show that there does not exist a recursive set $A \subseteq \mathbb{N}$ such that $\{\#\varphi : \Omega \vdash \varphi\} \subseteq A$ and $\{\#\varphi : \Omega \vdash \neg\varphi\} \subseteq \mathbb{N} \setminus A$.
8. Let $\Gamma$ be a set of $\mathcal{L}$-sentences. Show that if $\mathcal{N} \models \Gamma$, then $\mathrm{Cn}(\Gamma)^{\#}$ is not recursive.
9. Prove Corollary 6.5.13.
10. Prove Corollary 6.5.14.
11. Prove Corollary 6.5.15.
12. Let $\Gamma$ be a set of $\mathcal{L}$-sentences. Suppose that $\Omega \subseteq \Gamma$, $\Gamma^{\#}$ is recursive, and $\mathcal{N} \models \Gamma$. Show that if $\Gamma \vdash \exists s\psi_{\Gamma}(s, \overline{\#\sigma})$, then $\Gamma \vdash \sigma$.
13. Let $\mathbb{W}$ be the set of all logically valid $\mathcal{L}$-wffs. Show that $\mathbb{W}^{\#}$ is not recursive.

14. Let $\Gamma$ be a set of $\mathcal{L}$-sentences such that $\Omega \subseteq \Gamma$ and $\Gamma^{\#}$ is recursive. Let $\psi_\Gamma$ represent the relation $\mathbb{P}_\Gamma$ as defined in Definition 6.4.21 and let $\varphi$ be an $\mathcal{L}$-sentence. Suppose that $\mathcal{N} \models \Gamma$ and $\mathcal{N} \models \exists s \psi_\Gamma(s, \#\varphi)$. Show that $\mathcal{N} \models \varphi$.
15. Let $\psi_\Omega$ represent the relation $\mathbb{P}_\Omega$ as defined in Definition 6.4.21. By Lemma 6.5.1, let $\sigma$ be a sentence such that $\Omega \vdash \sigma \leftrightarrow \neg\exists s \psi_\Omega(s, \overline{\#\sigma})$. Thus, by Lemma 6.5.17, $\Omega \nvdash \sigma$. Show that $\mathcal{N} \models \sigma$. Conclude that $\Omega \nvdash \neg\sigma$.
16. Let $\Gamma$ be a set of $\mathcal{L}$-sentences such that $\Omega \subseteq \Gamma$ and $\Gamma^{\#}$ is recursive. Let $\psi_\Gamma$ represent the relation $\mathbb{P}_\Gamma$ as defined in Definition 6.4.21. Now, by Lemma 6.5.1, let $\sigma$ be a sentence such that $\Gamma \vdash \sigma \leftrightarrow \forall s \neg\psi_\Gamma(s, \overline{\#\sigma})$. Show that $\Gamma \vdash \neg\psi_\Gamma(\overline{n}, \overline{\#\sigma})$, for every $n \in \mathbb{N}$. Also show that $\Gamma \nvdash \forall s \neg\psi_\Gamma(s, \overline{\#\sigma})$.
17. Let $\Gamma$ be an extension of Peano arithmetic. Suppose that $\Gamma^{\#}$ is recursive. Show that if $\Gamma \vdash \mathrm{Con}(\Gamma)$, then $\Gamma$ is not consistent.

**Exercise Notes:** For Exercise 7, use proof by contradiction and Theorem 6.3.24. Then apply Lemma 6.5.1 with a negation. For Exercise 9, show that $\mathrm{Cn}(\mathrm{Cn}(\Omega)) = \mathrm{Cn}(\Omega)$. For Exercise 10, note that if $\Gamma \cup \Omega$ is consistent, then $\mathrm{Cn}(\Gamma) \cup \Omega$ is also consistent. For Exercise 12, see Lemma 6.5.16, its proof, and (6.46). For Exercise 14 and Exercise 15, see Exercise 7 on page 224. Exercise 16 provides an example of a deduction system in which the "$\omega$-rule" does not hold.

# Bibliography

[1] Cunningham DW. A logical introduction to proof. New York, NY: Springer; 2012.
[2] Enderton HB. A mathematical introduction to logic. 2nd ed. Burlington, MA: Harcourt/Academic Press; 2001.
[3] Cunningham DW. Set theory. A first course. Camb. Math. Textb. Cambridge: Cambridge University Press; 2016. Reprinted in 2022.
[4] Gödel K. Über formal unentscheidbare Sätze der Principia Mathematica und verwandter Systeme. I. Monatsh Math. 2006;149(1):1–30. Reprinted from Monatsh. Math. Phys. 1931;3(8):173–198 [MR1549910], With an introduction by Sy-David Friedman.
[5] Hinman PG. Fundamentals of mathematical logic. CRC Press; 2018. Available from: https://books.google.com/books?id=6UBZDwAAQBAJ.
[6] Chang CC, Keisler HJ. Model theory. 3rd edition. Dover Publications; 2013.
[7] Enderton HB. Computability theory. An introduction to recursion theory. Amsterdam: Elsevier/Academic Press; 2011.
[8] Barwise J (Ed). Handbook of mathematical logic. With the cooperation of Keisler HJ, Kunen K, Moschovakis YN and Troelstra AS. Reprint. vol. 90 of Stud. Logic Found. Math. Amsterdam: Elsevier; 1982.
[9] van Dalen D. Logic and structure. 5th ed. Universitext. London: Springer; 2013.
[10] Srivastava SM. A course on mathematical logic. 2nd ed. Universitext. New York, NY: Springer; 2013.

https://doi.org/10.1515/9783110782073-007

# Symbol Index



https://doi.org/10.1515/9783110782073-008

# Subject Index



https://doi.org/10.1515/9783110782073-009